# Biotechnology and Innovation Systems

# Biotechnology and Innovation Systems

## The Role of Public Policy

*Edited by*

Bo Göransson and Carl Magnus Pålsson

*Research Policy Institute, Lund University, Sweden*

**Edward Elgar**

Cheltenham, UK • Northampton, MA, USA

**International Development Research Centre**

Ottawa • Cairo • Dakar • Montevideo • Nairobi • New Delhi • Singapore

Published by
Edward Elgar Publishing Limited
The Lypiatts
15 Lansdown Road
Cheltenham
Glos GL50 2JA
UK

Edward Elgar Publishing, Inc.
William Pratt House
9 Dewey Court
Northampton
Massachusetts 01060
USA

International Development Research Centre
PO Box 8500
Ottawa, ON K1G 3H9
Canada
www.idrc.ca/info@idrc.ca
ISBN 978 1 55250 538 0 (IDRC e-book)

A catalogue record for this book
is available from the British Library

Library of Congress Control Number: 2011934818

MIX
Paper from responsible sources
FSC® C018575

ISBN 978 1 78100 138 7 (cased)

Typeset by Servis Filmsetting Ltd, Stockport, Cheshire
Printed and bound by MPG Books Group, UK

# Contents

# Figures

# Tables

# Boxes

# Contributors

**Anda Adamsone-Fiskovica** is a researcher and project manager at the Centre for Science and Technology Studies of the Latvian Academy of Sciences. She holds an MSc in Sociology from the University of Latvia and an MA in Science and Technology Studies from the Linkoping University (Sweden). She has also finished doctoral studies in sociology at the University of Latvia (2003–07). Her main research interests are related to R&D and innovation policy, development of higher education, and science–society relations in Latvia. She is the Latvian team leader in several international research projects and a member of a number of European innovation and science researchers' networks.

**Isabel Bortagaray** is Senior Program Officer with the International Development Research Centre, Ottawa, Canada. She holds a PhD in Public Policy (Science and Technology Policy, Georgia Institute of Technology). Her research has focused on science, technology and innovation studies in the context of developing countries, including work on Argentina, Brazil, Costa Rica, Colombia, New Zealand and Uruguay. More recently Dr Bortagaray has taken part in a project on the Distributional Assessment of Emerging Technologies led by Susan Cozzens at the Technology Policy and Assessment Center (Georgia Institute of Technology), as well as a project on The Evolving Role of Academic Institutions in Innovation Systems and Development in Uruguay coordinated by Judith Sutz and Rodrigo Arocena (University of the Republic, Uruguay).

**Tran Ngoc Ca** is Director of Secretariat for the National Council for Science and Technology Policy, and Deputy Director, National Institute for Science and Technology Policy and Strategy Studies, Hanoi, Vietnam. He has been working on projects on science, technology and innovation policy and management, information technology and electronics commerce for UNDP, UNIDO, World Bank, European Commission, CIDA, SIDA/SAREC, IDRC, among others. He is involved in various teaching and consulting works. Dr Ca gained an engineering degree at Moscow Mining University (former Soviet Union), a Master's at Lund University (Sweden) and a PhD at the University of Edinburgh (UK). He spent time in US universities such as UC Davis, UC Berkeley and Stanford as visiting Fulbright scholar. He has published many articles and chapters in books, and a book titled *Technological Capability and Learning in Firms: Vietnamese Industries in Transition* (Ashgate Publishing, UK). He is a member of several professional associations in Vietnam and abroad.

**José Eduardo Cassiolato** is Professor of Innovation Studies and Coordinator of RedeSist at the Economics Institute, Federal University of Rio de Janeiro, Brazil. He holds a DPhil from SPRU, the University of Sussex, UK. His main research activities are connected to science and technology policies and national and local systems of innovation in less developed countries. Such research focuses particularly on interactive learning and capacity building and in national and local processes of innovation-related development. Currently he is in charge of an international comparative study on the National Systems of Innovation of Brazil, Russia, India, China and South Africa.

**Le Van Chuong** has a first degree in mining engineering and is Head of Department for postgraduate training, National Institute for S&T Policy and Strategy Studies (NISTPASS), Vietnam.

**Bitrina D. Diyamett** is an Executive Director of the African Technology Policy Studies Network, Tanzania Chapter (ATPS-Tanzania), an NGO engaged in policy research in science, technology and innovation in Tanzania. She holds a PhD in development studies, University of Dar es Salaam; and a Master's degree in Science and Technology Policy from Lund University, Sweden. Much of her research work, including her PhD thesis, consultancies and publications, focuses on systems of innovation in the context of least developed countries.

**Bo Göransson** is Senior Research Fellow and coordinator of the LEAP4D research group at the Research Policy Institute (RPI), Lund University, Sweden. He holds a PhD in International Economics from Aalborg University, Denmark. His research work focuses on how knowledge and learning systems drive economic development and growth. Much of the research deals with issues related to innovation policies, capacity building and impact of new technologies in developing countries, particularly in the area of information and communication technologies (ICT). In a related line of research, he studies the role of universities in innovation systems and development.

**Birgitte Gregersen** is Head of Department at the Department of Business Studies, Aalborg University, Denmark. She is a member of the IKE-Research Group and has researched and published within the field of technical change and employment, IT in the public sector, public technology procurement, studies of national systems of innovation, university–industry linkages, innovation policy, and sustainable development. Her current research is centred around systems of innovation with a special focus on institutions and learning capabilities in a sustainable development perspective.

**Janis Kristapsons** is the founder (1991) and head of the Centre for Science and Technology Studies of the Latvian Academy of Sciences and advisor to the President of the Academy. His academic background is in nuclear and solid state physics and sociology. His main research interests currently cover research and innovation policies in the East and Central European countries, R&D and inventing activity, technology transfer and industry–academia linkages, as well as S&T indicators and research evaluation. Together with his Estonian and Lithuanian colleagues he has written a monograph on the transition of the Baltic R&D systems (2003).

**Ralf Lindner** is a Professor for Political Science at the Department for Politics and Public Affairs at the Quadriga Hochschule in Berlin and a senior researcher at the Fraunhofer Institute for Systems and Innovation Research (ISI) in Karlsruhe. He received his degree in Political Science and Economics from the University of Augsburg, completed graduate work at the University of British Columbia (Vancouver) and postgraduate studies at Carleton University (Ottawa). His doctoral dissertation focuses on the application and integration of digital networks in the communication strategies of intermediary organizations in North America. In addition to his research in the area of new media and society, Professor Lindner has specialized in science, technology and innovation policy analyses.

**Aija Lulle** has worked with the Centre for Science and Technology Studies of the Latvian Academy of Sciences since 2008. In 2001, she graduated the Master's programme in Sociology at the University of Latvia and is currently continuing her PhD studies in human geography at the same university. Since 2004 she has been working with several national and international research organizations and units on various mobility-related topics. Her main research interests are related to migration and transnationalism studies.

**Paula Macucule** is a junior researcher at the Biotechnology Center of the Eduardo Mondlane University (EMU), Mozambique. She holds a BSc in Biology from EMU. Her main research area is molecular parasitology and she is actively engaged in the characterization of drug resistance patterns in populations of African pathogenic trypanosome species in Mozambique.

**Nguyen Phuong Mai** has a background in Law studies, received an MSc from Maastricht University, and now works for the Department of Technology Market and Innovation, National Institute for S&T Policy and Strategy Studies (NISTPASS), Vietnam.

**Emmarold Mneney** is a Principal Agricultural Research Officer of the Mikocheni Agricultural Research Institute (MARI), Tanzania. He has a PhD in biotechnology from the University of London, UK and long-standing research interest in the field of agricultural biotechnology. His current research interests are molecular markers, tissue culture and genetic diversity and germplasm conservation of various tropical crops including cashew, sweet potatoes, sorghum, maize, cassava, banana and tomato. He also has experience in technology transfer technologies, biosafety and innovation systems. He has worked on several collaborative programmes and his current/recent research projects include: Genetic transformation for drought tolerant maize; development of Marker Assisted Selection (MAS) technologies to support breeding for tolerance to Aluminium (Al) toxicity in sorghum; and Product Development Partnership (PDP) for the appropriation of knowledge and innovation in biotechnology in Tanzania.

**Ana Maria da Graça Mondjana** is a Professor of Plant Pathology at the Faculty of Agronomy and Forestry Engineering, Eduardo Mondlane University, Mozambique. She holds a PhD in Plant Pathology. Although her research has been in a wide range of crops and diseases, her main focus has been on the epidemiology and management of viral diseases of vegetables and leguminous crops.

**Luis Félix Montalvo Arriete** is Senior Researcher of the Chair Science Technology Society and Innovation at the University of Havana, Cuba. He holds a PhD in Scientific and Technological Policy from the University of Campinas (UNICAMP), Brazil. Much of the research deals with issues related to science and technology policies, innovation policies and innovation systems. His research work focuses on the role of universities in innovation systems.

**Burton L.M. Mwamila** is an experienced engineering academic, researcher and consultant. He spearheaded the establishment of the College of Engineering and Technology, for which he served as its first Principal from 2005–09. He was the first Chairman of the National Council for Technical Education from 1999–07. He is currently the Chairman of the Tanzania Commission for Science and Technology (COSTECH) and the Tanzania Automotive Technology Centre (TATC). At regional level he has been a Member of the Governing Council of the African Network of Scientific and Technological Institutions (ANSTI) from 2000–09, and at continental level he has been Chairman of the Executive Board of the Pan-African Competitiveness Forum (PACF) since 2008. Since July 2009 he has been the founding Vice Chancellor of the Nelson Mandela African Institute of Science and Technology (NM AIST) in Arusha, Tanzania.

**Luis Neves** is a Senior Lecturer of Parasitology at the Veterinary Faculty, Eduardo Mondlane University, Mozambique and Director of the Biotechnology Center of the same university. He holds a PhD in Veterinary Parasitology from the University of Liverpool, UK. His main area of research is molecular protozoology with emphasis on the development of molecular diagnostic tools and vaccines. Since 2003 he has also been interested in the research of institutional models for science advancement in developing countries.

**Jorge Núñez Jover** studied chemistry, has a PhD in philosophy and is Director for Postgraduate Studies at the University of Havana. He is the coordinator of the Master's and Doctoral programmes on Science, Technology and Society. He holds the chair in science, technology, society and innovation. He has published 10 books and over 100 papers. His research is in the areas of education, science, technology and society; postgraduate studies evaluation and accreditation, as well as university science policy. He is Invited Professor in universities in Spain, Germany, Brazil, Mexico and other countries in Latin America and the Caribbean.

**Isarelis Pérez Ones** graduated in Sociology in 1997, and also with a Master's in Science, Technology and Society (STS) at the University of Havana where she is now an assistant professor. She was a graduate student at Roskilde University, Denmark (2002–03). In 2005 she took a training PhD Program, 'Globelics Academy', at the Technical University of Lisbon. She is finishing her PhD on STS at the University of Havana. Her main interests are related to the social studies of science and technology and higher education. Her current work has focused on the role of universities in the systems of innovation.

**Carl Magnus Pålsson** is Associate Professor at the Research Policy Institute (RPI), Lund University, Sweden. He holds a PhD in history of science and ideas. His research interests revolve around innovations systems and policies in the knowledge economy, and how organizations for higher education and research are transformed in historical, cultural and institutional dimensions.

**Tran Thi Phuong** has a background in natural and engineering sciences, received an MSc from Hue University and studied in Canada, is pursuing a PhD degree at the Asian Institute of Technology (Bangkok), and is a lecturer at Hue Agriculture and Forestry University (HUAF), Vietnam.

**Márcia Siqueira Rapini** is an associate researcher at CEDEPLAR – Federal University of Minas Gerais (UFMG) and a PhD student in Economics at the Economics Institute, Federal University of Rio de Janeiro, Brazil. She graduated from UFMG in 2000 and holds a Master's degree in Economics (UFRJ, 2004). Her research interests within industrial economics and economics of science of technology include: national systems innovation, university–industry interactions, and finance and innovation.

**Thomas Reiss** is head of the Competence Center 'Emerging Technologies' at Fraunhofer ISI. He holds a PhD in Molecular Biology from Freiburg University (1983) and joined Fraunhofer ISI in 1987. His current research is focusing on national and sectoral innovation systems; foresight and impact assessment of new technologies; and innovation policies. He is a member of the Management Committee of the European Techno-Economic Policy Support Network (ETEPS) and member of the editorial boards of the *International Journal of Biotechnology* and the *International Journal of Learning and Change*. At the Karlsruhe Institute of Technology (KIT) he lectures on the management of new technologies.

**Carlos Miguel Ribeiro** is a Senior Lecturer in Tropical Horticulture at the Faculty of Agronomy and Forestry Engineering, Eduardo Mondlane University, Mozambique. He holds an MSc in Crop Science from the University of Viçosa, Brazil. His main research interests are in the area of tropical fruits with emphasis on socio-economical factors associated with fruit production.

**Galina Sagieva**, Doctor in Economics, is the Head of Department for Studies of Intellectual Property and Technology Transfer of the Institute for Statistical Studies and Economics of Knowledge, State University – Higher School of Economics (SU-HSE), and a Senior Lecturer in Innovation Management. Her research interests refer to S&T statistics (methodology, analysis) and policies, which are reflected in over 100 publications. She has taken part in international projects initiated by the OECD, EC, World Bank, IIASA, IAIA, among others. She is a member of the national expert group of Russia's accession to the OECD on issues of S&T policy and development of biotechnology.

**Ulrich Schmoch** is private lecturer at the University of Karlsruhe (Germany), senior researcher at the Fraunhofer Institute for Systems and Innovation Research, and scientific manager at the office of the Expert Commission on Research and Innovation in Berlin (Germany).

He received a degree in mechanical engineering, a doctoral degree in social and political sciences at the University of Hanover (Germany), and a state doctoral degree in sociology of science and technology from the University of Karlsruhe (Germany). The focus of his research is systems of innovation, knowledge transfer, innovation indicators, and structures of scientific research.

**Sara Gonçalves Antunes de Souza** has been an Assistant Professor at Universidade Estadual de Montes Claros UNIMONTES since 1999 and a PhD student in Economics at Universidade Federal do Rio de Janeiro-UFRJ. She holds a Master's degree in Economics from Universidade Federal de Minas Gerais-UFMG. Her main research subjects in industrial economics and economics of technology include innovation systems and university–industry linkages, business incubators and biotechnology. During 2005–06 she was coordinator of the Center of Intellectual Property (PROTEGE/ÁGORA) at Universidade Estadual de Montes Claros-UNIMONTES.

**Judith Sutz** is the Academic Coordinator of the University Research Council of the Universidad de la República, Uruguay, and Full Professor of Science, Technology and Society. She holds a PhD in Socio-Economics of Development from the University of Paris-Sorbonne. Her research work focuses on innovation and the production and social use of knowledge in developing countries. She was part of the Task Force on Science, Technology and Innovation of the UN Millennium Project, and she is a fellow of WAAS and integrates the Scientific Board of Globelics.

**Erika Tjunina** is a senior researcher at the Centre for Science and Technology Studies of the Latvian Academy of Sciences. She received her diploma in Chemistry from the Riga Technical University and holds a doctoral degree in Engineering Sciences in polymer mechanics. Her research work at the Centre primarily deals with the scientometric analysis (quantitative S&T indicators) of the Latvian science and the studies of the inventions and inventors of Latvia. She has also been involved in a range of projects dealing with the role of academic institutions in the national R&D and innovation systems.

**Wang Haiyan** is Associate Research Professor at the Chinese Academy of Science and Technology for Development, Ministry of Science and Technology, PRC. Her research mainly focuses on regional development, National Innovation Systems, S&T Strategy and related policies. She has undertaken about 30 international, national and regional projects in the fields of science and technology strategy, regional development as research director or main researcher, and has published 5 books and about 30 articles in the fields of science and technology strategy, innovation policy research, national innovation systems and regional development.

**Zhou Yuan** has a PhD in regional development and is Research Professor and Deputy Director-General, The Administrative Center for China's Agenda 21 (ACCA21), Ministry of Science and Technology, PRC. He is Vice President of the China Association of High Technology Industry Development, and is International Advisory Board Member of the Center for Science, Technology and Innovation in China, The Levin Institute, The State University of New York. He is Professor of The University of Science and Technology of China, Xian Jiaotong University, The University of Science and Technology of Beijing, Institute of Nature Science History of the Chinese Academy of Science. He has undertaken over 50 international, national and regional projects as research director in the fields of science and technology strategy and regional development and has published 10 books and over 100 articles in Chinese and English in the fields of science and technology strategy and policy, and regional development.

**Graziela Ferrero Zucoloto** is a researcher at the Institute of Applied Economic Research (IPEA) and RedeSist and a PhD student in Economics at the Economics Institute, Federal University of Rio de Janeiro (UFRJ), Brazil. She graduated from the University of São Paulo (USP) in 1999 and holds a Master's degree in Economics (USP, 2004). Her main research interests in Industrial Economics and Economics of Technology include: industrial and innovation policy, innovation and the health sector, multinational enterprises and innovation, intellectual property rights (IPR), IPR and competition policy.

# Acknowledgements

The Developing Universities project (UniDev) from which this book emanates was made possible through the support of The International Development Research Centre, Canada, and the SAREC department within the Swedish International Development Cooperation Agency (SIDA). The editors would also like to extend their grateful thanks to Anna Ekermo, research assistant at the Research Policy Institute, Lund University, who assisted in vital phases of the editing process.

# 1. Strategies for appropriation of biotechnology

**Bo Göransson and Carl Magnus Pålsson**

## BACKGROUND

Biotechnology[1] has emerged in the last decades as a field perceived to be a crucial component in the knowledge economy. By its interdisciplinary nature biotech is boundary-crossing, and the expectations for renewal of academic research and industry production across sectors are mounting. In policies and implementation there is a consensus across the globe that biotech, as an academic as well as an industrial field, holds a revolutionary potential. Consequently, there are substantial issues, regarding what 'the revolution' might carry in its wake. Some of those questions relate mostly to the future, and are to some degree a matter for speculation, while others have clear social, economic and ethical ramifications for today's society.

Based on a new recombinant DNA technique introduced in the 1970s, the global biotech industry exhibited high growth rates over the following years in terms of sales and employment. However, the cost of performing biotech R&D has kept pace. Biotech is one of the most research-intensive industries in the world. In the United States, the largest biotech market in the world, US biotech companies spent USD25 billion (PPP) on research and development in 2006 (OECD, 2009, p. 24). The five largest biotech companies in this market invested an average of USD170 000 per employee in R&D in 2007 (Bio, 2008). These high costs have contributed to the heavy losses the industry has posted for much of its existence. Between 1994 and 2006, the US biotech market showed an average yearly net loss of USD4.8 billion (Bio, 2008). The recent global recession has further exacerbated this situation, albeit leaving some emerging economies such as China and India largely unaffected (Ernst & Young, 2010).

The rise of the biotech industry has also brought to the fore the long-standing discussion on the role of the state versus the market (cf. Harvey and McMeekin, 2007); how can or should the boundaries between public and private appropriation of biotechnology be delineated when much of the knowledge creation process takes place in research centres sponsored by public money while the profits of commercialization often accrue to the private sector?

In this book we will look at how actors in the biotech field organize themselves in order to facilitate cooperation and appropriation of biotech results as well as how these strategies translate into public policy. The case studies have been carried out by members of the network *Developing Universities – The Evolving Role of Academic Institutions in Innovation Systems and Development* (UniDev). The UniDev network was launched in 2005 as a joint undertaking by research partners in 12 countries. The research consortium consists of researchers from developed as well as developing countries. The overall objectives of the network are to contribute to a better understanding of the changing role of academic institutions in national contexts as well as to contribute to initiating a process of policy learning and exchange between countries in different stages of economic development. The findings of the research network have shed light on the changing role of academic institutions within the contexts of innovation and economic growth and development, and in a comparative, multi-stakeholder perspective (cf. Göransson and Brundenius, 2010; Science and Public Policy, 2009; Arocena et al., 2008).

In this book we will present case studies on private and public appropriation of biotechnology carried out by members of the UniDev network in 12 countries, ranging from low-income economies to highly developed countries. The case countries are Brazil, China, Cuba, Denmark, Germany, Latvia, Mozambique, Russia, Sweden, Tanzania, Uruguay and Vietnam. The diversity of the selected countries is in a way both a strength and a weakness of this book. On the one hand, it is clear that some of the experiences of Mozambique in biotech development are not readily transferable or even relevant for the policy-making process in, say, Germany. On the other hand, the mere fact that biotechnology is recognized by most countries as a crucial element in supporting domestic growth and development, opens up a considerable scope for policy dialogues and learning among different regions and countries.

The cases presented in the following chapters examine the status of biotechnology in each country and discuss how public and private partnerships in biotechnology can be formed, enhanced and sustained in support of national growth and development goals. The presentations aim at highlighting the processes driving successful examples of technological innovation, technology transfer and product delivery of biotechnology, particularly when the result has been products aimed at, and benefiting, the poor. In these processes academic institutions, such as universities and other research and development organizations, play a crucial role since they are currently the dominant players in biotechnology research in developing countries.

There is also a more implicit set of issues and questions emerging from within a project like this one. As a counterpoint theme there runs, or should run, a questioning of the very tenets behind the current bandwagon towards streamlining or aligning policies across nations. In an evolutionary perspective, central to, in principle at least, innovation system thinking, there should be room for many varieties in the ecology. The motives for implementing various policies are often surprisingly homogeneous in their expressions, in spite of the fact that the underlying fundamentals differ considerably between countries. In the long run, systemic dynamics and viability benefit from a multiplicity of voices and perspectives. Thus it is the purpose of an overview like this one to open up for broader discussions also on the ideological basis supporting the innovation discourses surrounding primarily biotech, but potentially other fields as well.

The varying experiences of the case countries have required an open-ended approach. Taking the present structure and areas of specialization in each country as the point of departure, the case studies explore how these structures have evolved and how the main actors have influenced the process. Thus the chapters focus on the policy level of biotech development, that is examining how different actors have influenced national policy in the field of biotech as well as how the resulting policy has contributed to the present specific form of the biotech industry in each country. Much of the data required for such analyses have been extracted from existing material, and have been complemented by interviews with key actors from policy-making bodies, academia and private firms. More specifically, each chapter provides an overview of the level of public and private appropriation of biotech as well as an overview of the policy 'landscape', and addresses the following research questions:

- What have been the key elements of the national policy for biotech?
- What was the rationale for developing the specific policy?
- Which actors were relevant for designing the strategy and what were the major steps in the development of biotech policy?
- What are the mechanisms for and modes of biotech transfer from the university to firms?
- What have been the main results of university–industry biotech partnerships and technology transfer in terms of knowledge accumulation?
- What are the policy implications for future development in the field, and what institutional learning may be derived and, subsequently, transferred from the cases?

The Brazilian strategy for enhancing capabilities in the field of biotech is discussed in Chapter 2. For the last two decades, Brazil has attached great importance to this field and considers the fostering of biotech capability one of the central tenets of the science and technology policies. It is clear from the chapter that the Brazilian policy has been quite ambitious and has resulted in a number of positive outcomes, not least an impressive increase in the size and complexity of biotech research in the agricultural sector and in health and biological sciences. Notwithstanding these accomplishments, the authors of the chapter are doubtful whether an innovative capacity in a broader sense has developed. Despite government efforts to stimulate spin-off firms from university research, to support venture capital initiatives and to encourage joint university–industry research projects, a fully-fledged production structure has not emerged. Consequently, there is a mismatch between the growing competence of the scientific community and the productive structure.

Using the example of the local biotech innovation system of Minas Gerais, the chapter addresses the evolution of this system, the role played by Biobrás, the pre-eminent Brazilian biotech firm before it was acquired by Novo Nordisk, as well as the linkages of local firms to university research. Based on data from surveys in 2001 and 2004, the impact of the acquisition of Biobrás by a multinational firm on the dynamics of the local innovation system and how other local biotech firms survived after this acquisition is examined. The results of the surveys indicate that despite three decades of government programmes, the number of surviving biotech firms is dismally low and their innovative capacity remains uncertain.

The innovation climate and decision-making processes in biotechnology firms in Uruguay is the topic of Chapter 3. The study is actor-oriented, focusing on conditions for three biotech companies. The authors analyse the interaction among established actors, and those that have entered the field more recently; the very mix of 'old' and 'new' is found to be crucial to innovation potential. Boundaries between 'internal' and 'external', relative to the firms, are problematized. The decision processes at firm level are analysed in terms of, on the one hand, resources internal to the companies, and on the other hand access to external resources. This analysis is carried out not least in terms of user–producer relationships. The degree to which 'external cognitive actors' can be enrolled in the firms' processes is found to be crucial for the success of innovation processes. The experiences from Uruguay are compared to and contrasted with innovation agendas and decision-making processes in biotechnology research centres in Cuba. The Cuban case highlights the vital role of policies in the creation of learning paths for research and entrepreneurship.

The next chapter delves deeper into the development of the biotechnology sector in Cuba. Cuba presents an interesting contrast to how the biotech sector has developed in most other countries. In Cuba, the state has assumed the sole responsibility for the building up of research competence as well as for production structures. The absence of a private sector, coupled with the trade barriers caused by the US embargo, has presented unique challenges to the Cuban innovation system. Chapter 4 examines these challenges and discusses how the policy-making process has evolved to deal with them in the area of biotech, a strategically important sector for meeting the twin fundamental objectives: fulfilling social needs, particularly through development of the health care system, and generating revenues from exports. Moreover, it shows the role of higher education in the development of the biotech industry and, through three success stories, indicates some of the progress made by the biotech industry and the S&T policy. Two of the case studies are drawn from the health care sector while the third deals with plant biotechnology in the agricultural sector. The progress Cuba has achieved over the last 15 years in building a sectoral innovation system in biotech has been considerable. However, the case studies show that the innovative capability has not been equally well developed throughout the sector. To find a remedy for the existing asymmetries in the priorities and consistencies of the policies governing the sector remains a major challenge for the S&T policy process.

As in most developing countries, modern biotechnology is fairly recent in Vietnam. Officially starting in 1994 with a government resolution, the development of the biotech sector has only recently started to take off. In two case studies of Hue University and Can Tho University respectively, Chapter 7 examines the effects of the implemented policies on the enhancement of research capabilities and finds that universities indeed play an important role in providing human resources for biotechnology development, but they also lack critical research competence for a sustainable development of the sector. The reason for this is partly found in the earlier, Soviet-style, division of labour where all research was carried out in research institutes and the universities were restricted mainly to training and educational activities. In the transition towards a more market-based innovation system, the universities are struggling to build up their research capacity, but the government resources allocated for biotech research are out of reach for most of them. It is only some of the bigger universities situated in the large metropolitan areas of Hanoi and Ho Chi Minh City that can lay claim to at least embryonic biotech research competence. Perhaps more worrying is that there is a general lack of infrastructure for commercializing biotech research results emanating from university research as well as from the R&D institute sector. One of the immediate challenges for the government is, thus, to tackle the underdeveloped infrastructure for supporting the effective appropriation of innovations in biotechnology, by putting in place a policy more conducive to biotechnology development.

China has also had to deal with the transition from a research and innovation system dominated by specialized government-funded R&D institutes towards a more decentralized and market-oriented system. As the presentation in Chapter 8 bears out, China appears to have advanced considerably further than Vietnam in this restructuring process. China started implementing governmental programmes on biotech in the late 1970s and has experienced a rapid growth in virtually all aspects of biotechnology. Today China accounts for a substantial share of the world's public research expenditure on agricultural biotechnology and has a rapidly growing biotech industry with total revenues from biotech products in China having increased 50 times during the last 10 years. However, imbalances remain in the distribution of research capacity. Although the university sector employs fewer researchers than the business sector (which includes both private and state-owned enterprises), the quality and level of research at the universities is in general higher than the enterprise sector is capable of producing.

The mismatch between the technology demanded by enterprises and the R&D efforts by universities indicates that the absorptive capacity in the biotech industry is not sufficiently developed to take advantage of the research results produced by university-based R&D. A low competence on the part of the receiver of new technology obviously has serious consequences for the linkages and interplay between universities and industry. The lack of absorptive capacity in the business sector is an urgent issue for the policy-makers to address in the development of the Chinese biotech sector.

Contrary to the case of the rapid development in China, modern biotechnology in Mozambique is in an embryonic state (Chapter 6). Industrial applications of biotech are mostly confined to breweries and dairies, applying traditional production techniques by utilizing various fungi or bacteria. The few examples of modern, science-based, biotechnology to be found are confined to the capital, Maputo, and to a small group of universities and public research centres. Moreover, these centres do not operate under a coherent strategic model with well-defined areas of responsibility, but function more under an ad hoc regime with uncoordinated relationships between them. The human resources are scarce; less than 100 researchers in Mozambique are engaged in biotech R&D, and about half of them hold a PhD degree. Thus, the present lack of manpower as well as the organizational model restrict the development of the biotech sector in Mozambique. Yet the potential application of biotech is recognized by the government as an important means to transform indigenous knowledge into new, localized, products and to support the socio-economic development of the country. However, most of the research carried out in Mozambique is commissioned research, aimed at immediate and marketable solutions to pressing needs. Such a narrow emphasis on short-term goals, coupled as it is with a heavy donor dependency, could prove unfavourable for the long-term capacity-building of research staff, as well as for the sustainability of the research environments.

The efforts of Tanzania to upgrade its performance in biotech, especially research and applications in the agricultural sector, is delineated in Chapter 5. All endeavours in the field should be appreciated in the context of the general goals of poverty reduction and sustainable economic growth. The focus is on the formation of partnerships to promote the efficiency and effectiveness of research and technological innovation. Among issues to confront are the creation of infrastructure for universities and research institutions, both for the immediate purpose of providing education and training, but also in order to counteract a significant and actual risk of brain drain. Only three out of ten universities have any operations in biotech, and at a rather low level at that; two have formal training programmes. Broadened involvement at more levels of academia and government is found to be an important policy goal, in order to exploit the full potential of human and natural resources. Wider participation can also be appreciated as a means of gaining control over the national agenda setting, currently too much in the hands of external actors. Ideally, the national resource allocation would suffice to improve conditions for research and education; however, that is a distant prospect. Instead, establishing partnerships is a crucial instrument for linking the Tanzanian development of biotech to the international community of researchers and to industry. A clearer articulation of purposes, with national and local needs in the foreground, is consequently a national priority, in order to bring human and natural resources to the best use.

A range of policy instruments are being formulated within the European Union, aiming at improving the overall performance of the EU, and also at exploiting and taking advantage of the niche competencies of the member states. In the broader international setting (e.g. within the OECD as well as through various UN agencies) countries are also comparing and aligning their policies. In Chapter 9 the science and biotech policies of five European countries (four EU member states: Denmark, Germany, Latvia and Sweden; one non-member: Russia) are given a broad, comparative and primarily statistical introduction. The data and material utilized for this overview is, in the case of the EU member states, produced for the very purpose of furthering Union-wide policy efforts. Historical trajectories and path dependencies are delineated, providing an introduction to the cases of the respective countries covered in Chapters 10–13.

In Chapter 10 the conditions for biotech appropriation in Denmark and Sweden are outlined, and are found to be in some crucial aspects rather similar in the two countries. Among general conditions we find the relative size of the economies, their dependence on export opportunities, and the ongoing transformation of them into advanced knowledge economies. Both countries find themselves, by most indicators, in the top league among OECD countries when it comes to performance in biotech. However, recent trends give cause for some concern; not least the fact that Sweden seems to be losing ground in terms of research impact. Sector-specific conditions for the biotech industry in both countries is characterized by a dominance by a relatively small number of large corporations; a general cross-sector phenomenon in Sweden, less so in Denmark. The dominant role of the large corporations contributes to path dependencies; a major ambition of policies of late has been to open up companies and academia to widened cooperation. Publicly funded programmes demanding matching funds from universities and corporations are used to influence actors regarding deeper collaboration. Local and national initiatives to form networks, clusters and 'triple helices' aim to strengthen regions. The cross-national 'Medicon Valley' initiative in the Øresund region, running since 1997, is a case in point, described as one of the most dense biotech regions in the world. However, the lack of consonance between the Danish and the Swedish side, for example in terms of policies, regulations and in terms of industry focus, has meant that substantial results from the cross-national linkages are still in the future.

Since Latvia's independence in 1991, the economy has faced considerable challenges. In Chapter 11 the focus is on four aspects of the development of the biotech sector in Latvia in the face of those challenges: the transition from a socialist to a capitalist economic system; interaction and power relations within the research system; outcomes and assessment of current policies; and a delineation of key factors facilitating or hindering the development of the biotech sector. Today's research organization traces its roots to the 1950s. Following independence, research organizations and industries faced drastic restructuring and downsizing; several indicators show that Latvia is performing below the EU average. However, conditions have improved to some degree in the last decade. Pharmaceutics is the biotech sector currently performing best. Among priority areas for biotech in Latvia we find general capacity building; the establishment of a niche for Latvian biotech specializations in relation to other nations; strengthening of links between academic research and industry, especially the absorptive capacity of the latter; and increased cooperation with international partners. On a political level, institutions for innovation, research and coordination are in a highly formative state, where regulative vs. laissez-faire policies are in the balance. Additionally, the Latvian economy is highly vulnerable to the developments of the international economy, apparent not least in the aftermath of the international financial crisis of 2008/2009.

In Chapter 12, the situation for German biotech is presented. The emphasis here is on the policies and performance of public research organizations (universities and research institutes) within the national innovation system. A range of quantitative measures for innovation activities and appropriation are utilized: availability of skilled human resources; productivity in the form of publications; specialization areas; and patent applications. Patenting statistics show that universities have an above average significance in knowledge intensive fields, such as biotech, compared to the average for all fields. The role of universities as partners for pharmaceutical firms is found to be crucial. However, the relative weight of education and research in the mix of university activities needs additional attention, not least as regards interdisciplinarity.

National innovation policies for the biotech field in Germany are focused on health biotechnology and generic biotechnology. In the chapter these policies are analysed in four dimensions: the development of the knowledge base; knowledge transmission and application; the market; industrial development. Policy gaps are identifed in, for example, the areas of appropriation and exploitation; education; and fiscal policies. Two shortcomings are found to be especially problematic: within the areas of human resources development, and the creation of a regulatory framework, respectively.

In Chapter 13 Russia's efforts to upgrade its biotech industry are analysed in the face of domestic and international economical and industry policy trends. Building upon quantitative data and an extensive interview survey among researchers and policy-makers, conclusions are presented on the performance of about 30 biotech sub-fields, as well as policy challenges to be confronted. Seen across fields, the picture, presented at a considerable level of detail, is mixed. Whereas Russian research is competitive in some areas, a general lack of resources is a problem, causing a lag, even a significant lag, in most of the research areas under consideration. As regards appropriation and commercialization there is a range of both practical and policy issues to handle in order to facilitate an increased interaction among publicly funded research organizations and industry. The handling of intellectual property is a case in point, patent applications being at a 'disastrously low' level. The main concern for the future is found to be legislation and the regulative framework, in its current form seriously impeding commercialization and efficient application of biotechnology in industrial production. There appears to be a general consensus among the respondents to the survey that the innovation climate of Russia needs considerable legal and policy support for the country to turn competitive.

In the following chapters we will present the case studies in more detail and discuss the driving forces for and consequences of appropriation of biotechnology in the 12 countries. The book is divided into four sections: we will first present evidence from Latin America, followed by case studies from Africa, Asia and Europe respectively. In the final chapter, implications for public policy and industry development will be discussed.

## NOTE

1. Where not elsewhere stated, in this book biotechnology is defined according to the 2002 OECD definition as 'the application of science and technology to living organisms, as well as parts, products and models thereof, to alter living or non-living materials for the production of knowledge, goods and services.' (OECD, 2009).

## REFERENCES

Arocena, R., I. Bortagaray and J. Sutz (2008), *Reforma Universitaria y Desarrollo*, Montevideo: Tradinco Publishers.

Bio (2008), *Guide to Biotechnology 2008*, Washington, DC: Biotechnology Industry Organization.

Ernst & Young (2010), *Beyond Borders: Global Biotechnology Report 2010*, Ernst & Young.

Göransson, B. and C. Brundenius (eds) (2011), *Developing Universities: The Changing Role of Academic Institutions in Development*, New York: Springer.

Göransson, B., R. Maharajh and U. Schmoch (eds) (2009), *Science and Public Policy*, **36**(2), March, special issue on the Third Mission of Universities.

Harvey, Mark and Andrew McMeekin (2007), *Public or Private Economies of Knowledge?*, Cheltenham, UK and Northampton, MA, USA: Edward Elgar.

Juma, Calestous and I. Serageldin (2007), *Freedom to Innovate. Biotechnology in Africa's Development*, Addis Abeba and Pretoria: AU and NEPAD.

OECD (2009), *Biotechnology Statistics 2009*, Paris: OECD.

# PART I

# Latin America

# 2. The recent evolution of the biotech local innovation system of Minas Gerais: university, local firms and transnational corporations

**José Eduardo Cassiolato, Graziela Ferrero Zucoloto, Márcia Siqueira Rapini and Sara Gonçalves Antunes de Souza**

## INTRODUCTION

Biotechnology refers to a broad and complex set of enabling technologies which have in common the use of biological cells or molecules applied in the production of products and services related to life sciences, biology, chemistry, medicine, agriculture and computation sciences (bioinformatics), among others. Biotechnology products have been increasingly used in vital areas such as health, agriculture and food production, environment, new materials and new sources of energy.

Brazil has been nurturing the development of biotechnology for more than 20 years. In fact, fostering biotechnology has been one of the central tenets of the Brazilian science and technology policies during this period. One of the main positive results of this policy is a significant increase in scientific and technological capabilities. Building on previous specialization in agricultural research and health and biological sciences, novel biotechnological research has flourished in Brazil. However, these capabilities have not been translated into a more complex productive structure, at least not to a degree comparable to scientific development. This is the more striking since a series of policy mechanisms have been put into practice since the mid-1980s to stimulate spin-off firms from university research, to support venture capital initiatives and to encourage joint university–industry research projects.

This chapter aims at discussing these problems using a case study to illustrate the Brazilian case. Biotechnology in Brazil has mostly been developed in areas related to the agro-industrial sector and to the health sector. More recently, taking advantage of the increasing specialization of the Brazilian productive structure in energy-related areas – bio-fuels, oil, and so on – biotechnology research has started to target energy-related activities. However, these efforts, albeit important, are still in their early stages, making an overall assessment difficult.

The case study chosen is the biotech local system in Minas Gerais. In that region, a successful local effort in building up biotechnology knowledge with the support of a strong local scientific base – particularly in molecular and cellular biology, animal genetics and biochemistry – involves four federal universities and three research institutes. Underpinning this development is also a favourable institutional environment that has emerged in the last 30 years. This effort has been supported by incentives from federal tax-breaks, financing of building facilities by local authorities, and entrepreneurial mobilization in creating seed money and venture capital.

Perhaps the most significant condition characterizing the evolution of the Minas Gerais local biotech system is the role played by Biobrás, a biotech firm specialized in the production of insulin. Arguably this firm has been the most important Brazilian biotech firm, its first and the only biotech firm that grew to a significant size. It developed as a spin-off from the Medical School of the Federal University of Minas Gerais in 1971 and began its industrial operations in 1976. The founder was a young biochemical PhD, who, since the beginnings of his academic career, had carried out research in the area of enzymatic protein that was potentially applicable to industrial use. Most of the biotech firms in Minas Gerais that are focused on human-health biotechnology have benefited in some way from Biobrás technology and market success.

Biobrás was the key enterprise of the local system. From the very start it was active in institutional building of the local system, particularly through the setting up of the Biominas Foundation and the Biotech Enterprise Incubator. Also, it attained a very high performance level, which had a positive effect on the local competitive environment, for the emergence of new biotech ventures.

In 2001, after several attempts, Biobrás was acquired by Novo Nordisk of Denmark, the world's largest producer of insulin. It is the contention of the authors of this study that this acquisition marked a significant shift in the evolution of the local innovation system. For example some studies (Judice and Soares, 2004; Judice and Vasconcelos, 2006), showed that the number of local biotech firms decreased from 2001 to 2004, as some firms disappeared. More important than the decrease in the number of firms, the quality of production also changed, with an impact on the organization and operation of the local innovation system.

This chapter addresses the evolution of this system, the role played by Biobrás and the linkages between this and other local firms to the university research; what the impact of the acquisition of Biobrás by Novo Nordisk is on the dynamics of the local innovation system and how other local biotech firms have survived after this acquisition.

The chapter is organized as follows. The second section will briefly present the evolution of the Brazilian scientific, technological and productive capabilities in the past 20 years. The third section will provide a historical account of the evolution of the world pharmaceutical industry and how biotechnology developments are changing some of its characteristics. The fourth section will present the evolution of the health biotech system of Minas Gerais since its inception in the early 2000s when Biobrás was acquired by Novo. The fifth section analyses the recent performance of the Biotech firms in Minas Gerais after Novo's acquisition of Biobrás. This analysis will benefit from new information gathered from two sets of novel empirical work. The first refers to a series of organized interviews in 19 firms in 2004. For the purpose of this study, 12 of these firms were again interviewed in December 2007 and January 2008. The methodology for this survey is the same as the one that RedeSist[1] has successfully used in Brazil since 1997 for the analysis of more than 80 local innovation systems in different parts of Brazil and in areas of manufacturing, agriculture and services.

## THE EVOLUTION OF BIOTECHNOLOGY IN BRAZIL, SCIENTIFIC CAPABILITIES AND FIRMS

### Scientific Efforts in Biotechnology

Table 2.1 presents the number of research groups in universities and other S&T organizations that were performing biotechnology research in Brazil during the 2002–2004 period.[2] The information refers to 18 selected biotechnology areas, according to a classification used by the Brazilian National S&T Development Council. The same table also presents some indicators of output of these research groups in terms of number of publications and patents per group. According to the information provided in Table 2.1 there are a large number of research groups doing S&T activities in biotechnology in Brazil during the period, and the number of research groups is growing rapidly. There were 1342 research groups in biotechnology in 2002 and 2013 in 2004. In plant genetics alone, 1158 groups were

*Table 2.1 Biotechnology research groups in Brazil, 2002–04: number of groups per selected biotechnology areas and biotechnology patents(*)*

| Research areas | 2002 | | | | | 2004 | | | | |
|---|---|---|---|---|---|---|---|---|---|---|
| | **Brazil** | South-east | SP | MG | RJ | **Brazil** | South-east | SP | MG | RJ |
| | **Total no** | % of total groups Brazil | | | | **Total no** | % of total groups Brazil | | | |
| Biotechnology (in general) | **1342** | 48.29 | 22.88 | 13.56 | 11.62 | **2013** | 48.09 | 23.35 | 13.02 | 11.48 |
| Genomics | **619** | 58.97 | 37.48 | 11.79 | 9.69 | **1026** | 56.34 | 34.60 | 10.62 | 10.53 |
| Proteomics | **49** | 67.35 | 44.90 | 8.16 | 14.29 | **188** | 63.83 | 27.13 | 14.89 | 21.81 |
| Pharmaceuticals | **614** | 54.07 | 34.53 | 9.93 | 9.45 | **975** | 54.77 | 32.10 | 8.21 | 14.15 |
| Pharmacogenomics | **2** | 100.00 | 50.00 | 50.00 | 0.00 | **26** | 50.00 | 7.69 | 23.08 | 19.23 |
| Pharmacogenetics | **19** | 63.16 | 26.32 | 26.32 | 10.53 | **45** | 60.00 | 26.67 | 22.22 | 11.11 |
| Biopharmaceuticals | **40** | 62.50 | 17.50 | 40.00 | 5.00 | **50** | 68.00 | 18.00 | 36.00 | 14.00 |
| Vaccines | **438** | 63.01 | 31.96 | 13.24 | 17.58 | **643** | 62.21 | 32.81 | 11.35 | 18.04 |
| Molecular diagnostics | **181** | 66.85 | 41.44 | 12.71 | 12.71 | **290** | 64.83 | 39.66 | 11.38 | 13.45 |
| Plant genetics | **1158** | 45.51 | 19.00 | 17.01 | 9.07 | **1693** | 44.12 | 20.32 | 15.30 | 7.97 |
| Biodiversity | **592** | 41.89 | 18.92 | 11.99 | 10.64 | **923** | 40.74 | 18.53 | 12.24 | 9.53 |
| Bioremediation | **154** | 52.60 | 23.38 | 11.04 | 18.18 | **295** | 56.61 | 24.75 | 13.22 | 18.31 |

| | | | | | | | | | | |
|---|---|---|---|---|---|---|---|---|---|---|
| Animal genetics | **880** | 48.18 | 23.98 | 15.80 | 7.95 | **1326** | 49.10 | 24.81 | 14.71 | 9.20 |
| Immunology | **735** | 61.36 | 28.98 | 14.42 | 17.82 | **1050** | 59.14 | 27.62 | 15.33 | 15.90 |
| Biomaterials | **263** | 69.58 | 44.87 | 15.21 | 8.75 | **454** | 65.86 | 43.83 | 13.44 | 8.15 |
| Toxinology | **37** | 81.08 | 62.16 | 5.41 | 13.51 | **62** | 67.74 | 48.39 | 11.29 | 8.06 |
| Recombinant vaccines | **178** | 65.17 | 29.21 | 20.79 | 15.17 | **268** | 65.67 | 31.34 | 22.01 | 12.31 |
| Stem cells | **116** | 70.69 | 50.00 | 5.17 | 15.52 | **339** | 66.67 | 46.61 | 7.08 | 12.68 |
| Biodiesel | **49** | 40.82 | 20.41 | 0.00 | 20.41 | **221** | 46.61 | 18.55 | 5.88 | 22.17 |
| International publication biotechnology | **1216** | 49.51 | 24.01 | 13.49 | 11.84 | **1857** | 49.17 | 24.02 | 13.30 | 11.58 |
| Patents and product registration biotechnology | **195** | 64.62 | 29.74 | 17.95 | 16.92 | **426** | 58.45 | 28.17 | 18.31 | 11.97 |

*Notes:*
(*) Number of patents and/or product registration per research group
SP = São Paulo
MG = Minas Gerais
RJ = Rio de Janeiro

*Source:* CNPq Research Groups Database, 2007.

active in 2002 and 1693 in 2004. Animal genetics, immunology, biodiversity, genomics, pharmaceuticals, vaccines and biomaterials are other important areas for Brazilian biotechnology. Emerging and fast-growing areas in 2004 were genomics, proteomics and stem cell research. These groups published 1216 papers in internationally recognized publications in 2002 and 1857 in 2004. Patent applications reached 195 in 2002 and 426 in 2004. The University of Campinas, in São Paulo, is responsible for the largest number of biotech patents in Brazil (Milan, 2006).

Biotechnology research is highly concentrated in the richest south-east region, particularly in the states of São Paulo, Minas Gerais and Rio de Janeiro. According to the data base of the National S&T Council, the south-east region is responsible for no less than 40 per cent of research groups in biotechnology in 2002 and 2004. It accounts for more than 60 per cent of research groups in 11 of the biotechnology areas in 2002[3] and in 10 areas in 2004.[4] The 2004 data show a small de-concentration. Although not shown in Table 2.1, this de-concentration, in fact, refers to the growing importance of the state of Rio Grande do Sul and of the federal district, Brasília, both of them hosts to important federal universities. The less concentrated biotechnology research areas are in agriculture biotechnology (plant and animal genetics), a fact that is related to the spatial organization of EMBRAPA, the Brazilian Enterprise for Agricultural Research, a government-owned research institution that has facilities in different parts of the country and which conducts research associated with the six different biomes that characterize the Brazilian ecology.

The state of São Paulo – Brazil's richest – has the main biotechnology research agglomeration. In 2002, in 11 research areas,[5] São Paulo accounted for 30 to 60 per cent of the total number of research groups in Brazil, and in 2004 São Paulo had from 30 to 50 per cent of biotech research groups in eight areas.[6]

The states of Rio de Janeiro and Minas Gerais share second place after São Paulo, ranging from 8 to 10 per cent of the biotechnology research groups in Brazil in the selected areas. The state of Minas Gerais has its main strengths in biopharmaceuticals and pharmacogenomics (which were both emergent research areas in 2002 to 2004, with very few research groups). The state of Rio de Janeiro, in its turn, is strong in biodiesel research and proteomics (which were also emergent research areas in 2002 to 2004).

The number of indexed publications in health biotechnology in Brazil jumped from 96 in 1998 to 179 in 2001 (Ferrer et al., 2004). Universities in Brazil publish 80 per cent of all Brazilian health biotechnology papers in international peer-reviewed journals. According to Ferrer et al. (2004) the largest contributor is the University of São Paulo, publishing almost

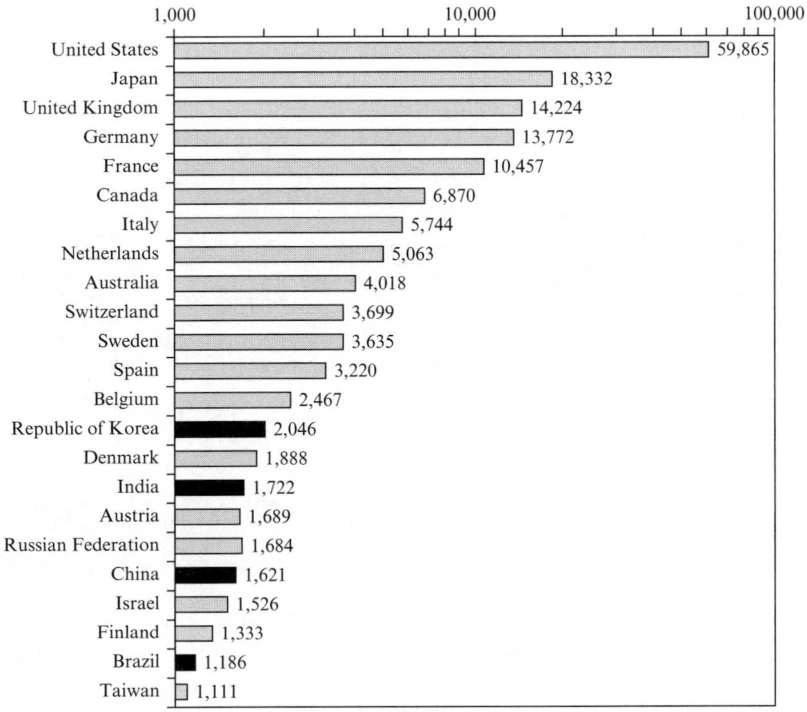


*Source:* Thorsteinsdóttir et al. (2004)

*Figure 2.1 Total number of published papers in health biotechnology, 1991–2002 (top 23 countries)*

26 per cent of this type of article. Both the Federal University of Rio de Janeiro and the Federal University of Minas Gerais are also publishing substantially in this field, with around 11 per cent of the papers each. The universities are heavily involved in domestic collaboration, which has increased extensively since 1991. The Ludwig Institute of Cancer Research collaborates extensively with the Federal University of Minas Gerais, and both the Federal University of Rio de Janeiro and the Federal University of Minas Gerais collaborate considerably with FIOCRUZ (Ferrer et al., 2004).

In terms of international comparison, Figure 2.1 presents the top 23 countries that publish papers in health biotechnology in the world during the 1991–2002 period. Compared to the leading countries in the world, the contribution of Brazil (1186 papers) and other developing countries is modest. Figure 2.2 shows that some of the developing

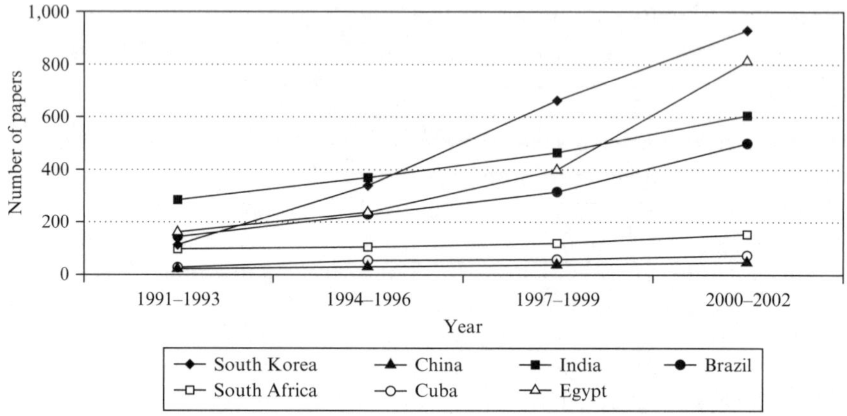


*Source:* Thorsteinsdóttir et al. (2004)

*Figure 2.2 Number of published papers by selected developing countries, 1991–2002*

countries are approaching and/or overtaking smaller developed countries in their health biotechnology publications. But if the total number of publications is still relatively small, one should note that, together with other developing countries, Brazil is improving its performance significantly. Figure 2.2 shows the evolution of health biotechnology publications in China, India, Brazil, South Africa, Cuba, Egypt and South Korea, from 1991 to 2002. In the early 1990s, the total number of health biotechnology papers was very small in all the countries and only India had an early start in supporting research in the field, producing more than 200 papers for the three-year period 1991–93. Since then, the total number of health biotechnology papers published in the world was stable during the period of 1991 to 2002, increasing from about 10000 in 1991 to about 12000 in 2002. The proportion of health biotechnology papers of these countries dropped slightly during this period, from 1.8 per cent in 1991 to 1.6 per cent of total papers in 2002. Around the mid-1990s, publications in health biotechnology started to increase significantly in almost all of these countries, particularly South Korea, India, China and Brazil.

Scientific, technological and productive activities of biotechnology in Brazil are heavily concentrated in the south-east, the richest part of Brazil where also the best universities and research institutions are located. As might be expected, the evolution of biotechnology research and productive activities in the three most important states – São Paulo (SP), Minas

Gerais (MG) and Rio de Janeiro (RJ) – is deeply influenced by the institutional development in these states, and institutional arrangements are crucial to explaining their differences.

In São Paulo, a state with one of the oldest traditions of research in the area (particularly in biology), two recent programmes have been instrumental in organizing research capabilities and connecting with global research: the genomic research programme of São Paulo, FAPESP – Genome Project,[7] and the Brazilian Genome Program, implemented by the Ministry of Science and Technology (MCT) through its National Research Council (CNPq). FAPESP's Genome project initiated the Brazilian genomic research, in 1997, on *Xylella fastidiosa*,[8] a functional genome project that got a cover article in *Nature* (Simpson et al., 2004). This experience involved 65 labs in sequencing learning and practice, with a total of 450 trained researchers (GERAQUE, 2003). The Xylella genomic project has set up the genomic research model in Brazil. It was organized through a large research network involving public and private institutions, the Sequencing and Nucleotide Analysis Organization (ONSA). Besides academic institutions, the genome project included several private firms in sectors such as biofuels, paper and pulp. Several international research groups also participated in the project.

In Rio de Janeiro the evolution of health biotechnology has largely been associated with the history of the Oswaldo Cruz Foundation (Fiocruz). Fiocruz is a federal institute, founded in 1900 in Rio de Janeiro with the objective of controlling bubonic plague, yellow fever and smallpox. It was constituted outside the university, inspired by the Pasteur Institute. However, in contrast to Pasteur, Fiocruz was connected to the public health system. As Gadelha (2002, p. 74) points out, Fiocruz, as the Pasteur Institute, is in transition to change 'from little science to big science'. This institution, affiliated to Brazil's Ministry of Health, has become a strong producer of new knowledge in the health biotechnology field in Brazil, and its scientists published more than 16 per cent of the Brazilian papers in the international peer-reviewed literature from 1991 to 2002 (Ferrer et al., 2004). Additionally, Fiocruz provides training on a variety of topics related to the biomedical field. The institute also runs two R&D units and manufacturing plants, Bio-Manguinhos, which has vaccine and diagnostics production facilities (it is the largest Brazilian manufacturer of vaccines), and Far-Manguinhos, which produces pharmaceuticals.

The third case is the one that will be analysed in more detail in this chapter, the health biotech system of Belo Horizonte, the capital of Minas Gerais State. It has been considered the most advanced innovation biotechnology system in Latin America (Biominas, 2001).

## Biotechnology Firms in Brazil

Although Brazil has been implementing policies for biotechnology since the mid-1980s, the first systematic attempt to identify the Brazilian private biotechnology sector was made in 2001. Using a broad definition of biotechnology – it included all life science firms – this study identified the existence of 304 biotechnology companies in Brazil (Biominas, 2001). Eighty-one per cent of these firms were located in the three largest states: 129 (42 per cent) in São Paulo, 89 (29 per cent) in Minas Gerais and 28 (9 per cent) in Rio de Janeiro. Approximately 70 per cent of these firms were locally owned, 25 per cent were multinational subsidiaries and 5 per cent were state-owned firms. Of their combined products, 26 per cent were directed at the health care market and the rest at agricultural, environmental and other industrial fields.

A more recent survey of Brazilian biotech firms (including all bioscience firms) identified 182 firms in 2007. Seventy-nine per cent (143) of these firms were based in the three largest states – São Paulo, Rio de Janeiro and Minas Gerais. There were only two firms in the Amazon and 10 in the Brazilian north-east, the least developed states. Sector-wise, more than 50 per cent of these firms were in health sectors (36.9 per cent in human health and 18.3 per cent in animal health). Agriculture, with 22.5 per cent of firms, ranked second (Figure 2.3). Worth noting is the fact that though the numbers are low, firms are starting to be set up in areas that are specifically important to the Brazilian national innovation system, such as bioenergy (4.2 per cent of total firms) and environment (14.1 per cent of firms).

Seventy-one of these firms were engaged in what could be termed as 'modern biotechnology'. Most of these firms are relatively new: 75 per cent have been in operation for less than ten years and 25 per cent were set up since 2005. Seventy-five per cent are micro and small firms with a maximum turnover of USD500 000 per year. They are also heavily concentrated in São Paulo (42.3 per cent of total), Minas Gerais (29.6 per cent) and Rio de Janeiro (8.5 per cent of total). A significant number of modern biotechnology firms (32 per cent) are part of incubators and science parks. Only 11 firms deposited a total of 17 patents. This means that 85 per cent of these biotechnology firms do not have any patents, 10 per cent have only one patent and only 5 per cent have two or three patents.

From these studies one may infer:

- Although there are several government programmes attempting to foster biotech firms, they have not been very successful: the total number of biotech firms in Brazil has decreased in the 2000s; it could be argued that the number of surviving firms is not necessarily

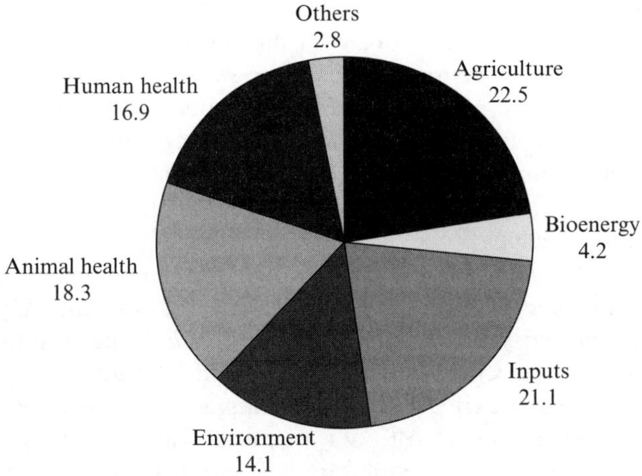


*Source:* Fundação Biuominas (2007)

*Figure 2.3 Brazil: distribution of biotechnology firms by sector of activity, 2007*

evidence of policy failure, as the focus of a good policy should not be the increase of the number of firms in a particular sector. However, the Brazilian policy has historically been targeting the setting up of new firms, and in this sense one may argue that its focus has proved both not to be working and to be wrong. It could also be argued that firms that disappear were acquired by larger firms and positively inserted in the production chain. As will be shown below, the scattered evidence suggests otherwise, as some of the small biotech firms in Belo Horizonte were acquired to perform routine, non-technology-intensive activities.

- These firms have a low propensity to patent: Brazilian biotech firms patent much less than comparable ones in other developing countries, such as India and China.
- Firms obviously find it difficult to grow; some authors (Fonseca, 2009; Judice and Vedovello, 2008) suggest that, in the process of growth, small biotech firms in Brazil tend to become more specialized in niche markets, which intrinsically limits their growth perspective. Unless they have a captive market – such as public health – firms define their specialization pattern from their own scientific technological base (Fonseca, 2009).

As these firms succeed, their intangible assets become valuable to the market. This is, at present, when large transnational corporations (TNCs), particularly in pharmaceuticals and in the seeds sector, turn their attention to them and propose strategic alliances, contractual partnerships and eventually totally control their capital. The strategy of these TNCs is very simple: acquire knowledge assets.

**Brazilian Biotechnology Policy**

The Brazilian government started to design and implement policies in the area in the early 1980s when CNPq (the Brazilian National S&T Council) set up a specific research programme – *Programa Nacional de Biotecnologia* (PRONAB/CNPq) – to stimulate biotech research. Some years later, after a Ministry of S&T was established in Brazil in 1984, a special Secretariat for Biotechnology was created with the objective of defining and implementing a national policy for biotechnology.

One of the main actions of the Secretariat was the setting up in 1987 of a special programme for human resources in biotechnology – Programa RHAE-MCT – which lasted until 1997. One of the novelties of the RHAE programme was that firms could get financial resources to attract human resources through research projects in partnership with universities.

Throughout the 1990s, with Brazil in a deep economic crisis, public support for biotechnology (and for S&T as a whole) declined. However, in 2000 a new programme for biotetechnology, genetic resources and genoma – PBRG-G – was launched. Under this programme more than USD120 million for a three-year period was allocated to the development of biotechnology products and processes relevant to industrial production. This new programme was part of major changes in the Brazilian S&T policy that concentrated on innovations.

Under PBRG-G there was an intention to create competence in connected areas such as bioelectronics, conventional genetics and physiology, with the explicit aim of attracting new foreign direct investment. The main actions of PBRG-G targeted the characterization and conservation of genetic resources (including the creation of a germplasm bank and culture collections), support to conservation and to the sustainable use of biodiversity and biosecurity. The setting up of regional research centres in biotechnology was also stimulated.

Arguably the most important action of PBRG-G was the setting up of a Genoma research network, initially consisting of 26 research labs, to sequence *Chromobacterium violaceum* and to implement projects to support collections of microbean cultures and human cells.

But it was the setting up of sector funds for different research areas that constituted the most significant change in the Brazilian S&T policy of the last 10 years, in the sense that they guarantee resources to an area that traditionally was of marginal importance for the Brazilian government. These funds receive financial resources from several sources that are used to stimulate university–industry partnerships.

The biotechnology sector fund was set up basically to strengthen the old PBRG-G. But its implementation lacked focus and one could witness a significant dispersion of limited resources into a vast number of projects, mostly R&D projects. In 2007, for example, more than 25 projects were approved, the majority with resources amounting to less than USD50 000.

Recent policy mechanisms have been introduced by the Ministry of Science and Technology and its agencies, which have a significant impact on biotech innovation. One of them is the *Subvenção Econômica* – literally economic subvention – which is a non-reimbursable grant given to firms (via public tenders) to perform innovation-related activities, basically joint R&D projects with universities and other R&D institutions. Between 2006 and 2008, biotechnology projects amounting to approximately BRL139 million (around USD60 million) were selected and supported. Also worth mentioning is the Program Inova Brasil (Brazil Innovates), created in 2008 with the aim of financing innovation in Brazilian firms, with negative real interest rates for firms in strategic areas of the Brazilian industrial policy (among which biotechnology) and the Prime Program,[9] which targets spin-offs from universities and research institutes in partnership with a network of Brazilian incubators. As these programmes were launched recently, it has not yet been possible to evaluate their impact.

Finally, it is worth mentioning the institutional support for promoting risk capital in small high-tech firms. This is specifically being done by Finep – the financial arm of the Ministry of Science and Technology – with the Inovar Project. This is a project that has been operating for more than 10 years, but with paltry results.

## THE PHARMACEUTICAL INDUSTRY AND BIOTECHNOLOGY

Perhaps the most significant fact that characterizes the evolution of the world's pharmaceutical industry is the remarkable historical register, across two centuries, of a highly innovative industry (Achilladelis and Antonakis, 2001). In fact, throughout this period, industry has played an important role for the evolution of scientific knowledge in the area. At the same time the industrial structure of the pharmaceutical industry remained practically unchanged. Such firms have been able not only to cope with the successive technological changes that affected the industry, but even, given their political and oligopolistic powers, to shape these transformations.

This is a highly profitable industry. Based on information provided by Fortune 500, Love (2001) reveals that in the 1970s the profitability of the pharmaceutical industry was twice the average of all firms in the ranking. In 1980 this proportion increased to three times, and in the 1990s it was four times the average of Fortune 500 firms. This trend was observed not only in the US. The Indian pharmaceutical industry, for example, has shown higher profitability than all other sectors of the Indian economy from 1970 to 2000.

Together with innovation, marketing activities are the main competitive weapons of the industry. In fact, although there is a much publicized idea that firms in the sector spend a lot on innovation, Angell (2004) reveals that R&D is a relatively small part of the budget of pharmaceutical firms, compared with administrative and marketing expenditures. It is estimated that marketing expenditures represent twice as much as R&D expenditures of large pharmaceutical firms. According to *The Economist* (2005) 33 per cent of the sales of Novartis were destined to the promotion of drugs and only 19 per cent to R&D.

In fact there is significant controversy in the classification used by the industry to register R&D expenditures, making it hard to determine precisely the amount spent on R&D. What several studies have demonstrated, however, is that the state increasingly pays for most of the total R&D costs in the development of new drugs. According to Bond and Glynn (1995), in 1993 the US federal government was responsible for approximately 39 per cent of a total of USD30 billion of R&D expenditures in the biomedical sector. State and municipal government were responsible for 7 per cent, non-profit organizations for 4 per cent and the productive sector was responsible for the remaining 50 per cent. According to the same authors, in 1940 US firms spent USD25 million, non-profit organizations USD17 million, and the federal government only USD3 million (7 per cent of total).

In fact, possibly the most significant characteristic of the industry is its structural relation with the state. Several characteristics of the health sector are responsible for the fact that this segment is the one where the governmental intervention has been, historically, the most intense. In the area of health services as a whole, and in the production and consumption of drugs in particular, government intervention continues to characterize the dynamic of the industry from all points of view: with regards to the regulation of the activities, to the stimulus to S&T development, and to the demand for drugs, via procurement.

McKelvey and Orsenigo (2001) propose that the history of the pharmaceutical industry as an evolutionary process of adaptation to major technological and institutional 'shocks', can be divided into three major periods. The first period, between 1850 and the end of the Second World War (1945) is the time when drugs were closely related to chemicals. According to the authors, this period was one in which hardly any new drug development occurred, with minimal research based on relatively primitive methods being conducted. The pharmaceutical industry was not tightly linked to formal science nor was it characterized by extensive in-house R&D for new drugs.

A second period, a golden age for the pharmaceutical industry, followed. This period, which started with the end of the Second World War and finished roughly in the early 1980s, was characterized by very high R&D spending, leading to a steady flow of new drugs. This was a time when large MNCs in pharmaceutics enjoyed very high rates of growth in earnings and return-on-equity. Although this period started after the war it profited from the war effort in the sense that the US and British governments organized a massive research and production effort that focused on commercial production techniques and chemical structure analysis. Due to the technical experience and organizational capabilities accumulated through the intense wartime effort to develop penicillin, as well as to the recognition that drug development could be highly profitable, pharmaceutical companies embarked on a period of massive investment in R&D.

At the same time there was a very significant shift in the institutional structure surrounding the industry. First, whereas before the war public support for health-related research had been quite modest, it boomed to unprecedented levels after the war. Thus, science push and science connections began in earnest. Second, the development of the Welfare State – especially of national healthcare systems – provided a rich, 'organized' and regulated market for drugs, even if the features obviously varied drastically across countries.

Although the intensity of R&D was high, the approach to research – the so-called 'random screening' – was not so advanced, as natural and chemically derived compounds are randomly screened in test tube experiments and laboratory animals for potential therapeutic activity. However, the net result was positive, as several important classes of drug were discovered and several hundred new chemical entities (NCEs) were introduced in the 1950s and 1960s.

Nevertheless, the search process itself was rather inefficient, and so the successful introduction of NCEs has to be considered to be quite a rare event. Estimates suggest that, out of all new compounds that were tested, only one in 5000 reached the market.

In short, innovative new drugs arrived quite rarely, but after the introduction they experienced extremely high rates of market growth. A few 'blockbusters' dominated the product range of all major firms (Matraves, 1999, p. 180; Sutton, 1998) and they were dependent on these singularly successful products, which also experienced rapid growth.

The success of this way of organization of the innovation process led to a favouring of certain types of innovation, which were reinforced by mechanisms of appropriability of the potential profits deriving from innovation. In fact, the pharmaceutical sector is one of the few industries where patents provide solid protection against imitation (Klevorick et al., 1987).

Some associated fundamental changes also characterized this second period: increasing stringency of the procedures for product approval in most countries, development of health care systems, and finally, one that would fundamentally affect the transformation to the third period, the relation between public-funded basic research performed by universities and their linkages with industry.

This third period, which started around the early 1980s, refers to the impact of the scientific revolution on the dynamics of the industry, as substantial advances in publicly funded research in areas such as physiology, pharmacology, enzymology and cell biology led to enormous progress in the ability to understand the mechanism of action of some existing drugs as well as the biochemical and molecular roots of many diseases. This had a profound impact on the process of the discovery of new drugs within pharmaceutical firms, as it introduced a significantly more effective way to screen compounds. As pointed out by McKelvey and Orsenigo (2001), the availability of drugs whose mechanisms of action were well known made possible significant advances in the medical understanding of a number of key diseases. Such 'guided search' combined medical understanding with an understanding of disease and drug action, and enabled the firms to concentrate on areas likely to give further returns.

However, the shift in the technology of drug research from random screening to one of guided discovery or 'drug discovery by design' was critically dependent on the ability to take advantage of publicly generated knowledge (Gambardella, 1995; Cockburn and Henderson, 1996).

What is crucial for the purposes of this study is that (i) this transition towards new techniques of drug discovery was in mid-course when molecular genetics and rDNA technologies opened an entirely new frontier for pharmaceutical innovation; and (ii) the advent of modern biotechnology has had a significant impact on both the organizational competencies required to be a successful player in the pharmaceutical industry, and on the industry structure in general.

The most noticeable manifestation of the transformations occurring in the pharmaceutical innovation system has been the appearance of a new breed of actors, that is new specialized biotechnology firms (NBFs).

While biotechnology-related products became integrated with pharmaceuticals, the large majority of these new companies never managed to become a fully integrated drug producer. The growth of NBFs as pharmaceutical companies was constrained, as these companies often lacked competencies in other areas of the innovative process: in particular, knowledge and experience of clinical testing and other procedures related to product approval on the one hand and marketing on the other. Some, like Genentech, worked to hire a broader range of people with appropriate skills, while others remained more specialized in their activities. Thus, many of these NBFs have exploited their basic competence and acted primarily as research companies and specialized suppliers of high technology intermediate products, performing contract research for and in collaboration with established pharmaceutical corporations.

Many NBFs are founded with the aim of becoming large integrated pharmaceutical producers. But, given the organizational and financial constraints, they often become specialized suppliers of specific techniques and research projects. In the process of experiencing this transformation, the principles on which projects are selected and developed, the financial strategy and so on, obviously change as well. Large corporations internalize some basic principles of academic research into the organization of their laboratories and in the incentive structure of their researchers.

As mentioned, considering that in the majority of cases it is the government that pays for the drugs, in several industry segments the price mechanisms established by sellers and buyers become irrelevant. In Brazil, this is the case, for example, for AIDS drugs that have to be freely supplied to individuals who are affected by the disease. With the state acting as a monopsony, the role of procurement is central as a stimulus to investment, production and innovation in the sector.

In recent decades the industry has undergone dramatic structural changes, with the rise of the biotechnology sector and substitution away from other therapeutic modalities such as surgery, and increased competition from globally active generic manufacturers. Some authors suggest that the industry is facing a crisis that threatens the established model. Factors such as a decrease in R&D productivity, associated with a growing number of patent expirations of blockbusters, and a growing pressure from consumers for reduction of health expenditures are forcing the industry to make adjustments to the established patterns.

These pressures for cost reductions have led to some degree of disintegration and geographic dispersion. Innovative activity is nonetheless highly geographically concentrated, reflecting the economic significance of factors such as localized knowledge spillovers and the strength of patent protection, as well as the influence of government policies such as price regulation, state procurement of drugs, and health and safety regulation. Rising R&D expenditures in the face of health care cost containment pressurize and apparently slow down research productivity, giving pharmaceutical companies a powerful incentive to seek out cost savings and new models for production and innovation.

Besides an emphasis on concentrating the production on a reduced number of more profitable drugs, and a diversification towards activities of lower risk (as for instance generic drugs), the new strategy has privileged two correlated processes:

- a significant increase in technological and commercial partnerships with small biotechnological firms;
- a growing relocation of some technological effort to different countries, particularly China, and, on a smaller scale, India.

Large pharmaceutical firms are directing most of their externalized R&D to small biotechnology firms, which accounted for approximately 20 per cent of their total R&D budget in 2004. In that year more than 500 new agreements between these large pharmaceutical firms and small biotech firms were signed, a much higher number than the 230 signed in 1999 and less than 70 in 1993.

Although representing a general characteristic of the industry, this type of strategy is more accentuated in medium-sized global pharmaceuticals. Without the power to compete directly with the largest firms, these companies concentrate in therapeutic niches. Such a strategy is conditioned by a specialization in specific therapeutic classes. In this case the coordination of global networks is of paramount importance. The main medium-sized groups defend their competitive position in specialized chains: for example Novo Nordisk in diabetes and biotechnology, Schering in gynaecology and medical imagery, Baxter in blood derivatives and so on.

The other particularly important trend was the setting up of R&D facilities in developing countries, particularly China. For instance, in the last few years global pharmaceutical firms set up research centres in Shanghai, Beijing and other Chinese cities. China is attractive both for its buoyant markets, and for its excellent S&T infrastructure, with research institutes producing prominent contributions to pharmaceutical research. However, although some evidence of cost-driven geographic redistribution of R&D into new low-cost locations is found (Table 2.2), this process has thus far been limited (Cockburn, 2008).

In fact, some authors (Cockburn, 2008) suggest there are limits to the types of activities that are likely to be relocated outside the industry's traditional locations. Cockburn points out that

*Table 2.2 Estimated R&D expenditures in pharmaceutical industry by country, 1990–2004*

| | 1990 | 1995 | 2000 | 2004 |
|---|---|---|---|---|
| Total BERD at PPP (current million $), of which: | 16853 | 24587 | 33781 | 46216 |
| USA | 37.3% | 41.5% | 38.3% | 36.5% |
| EU15 | 39.8% | 36.3% | 40.4% | 39.0% |
| UK | 12.1% | 11.8% | 13.3% | 11.1% |
| France | 6.4% | 8.5% | 7.8% | 7.6% |
| Germany | 8.1% | 5.0% | 6.7% | 7.5% |
| Italy | 5.5% | 2.5% | 1.9% | 1.5% |
| Sweden | 2.1% | 2.7% | 3.7% | 3.6% |
| Japan | 16.2% | 14.9% | 14.3% | 14.8% |
| Other developed countries[a] | 6.7% | 6.3% | 5.8% | 8.0% |
| 'New Europe'[b] | – | 0.8% | 0.9% | 1.2% |
| Other emerging economies[c] | – | 0.1% | 0.4% | 0.6% |

*Notes:*
a. Australia, Canada, Iceland, Korea, Norway, Singapore and Switzerland.
b. Czech Republic, Hungary, Poland and Slovenia.
c. Taiwan, Mexico and Turkey.

*Source:* Cockburn (2008)

> substantial offshoring of R&D activities is most likely to occur where the research activity is relatively routinized, uses large amounts of relatively low-skilled labor, and does not need to be tightly integrated or co-located with other R&D activities [. . .] large-scale, late-stage clinical trials with 'low-tech' endpoints (such as measuring blood pressure) are examples of this kind of activity, and indeed the global allocation of research effort in these areas shows signs of a significant response to cost differences across countries (Cockburn, 2008, p. 287).

Clinical tests, for example, are important R&D activities in the pharmaceutical industry given their relevance to the approval process of new drugs. Such activities are relatively easy and cheap to transfer to other parts of the world (Table 2.3). As they are highly labour-intensive, it is possible to reduce R&D costs by transferring them to regions where labour is cheaper than it is in countries where MNCs' headquarters are located. In China, for example, clinical tests are around 30 per cent of the cost in the US.

*Table 2.3 Regional share of worldwide clinical trial sites (per cent)*

| Year | 2002 | 2003 | 2004 | 2005 | 2006 |
|---|---|---|---|---|---|
| North America | 58.2 | 54.1 | 59.8 | 54.3 | 49.5 |
| Western Europe | 30.6 | 26.7 | 24.1 | 27.4 | 27.6 |
| Oceania | 3.3 | 4.0 | 3.2 | 4.5 | 4.6 |
| Latin America | 1.7 | 3.5 | 3.7 | 3.3 | 4.3 |
| Eastern Europe | 3.8 | 7.4 | 4.9 | 5.9 | 8.1 |
| Asia | 1.1 | 3.1 | 2.5 | 3.2 | 4.0 |
| Middle East | 0.3 | 0.3 | 0.7 | 0.7 | 0.6 |
| Africa | 1.0 | 0.8 | 0.9 | 1.0 | 1.0 |
| Developed countries | 92.4 | 85.0 | 87.4 | 86.4 | 82.0 |
| 'Emerging' countries | 7.1 | 14.2 | 11.8 | 12.6 | 17.1 |
| Others | 0.5 | 0.9 | 0.9 | 1.0 | 0.9 |

*Source:* Cockburn (2008)

Using several proxy indicators such as patent applications, academic publications, or databases that document clinical trial sites, Cockburn (2008) found no evidence of globalization of R&D in terms of location of drug discovery activity. In fact the author claims that

> less routinized, more science-intensive [activities] are much less likely to relocate to low-cost locations. Decisions about where to locate science-intensive drug discovery appear to be much less sensitive to labor costs and may be driven primarily by factors such as proximity to leading-edge academic research and . . . [other] externalities [mostly found in developed countries] (Cockburn, 2008, pp. 227–8).

Table 2.4 presents evidence on R&D spending abroad by US-based pharmaceutical MNCs, members of PhRMA, the Pharmaceutical Research and Manufacturers of America.[10] In 2005, these firms spent around $9 billion outside the United States (21.5 per cent of their total R&D spending). Most of these resources were located in Western Europe (73.4 per cent) and Japan (11.5 per cent). All other countries were recipients of very low amounts of R&D expenditure, and even India and China accounted for a negligible share.

In short, there is an international division of labour in the way the industry is organized globally. The production of drugs involves four basic stages: R&D of new drugs; industrial manufacturing of pharmaceutical inputs; formulation and final processing of drugs; and distribution by retailers and health service providers. Activities of higher technological

*Table 2.4 R&D expenditures abroad by US pharmaceutical MNCs (PhRMA members), 2005*

| Area | R&D expenditures (USD million) | Share (%) |
|---|---|---|
| Africa | 28.0 | 0.3 |
| Canada | 479.3 | 5.4 |
| Latin America and Caribbean | 174.9 | 2.0 |
| Asia-Pacific (except Japan) | 117.5 | 1.3 |
| India and Pakistan | 10.9 | 0.1 |
| Japan | 1025.4 | 11.5 |
| Australia and New Zealand | 144.6 | 1.6 |
| Europe | 6524.7 | 73.4 |
| Central and Eastern European nations, including Russia | 244.6 | 2.8 |
| Middle East | 37.7 | 0.4 |
| Uncategorized | 101.3 | 1.1 |
| Total | 8888.9 | 100.0 |

*Source:* Cockburn (2008)

intensity tend to be concentrated in developed countries, while subsidiaries of large pharmaceutical MNCs concentrate on the formulation/production of goods in some less relevant technological activities – such as clinical tests – and the search for localized knowledge such as those related to biodiversity (Gadelha, 2003).

## THE MINAS GERAIS HEALTH BIOTECHNOLOGY LOCAL SYSTEM: THE FIRST 20 YEARS

Several studies have pointed out that, in different parts of the world, biotechnology production tend to occur as spin-offs of high-quality academic research, and start-up firms locate close to universities that contain such capabilities. The same process has happened in Minas Gerais. At the end of the 1990s, in the metropolitan area of Belo Horizonte, Minas Gerais capital, and in smaller towns of Minas Gerais, there were more than 35 biotech firms. These firms specialized in the areas of health and agriculture, each one with different types of industrial organization and patterns of competition. Health biotech firms were small and medium-sized, locally-owned firms, while most agro-biotech firms belonged to multinational chemical-pharmaceutical firms, such as Monsanto and Novartis.

These two specialization areas comprise different institutional arrangements, one of them being led by new biotech firms set up by university scientists and technologists, and the other by global leaders in the agro-biotech markets, linked to strong public research centres of EMBRAPA, the Brazilian national agricultural research enterprise. The health biotech institutional arrangement has evolved in a quite different pattern from the demand-induced nature of the local agro-biotechnology activities. Scientific knowledge supplied by biology and biochemistry departments of local universities played a major role in the emergence of local health biotech firms.

In 1999 the health biotech institutional arrangement comprised 30 firms, divided into 17 small and medium-sized established firms (Table 2.5) and 14 incubated enterprises (Table 2.6). Most of them are located in the Belo Horizonte Metropolitan Area and affiliated to the Biominas Foundation, which has taken an active role in promoting local biotechnological development. Specifically, it promoted the creation of the Biotechnology Enterprise Incubator of Minas Gerais – IEB-MG.

The majority of these firms focused on human-health biotechnology, which has benefited to some degree from Biobrás market success. The range of products included insulin, enzymes, diagnostic kits, heart bioprotheses, chemotherapy products, heart valves, heart surgery equipment and health-applied software. Other specialization areas were animal health and agri-biotechnology, fine chemicals and environmental biology.

The local scientific and lab base in molecular and cellular biology, animal genetics and biochemistry involves four federal universities and three institutes, which have been crucial to the development of this biotech agglomeration. The Federal University of Minas Gerais (UFMG) has been the main source of enterprise spin-offs, pioneered by Biobrás in the 1970s from the Biochemistry Department (see below). The Federal University of Juiz de Fora (UFJF) has also contributed to the biochemical capability with Quiral, a specialized chemotherapy drug firm, being its main spin-off (Biominas, 2001).

The Federal University of Viçosa (UFV) also contributed significantly to biotechnology innovations in Minas Gerais, especially in plant and animal genetics. Its Biotechnology Department, for example, developed the first pure lineages of pigs in Brazil in the early 1990s (Lemos, 1992). This department's Centre for Applied Agriculture (Biagro) has 27 laboratories, 350 scientists and technical personnel and USD5 million applied to specific projects (Silva, 2004, p. 15). Two smaller universities, the Federal University of Lavras (UFLA) and the Federal University of Ouro Preto (UFOP) are also involved in biotechnological research, the former in agronomy and the latter in minerals.

*Table 2.5 Established enterprises in health biotechnology, Belo Horizonte, 1999*

| Enterprise | Specialization area | Product line |
|---|---|---|
| Biobrás S.A. | Human health – biochemistry | Insulins |
| Cepe Ltda | Laboratorial services | Bromatologic, pharmaceutical and environmental analysis |
| Conap Ltda | Human health – biochemistry | Apitoxin and derivatives |
| Diamed Latino America | Human health – immune haematology | Products for diagnosis in immune haematology |
| DMG Ltda | Medical equipment | Cardiac surgery equipment |
| Gene Ltda | Human health – genetic engineer | Molecular diagnosis *S. mansoni* genoma |
| In Vitro | Human health | Diagnostics kits |
| Labcor Ltda | Human health | Cardiac bioprosthesis and arterial graft |
| Labtest S.A. | Human and animal health – biochemistry | Reagents and kits for diagnosis |
| Lema Ltda | Animal health | Vaccines, drugs and sera |
| Quiral Ltda | Human health – antiblastic chemotherapy | Anti-cancer products |
| Redelab | Human-health services | Diagnosis and analysis |
| RPI Ltda | Human health – information technology | Software |
| Tri Technologies | Human health – bioprosthesis | Cardiac valves |
| Vallée S.A. | Animal health – genetic engineer and immunology | Vaccines and anti-parasites |
| Zammi Ltda | Human health – medical equipment | Cardiac surgery equipment |

*Source:* Biominas (2001)

Local research institutes also have laboratories and scientific biotechnological capabilities. The Centro de Pesquisas Rene Rachou, a local affiliate of the national health research institute (FIOCRUZ), has a noted molecular biology laboratory. Two state government institutes have more specific capabilities, the Ezequiel Dias Foundation (FUNED) focused on vaccine research and industrial production and the Technological Centre (CETEC) on bio-metallurgy. Finally, in agro-biotechnology the National Research Centre for Maize and Sorghum (CNPMS), an affiliate of EMBRAPA in the town of Sete Lagoas, has recently created a specialized department in applied molecular biology (Silva, 2004, pp. 15–16).

*Table 2.6 Incubated enterprise of Biominas Foundation, 1999*

| Enterprise | Specialization area | Product line | Spin-off from |
|---|---|---|---|
| Biobrás Software | Applied information technology | Bio-industry software | Biobrás |
| Biocarbo Ltda | Fine Chemistry | Tar by-products | Acesita |
| Bioconsult Ltda | Consulting services on biology | Aquaculture | – |
| Biológica Ltda | Environmental microbiology for industrial use | Environmental control projects | UFMG |
| Bion Consultoria | Immunology | Polyconal antibodies | UFMG |
| Biorganica | Human health – genetic engineer | Anti-cancer from natural sources | Quiral |
| Cepa Biotecnologia Ltda | Human health – microbiology and genetics | Diagnostic kits | UFMG/Cuba partnership |
| Empack Ltda | Services | Packing and quality control | Katal |
| Greentech Ltda | Agri-biotechnology | Cloning of ornamental plants | UFMG |
| Hereditas Ltda | Human health | Molecular diagnosis of infectious diseases | UFMG |
| Katal Ltda | Human health | Kits and reagents for diagnosis | UFMG |
| Laboratorios JHS Ltda | Human and animal health | Bioceramics for bone rebuilding | UFMG |
| Phoneutria Biotecnologia | Human health | Bio-pharmaceuticals and restriction enzymes | UFMG |

*Source:* Biominas (2001)

The history of the Minas Gerais bioindustry is intertwined with the history of Biobrás. In fact, two factors are usually associated among relevant conditions that have favoured the geographic concentration of Brazilian bioindustry in the state of Minas Gerais: first, this state historically has developed competencies in medicine and has a tradition of pharmaceutical teaching and training (Judice and Baêta, 2005). Second, entrepreneurial developments have given rise to the first biotechnology company in Brazil, Biobrás. As Mytelka (1999, p. 37) observed:

> Biobrás, a large Brazilian-owned pharmaceutical firm has been a Third World firm pioneering North–South partnerships. Biobrás has substantial experience in research, development and production of enzymes, insulin crystals, human insulin and diagnostic kits. As a result of its strong research capabilities, it was able to form a joint venture with Eli Lilly for the production of human insulin by biotechnological methods. As the Brazilian biotechnology industry grows, networking among Brazilian firms is beginning to develop.

Biobrás was set up in the early 1970s as a spin-off from the biochemistry department of the Federal Univerisity of Minas Gerais. The founder was a scientist and entrepreneur who had returned from his PhD studies in the USA in the 1970s and received a Rockfeller Foundation grant to furnish and equip a laboratory of enzymes at the faculty of Medicine at UFMG. He started training students in a biochemistry master of sciences course which particularly attracted chemical engineers. They started research in fermentative processes and enzymes (pancreatine, bromeline and curd).

Biobrás emerged as a project with three partners. Financing was sought at FINEP (the Brazilian Innovation Agency) which approved the technical project for a pilot scaling-up process. However, the founding partners were unable to provide collateral that would cover FINEP's investment losses in case of failure; thus the project was not approved.

In their knowledge search, the partners then went to the New England Enzyme Center in Tufts University in the USA. There they started their learning process, a journey to an increasing international and national networking process to source knowledge and technology wherever it was possible and available. Two master's students were sent out to train at Tufts, later to return to Brazil, but the process had not resulted in the ideal scaling up. At this time, the partners went to the Engineering School of UFMG and through their professional contacts they found a private partner, an engineering company (Hidroservice). Biobrás and Hidroservice worked out together and partially succeeded in a precarious scaling-up process. Biobrás, by then, was a typical biotechnology start-up, except it was off-campus. It was run by the three partners and by seven master's students, and operated from a rented house in the central area of Belo Horizonte (the capital of Minas Gerais).

At this point their next step was to search for funding at SUDENE in order to build up an industrial plant.[11] This organization offered special funding conditions for the north-east area of Brazil, and the northern area of the state of Minas Gerais was eligible for these special conditions. Thus, Biobrás moved to the northern part of Minas Gerais State, and the industrial plant was built in Montes Claros.

The following steps were managerial initiatives to market prospecting. The consulting group Arthur Little was contracted to analyse the enzyme market, and a headhunter company was asked to contract a North American expert to be the industrial director of the Biobrás plant. Such an expert was found – a specialist in pharmaceutical enzymes (bromeline) – and contracted as director. As the industrial plant in Montes Claros went into operation, Biobrás moved towards the second learning curve which is considered much harder than the first one, the market domain and the commercializing of 'import substitutes'. Biobrás was, at this time, an enzymes producer, and its main clients were found in the national and international pharmaceutical industry.

The technological evolution of Biobrás involved upgrading its capability from a pharmaceutical enzyme producer to a high-grade purifier of protein, especially insulin formulates. This evolution is characterized by bottlenecks in access to market niches, a major constraint for science-based enterprises in a peripheral environment, even if they are benefiting from protective barriers. To a large extent it seems that this market-access challenge has been a decisive factor in bringing about Biobrás' technological evolution. Protection of domestic markets by import-substitution policy provided two important leverages for the firm's growth. On the demand side, protection from competitive pressure allowed time for learning and the building up of technological capability on new product lines. On the supply side, protective barriers to international competitors opened up opportunities for technological joint ventures and, accordingly, learning from imitation.

The first challenge was to transform enzyme production from laboratory level to pilot and later industrial scales. Since the company already had the technical capability, the lack of seed money became the main barrier. Subsidized money from the Federal Agency for Development of the Northeast Region (SUDENE) supplied the funding needed to accomplish this challenge. This resulted in the location of the industrial plant in the middle-sized city of Montes Claros, situated in a poor and relatively isolated region of Minas Gerais. Learning to operate on an industrial scale was acquired both by training two shareholder engineers in the USA and hiring a foreign specialist as industrial director. Pancreatine,[12] which up to that point had been an imported pharmaceutical input, was the first industrial enzyme produced. As soon as industrial production started, the enterprise immediately benefited from the tariff barrier by the protectionist national-similar law, favouring comparable locally produced products. Despite this apparent monopolist position, the product was an input of the value chain of Hoechst, which exerted its monopsonistic power to squeeze supply out of the new entrant.

The second attempt to produce a marketable product involved coagulum enzyme for the dairy industry, using a similar production technique.

The rival importer was Hala, a Danish firm with 60 years of experience in the market, and with a countrywide distribution network in Brazil. Once more the national-similar law imposed import tariffs from 20 to 80 per cent and virtually guaranteed a captive domestic market. However, the unexpected reaction by Hala was to open a plant in Brazil and to become a national manufacturer of a similar product, a move that put Biobrás out of the market.[13]

This second market failure led the enterprise to seek outlets for its excess production in the export market, which gave it its first experience of export trading, particularly to the USA. This first foray into the American market provided experience used in the technological upgrading of Biobrás' product line in 1979, from enzymes to hormones, that is, insulin crystals. This upgrading was facilitated by technical synergies in pancreas use, although difficulties were encountered both in reaching a satisfactory technical level of insulin purification and in finding markets abroad that would allow the plant to operate at minimum efficient scale, equivalent to three times internal demand. In this case the adoption of a protected-market strategy was considered to be useless, and the strategy employed was to form a joint venture with Eli Lilly, one of the world's two leading insulin producers.[14]

The joint venture with Eli Lilly provided Biobrás with access not only to insulin-purifying techniques but also to the international market. That is, this association made possible an effective learning by technology transfer of the world best practice, since transfer management and quality control were retained by the R&D department of Biobrás.[15] At the same time it paved the way for a distribution channel in the USA, considered to be a decisive competitive step in the pharmaceutical industry.[16] Between 1979 and 1983 Biobrás was able to monopolize the domestic insulin market. It received substantial benefits from import-substitution policies: a long-term low-interest loan by the National Development Bank (BNDES), which provided 70 per cent of the funds needed for the new plant; the national-similarity approval by the Brazilian regulatory agency (CACEX) that prevented imports; and a guaranteed demand for 80 per cent of its internal sales by the federal government, via the National Drug Agency (CEME).

Vertical integration to the final product, insulin formulates, was made possible in 1983 thanks to several favourable factors, which enabled Biobrás to terminate the joint venture with Eli Lilly. On the external front, Novo Nordisk was able to increase its market share in the American market from 4 to 20 per cent by a joint venture with Aquibb. This resulted in excess capacity at Eli Lilly's internal plants and the need to assume part of the production of insulin crystals at that time imported from the Brazilian associate. On the internal front, protectionist policy favoured a production verticalization strategy to take advantage of knowledge acquired from Eli Lilly on insulin purification. This upgrade from production of biochemical inputs to final formulates led to further improvements in the firm's technological capability, allowing it to shift its core competence towards high-grade protein purification. The new core business on final pharmaceutical products resulted in changes in Biobrás' competitive strategies. Sales efforts to build up trademark recognition and capture consumer loyalty constituted important improvements to the firm's distribution network.

Biobrás had an annual growth rate of more than 10 per cent until the early years of the 2000–2010 decade when Novo Nordisk of Denmark eventually acquired it. This performance reflects an upgrade in its product line to higher value-added products. Biobrás was an internal market-driven enterprise, whose exports constituted around 12 per cent of sales in its final years. Although organized as a capital-equity undertaking, it was always a family enterprise controlled by the family holding BIOPART Ltda, which owns 77 per cent of its shares.

By the 1990s Biobrás was spending around 7 per cent of sales on R&D and had started knowledge networking and sourcing oriented towards technology development and production internalizing in human recombinant insulin. Animal pancreas supplies fell short in the market and new technological and strategic trajectories were needed in order to produce insulin of better quality and market competitiveness.

Biobrás had been granted a USPTO patent in 1998, the fourth recombinant insulin patent in the world. Table 2.7 shows the institutions, universities, research institutes and private companies with which Biobrás established formal and informal links, technological, managerial and strategic collaboration and partnerships, in order to develop human recombinant insulin.

Biobrás experienced a remarkable technological evolution since it was created. Table 2.8 shows the main technological breakthroughs in the general history of the evolution of insulin, and when Biobrás introduced these technologies. It is important to point out that as a result of its commitment to technology and innovation, time lags between Biobrás' achievements and the international research frontier substantially diminished over time. Although having shown specific technical capability, Biobrás did not have the size to compete internationally with large MNCs. Before reaching maturity, its cumulative technological capability came from heuristic market searches of product-line niches that might provide opportunities for its growth under a relatively stable competitive environment.

At the end of the 1990s, Biobrás was part of a select team of firms

*Table 2.7 Biobrás' collaborative and partnership relations and network for human recombinant insulin development, 1991–98*

| Universities | Private companies |
|---|---|
| **Knowledge, research and technology** | **Purification processes, chromatography, customized engineering:** |
| 1. UFMG (MG- Brazil) | Pharmacia/Amersham (Sweden) |
| 2. UNB (Brasília - DF) | |
| 3. USP Ribeirão Preto (SP-Brazil) | **Stakeholder in the USA, training and technology transfer:** |
| 4. Miami University (USA) | |
| 5. Southampton University (UK) | LIPHA (USA) |
| 6. Karolynskas Institute (Sweden) | |
| 7. Shemiakin Institute (Russia) | **Water treatment** |
| 8. Pennsylvania State University (USA) | French company |
| 9. University of Columbia (USA) | **Lab and industry supplies** |
| | Norwegian company |
| **Management learning** | **Production of pills** |
| | Irish company |
| 1. INSEAD (France) | |
| 2. Fundação Dom Cabral (MG -Brazil) | **Packaging** |
| | São Paulo company |

*Source:* Lemos (2000); Judice and Baêta (2005)

producing insulin through recombinant DNA: it was the world's fourth largest producer after Eli Lilly, NovoNordisk and Aventis. In 2000, it was responsible for 80 per cent of the Brazilian market (Sutz, 2005), exported to 12 countries and received its first international patent. As mentioned above, this technological evolution of Biobrás was accompanied by the setting up of several small biotech firms, as offsprings: Biofar, Bioferm, In Vitro, Biobrás Software, Dialab, Biomm Inc. (in Miami), Biomm SA (in Brazil). Most of these firms have benefited in some way from Biobrás' technology and market successes.

In 2001, after losing a large public tender for the supply of insulin to the Brazilian Health Minister, Biobrás went into a severe financial crisis and was acquired by Novo Nordisk. The results of this public tender were heavily contested. According to one newspaper:

> The (Health) ministry, as subsequently recognized by justice, disregarded the legislation that force the government to give equal tributary treatment to Brazilian and foreign firms and the (legislation) that determines how ICMS is

*Table 2.8 Technological progress of Biobrás in insulin production*

| World frontier | Year | Biobrás |
|---|---|---|
| Insulin discovery | 1921 | |
| Beginning of insulin commercialization | 1923 | |
| Introduction of PZI formulation | 1930s | |
| Introduction of NPH formulation | 1940s | |
| Introduction of Lent formulation | 1950s | |
| Introduction of Single-peak formulation (1) | 1972 | |
| | 1976 | Beginning of Biobrás industrial activities with enzyme production |
| | 1979 | Beginning of production of insulin crystals based on single-peak grade |
| Introduction of single-component insulin, highly-purified insulin and monocomponent insulin (2) | 1980 | |
| Introduction of human insulin based on biosynthetic and semi-synthetic products | 1984 | Beginning of production of highly-purified insulin |
| | 1988 | Beginning of production of human insulin |
| Introduction of human insulin by recombinant DNA | 1994 | |
| | 1998 | Beginning of production of human insulin by recombinant DNA |
| | 2001 | Brazilian patent of human insulin by recombinant DNA<br>Biomm as a spin-off |
| | 2003 | Government authorizes sale of Biobrás to Novo Nordisk |

*Source:* Lemos (2000); Judice and Baêta (2005)

calculated . . . for the multinational company to win, the commission in charge of the tender of the ministry did not follow the law, and altered in cents the final price offered by Novo Nordisk. (Cruvinel, 2004)

In fact dumping practices by Novo Nordisk had been discussed in Brazil since the late 1990s. In 2000, the Ministry of Development examined insulin imports between January 1998 and July 1999 and found evidence of dumping practices in insulin imported from Denmark, USA and France. The same organization also verified that the Brazilian industry was damaged by such practices and concluded that Biobrás lost at least 20 per cent of its participation in the sales to the private market to foreign

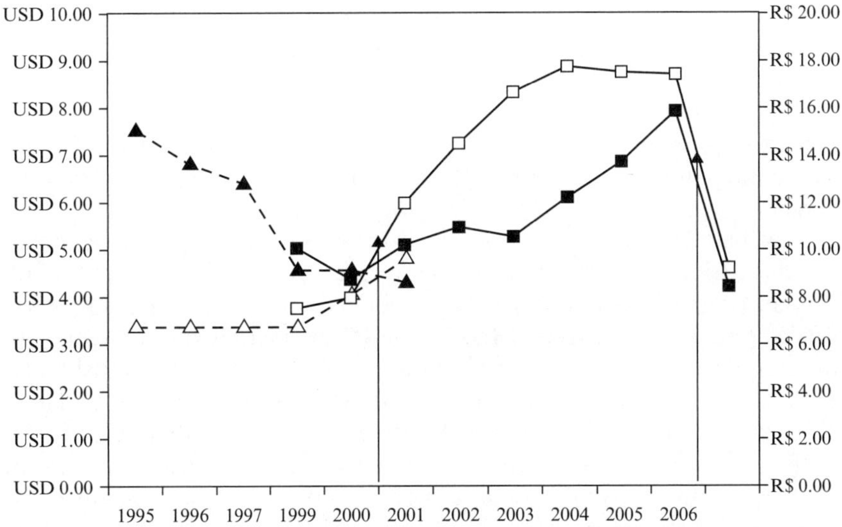


*Notes:*
2000–2001 – acquisition of Biobrás by Novo Industries;
2006 – announcement of agreement between Fiocruz and Indar to start internal production of insulin
▲ – Animal insulin USD; △ – Animal insulin R$; ■ – Human insulin USD; □ – Human insulin R$

*Source:* Costa (2008)

*Figure 2.4 Brazil: unit price paid by the Brazilian government for insulin (human and animal), 1995–2007*

competitors due to their behaviour. The two public tenders by the Ministry of Health in 2000 were won by Novo Nordisk, Denmark (Batista, 2002).

After a short legal battle at the beginning of 2001 between Biobrás and Novo Nordisk the major partners of Biobrás sold their shares (32 per cent of voting capital) to the Danish firm. One of the Brazilian partners stated to a journalist:

> we competed with high tech firms that were at least 50 times larger than us. If we did not have a minimum of support from the Brazilian government there were not too many options: we either had to sell our stakes or passively wait for the competition to break us (Cruvinel, 2004).

With a monopoly position, Novo Nordisk started to charge prices accordingly. Figure 2.4 shows unit prices paid by the Brazilian government from 1995 to 2007 (Costa, 2006). In the 1995–2001 period there was a substantial decrease in prices paid for animal insulin, which was the main product supplied by Biobrás: prices fell from around USD8 to USD4. After Biobrás was taken over by Novo Nordisk, the monopolist firm substantially increased its price from USD4 in 2001 to USD9 in 2005, only to fall drastically again in 2006 after the government announced it was going to stimulate the production of insulin in one of its own laboratories. The same pattern is shown in figures regarding the price paid for human insulin.

It is true that less than two years before being acquired by Novo Nordisk, Biobrás, jointly with the University of Brasília, patented a biosynthetic human insulin. After that Biobrás received public resources, via funding mechanisms, from FINEP, the Brazilian Innovation Agency of the Ministry of Science and Technology to produce this type of insulin (O Estado de São Paulo, 2001). Subsequently Biobrás was acquired by Novo Nordisk and, as several authors (Judice and Baêta) pointed out, at least in theory, part of the agreement between these two included a clause that allowed a spin-off from Biobrás (called Biomm) to keep the patent rights and its old technical team.

Silva (2004) also interviewed former managers of Biobrás on its acquisition by Novo Nordisk about the patent rights of the biosynthetic human insulin. Their response was that technological development would not be affected by the sale, precisely because, as mentioned above, patents were kept by Biomm.

However, although obtaining the patent was heavily publicized (it was announced as the fourth patent of this kind that was given in the whole world), it was never used by the new firm. Biomm never acted as a manufacturing firm; up to the moment this chapter was written they only worked as a consulting firm, that is, a firm that supposedly sells technical services. It never entered into production, even if public resources by Finep were spared for that. The only effect was that prices paid for insulin by the government escalated.

Then, even if the above-mentioned reaction of former managers of Biobrás could suggest an extension of its technological commitments, the fact is that the country lost accumulated expertise in producing insulin and developing technologies associated with the production, lost know-how and know-what. Other interviewees in Silva's (2004) work raised important, related issues. One local institutional officer pointed out that Biobrás became another case of a long list of high-tech firms acquired by foreign MNCs that do not keep local decision power and do not invest in real technological development. According to this officer, with its success Biobrás diffused the idea that by investing in R&D it is possible to succeed in the business world. With its acquisition, there was an inverse effect: to confirm the widespread idea that there is a limit for the local entrepreneur.

## THE LOCAL INNOVATION SYSTEM AFTER BIOBRÁS WAS ACQUIRED BY NOVO NORDISK

Sutz (2005) points out that knowledge accumulation is a difficult endeavour in Latin America as repeated processes of un-learning inhibit social learning capabilities. In Brazil several of these processes have happened as a consequence of government policies. The analysis of the impacts of Biobrás' acquisition by Novo Nordisk in the local biotech system of Minas Gerais is the subject of this section.

The biotech local system of Minas Gerais has been widely studied in the last 15 years. Here, the analysis of the characteristics of the local biotech system in Minas Gerais will be made using two blocks of information: first, information given by academic studies that analysed the system emphasizing the innovation processes; second, two surveys made by RedeSist researchers: one in 2004 and the other in December 2007 and January 2008.

Some of the studies that analysed the system before the acquisition of Biobrás by Novo Nordisk demonstrate a series of positive indications regarding its evolution. For example, a substantial degree of interaction between local firms and universities appears in the works of Souza (2001) and Fajnzylber (2002), which interviewed firms in 2000 and 2001. One of these positive results is the setting up of spin-offs: among the firms researched by Souza, 33 per cent were founded by lecturers and researchers of local universities who maintained their affiliation with them after they set up these firms. From the total of firms studied by Souza (2001), 67 per cent pointed out that they had some kind of connections with universities. Some said that they financially supported research done at universities (33 per cent), others asserted that they used university personnel in their own research (17 per cent), while others stated that they used equipment, labs and personnel from the university in a R&D partnership. The importance of such cooperation and exchange of information and personnel with universities and public labs was also confirmed with a higher intensity by the study of Fajnzylber (2002) where 81 per cent of firms pointed out that they used these forms of collaboration for their R&D.

The same pattern occurred with respect to inter-firm interactions. In fact both studies acknowledged the importance of these connections: in the study of Souza 40.6 per cent confirmed that they have interaction with other firms for R&D, while in the study of Fajnzylber 69 per cent of firms affirmed that they had this type of relationship. Although Biobrás was a single-product firm (insulin) it had a relatively large R&D lab that was shared by the smaller biotech firms in the local system (Lemos, 2000), and in this sense it acted as the focal point for sharing both tacit and codified knowledge.

Another impressive positive result of the 1990s refers to internal capabilities of firms: 72 per cent of firms interviewed by Fajnzylber and 83 per cent of those approached by Souza confirmed that they had an internal R&D infrastructure. This was an impressive result, knowing the difficulties small firms have in maintaining such internal activities.

The role of Biobrás was of paramount importance for this pattern of behaviour. Biobrás played a pivotal role in the local system. As the oldest and largest firm it had also the largest R&D lab. These studies confirm that the smaller biotech firm started using Biobrás' R&D facilities and gradually felt the need to have their own small facilities in order to better use the whole local infrastructure.

Obviously there were problems in the local system as well. With more than a decade of strong support by local, state and federal governments there was a relatively small number of firms that were able to survive. Although there is not any systematic attempt to identify the rate of survival of local small biotech firms, indirect evidence in these studies suggests that most firms that were set up in the 1980s did not subsist for a long period.

All the studies that were made after Biobrás was acquired by Novo Nordisk – including the survey specifically done for this study – showed significant changes in the system. The interviews of Silva also indicate that, in 2004, local biotech firms had a very low level of integration with the international market, which was also a feature of the period before Biobrás was acquired by Novo Nordisk. From those interviewed by Silva only one – a manufacturer of diagnostic kits – pointed out that it exported. However, only 20 per cent of its sales came from exports.

Most of these findings were confirmed by our own research. In RedeSist's research of 2004, 19 firms in human health were interviewed. At that time, organizations that supported firms, including Fundação Biominas (incubator) and SEBRAE (Brazilian Organization to Support SMEs) were also interviewed. Among these firms, 17 were controlled by local capital while the other two were controlled by both Brazilian and foreign capital. In 2007/2008, 11 of these firms were interviewed again. Nine were Brazilian owned, one foreign owned and one had mixed control.

A first set of questions put to these firms referred to the main characteristics. The questions asked: (1) whether firms introduced new products/processes in the three years before the survey and if these products/processes represented a novelty at the level of the firm, for the national market or for the global market; (2) which were the main sources of information used by these firms in their search; (3) the importance of innovation in the competitiveness of these firms.

In the 2004 survey, 47.4 per cent developed a product new to the firm but already in the market, 37 per cent developed a product that was new to the Brazilian market and 15.7 per cent a new global product. Regarding process innovations, approximately 58 per cent introduced a new process (31.6 per cent of interviewed firms introduced a process new to the firm and 26.3 per cent introduced a process new to the sector). These findings coincide with Silva's (2004) analysis. Based on qualitative information obtained through interviews, Silva pointed out that the vast majority of surviving biotech firms in Belo Horizonte were hardly innovative. They were in fact subcontracted by larger firms from other parts of the country to perform specific tasks that are more connected to production than to R&D. For example, one of the interviewed firms, which was labelled as a biotech firm, in fact performed quality control for larger textiles, chemicals and plastic firms in the states of São Paulo and Rio Grande do Sul. Another was a passive developer: it produced from specifications set up by its client, with the aim of stimulating the use of its raw material. Another was subcontracted by a larger firm in the same sector, but did not share information or anything else with this client: the contracting part determines the production methodology, the equipments to be used, and so on (Silva, 2004). In this case subcontracting is a contract to rent equipment, facilities and personnel, with no discernible technological work and relationship between the parts.

Regarding sources of information for innovation, Table 2.9 presents the importance attributed by interviewed firms to different sources of information for innovation in both the 2004 and the 2008 surveys. The 2004 survey stressed the importance of internal sources (R&D department, production, sales and marketing) for small enterprises, while R&D and production were more important for micro firms. In general such information suggests that as firms grow, market signals become more important. Regarding external sources, micro enterprises depend more on information from universities, research institutes and the Internet. Information from suppliers and clients, although relevant, were less important. Regarding small firms, the situation is inverted, with information from the Internet, participation in fairs and exhibitions, clients and suppliers, in that order, representing the most important sources. The importance of universities and research institutions declines as firms grow, which suggests a decoupling of biotech firms from academic sources of information.

A similar picture emerged from the 2008 survey. Internal sources were again considered as the most important source of information. Micro firms attribute the highest importance to production, while small firms

*Table 2.9 Importance attributed to sources of information for innovation by micro and small firms in the local biotech system of Belo Horizonte, 2008*

| Sources of information | Importance index* | | | |
|---|---|---|---|---|
| | 2000–02 | | 2005–07 | |
| | Micro | Small | Micro | Small |
| **Internal sources** | | | | |
| R&D department | 0.86 | 1.00 | 0.72 | 0.84 |
| Production | 0.85 | 1.00 | 0.80 | 0.72 |
| Sales and marketing | 0.76 | 1.00 | 0.72 | 0.92 |
| Others | 0.00 | 0.00 | 0.00 | 0.00 |
| **External sources** | | | | |
| Other firms of the group | 0.00 | 0.00 | 0.12 | 0.18 |
| Associated firms (joint venture) | 0.00 | 0.00 | 0.12 | 0.32 |
| Input suppliers | 0.67 | 0.80 | 0.76 | 0.52 |
| Clients | 0.75 | 0.80 | 0.75 | 1.00 |
| Competitors | 0.35 | 0.33 | 0.44 | 0.80 |
| Other firms in the sector | 0.57 | 0.65 | 0.15 | 0.36 |
| Consulting firms | 0.29 | 0.33 | 0.65 | 0.5 |
| **Universities and research institutes** | | | | |
| Universities | 0.82 | 0.63 | 0.64 | 0.56 |
| Research institutes | 0.82 | 0.48 | 0.70 | 0.64 |
| Centres for training human resources | 0.43 | 0.67 | 0.65 | 0.32 |
| Organizations for testing and certification | 0.59 | 0.65 | 0.52 | 0.70 |
| **Other sources** | | | | |
| Licences, patents and know-how | 0.58 | 0.50 | 0.65 | 0.26 |
| Conferences, seminars, etc. | 0.75 | 0.83 | 0.84 | 0.84 |
| Fairs and exhibitions | 0.51 | 0.90 | 0.50 | 0.84 |
| Leisure meetings | 0.30 | 0.45 | 0.30 | 0.30 |
| Local entreprenurial associations | 0.60 | 0.65 | 0.58 | 0.56 |
| Internet | 0.81 | 1.00 | 0.80 | 0.78 |

*Notes:* * Indices with value from 0 to 1, according to the following weighted average: (0* n. of 'no' answers) + (0.3* n. of 'low' answers) + (0.6* n. of 'average' answers)+ (1* n. of 'high' answers). The closer the index is to 1 the highest the importance attributed by firms to the particular source of information.

*Source:* RedeSist's surveys

emphasize sales and marketing, followed by internal R&D. External sources are much less important, with the exception of input suppliers and clients for micro firms and only clients and competitors for small firms. Universities and research institutes do not seem to be that important either, but it is worth noting that organizations for testing and quality certification have some importance for small firms. Finally, again, fairs and exhibitions, conferences and the Internet are mentioned by firms as important sources for getting new information.

However, what is more remarkable to observe from the data presented in Table 2.9 is that the importance attributed to the majority of sources of information for innovation has decreased from the time of the first survey – which captured the period 2000–02, when Biobrás was still present – to the second survey – which captured the 2005–07 period. In fact, all internal sources of information were downgraded, even if they still continued to be very important. For external sources that are linked to other firms the picture is more blurred. Other firms of the group and associated firms that were not important at all during the first survey, started to be mentioned by some firms, even if the scale was near to nothing. The importance of clients, competitors and consulting firms also increased for both micro and small firms, and achieved a higher relevance. However, other fims in the sector and input suppliers became less important. Regarding universities and teaching and research institutes, micro firms attributed much less importance in 2008 to universities and research institutes and slightly higher importance to centres for training human resources and organizations for testing and certification than had been indicated in 2004. For small firms there was a decreasing importance of universities and centres for training human resources, while research institutes and organizations for testing and certification acquired higher importance.

The discussion on sources of information for innovation should be performed with due regard to the available information regarding strategies and types of cooperation made by firms. When asked about cooperative activities, 74 per cent of firms affirmed having been engaged in some type of partnership. Although this general figure might suggest a positive outcome regarding partnerships, a more detailed view about types of cooperation indicates a low level of cooperation directed towards more technology-intensive activities (Table 2.9).

Although micro firms tend to use laboratories and equipment belonging to the incubator organization, the information presented in Table 2.10 suggests a low level of cooperation particularly for the development of products and processes and for human resources training and capacity building. Although somewhat higher for small firms, this kind of

*Table 2.10 Importance attributed to forms of cooperation by micro and small firms in the local biotech system of Belo Horizonte*

| Activities | Importance index* | |
|---|---|---|
| | Micro | Small |
| Acquisition of inputs and equipment | 0.21 | 0.50 |
| Joint sales of products | 0.13 | 0.25 |
| Development of products and processes | 0.25 | 0.58 |
| Design | 0.08 | 0.23 |
| Human resources capacity building | 0.25 | 0.50 |
| Financing | 0.17 | 0.58 |
| Participation in fairs and exhibitions | 0.23 | 0.48 |

*Notes:* * Indices with value from 0 to 1, according to the following weighted average: (0* n. of 'no' answers) + (0.3* n. of 'low' answers) + (0.6* n. of 'average' answers)+ (1* n. of 'high' answers). The closer the index is to 1 the highest the importance attributed by firms to the particular cooperation form.

*Source:* RedeSist's surveys

cooperation was also modest for them. In general, more important partnerships are those for getting funding and for the acquisition of inputs and equipment.

The qualitative information obtained in interviews explicitly demonstrates that the level of partnership relations between these firms and universities and research institutions has been particularly small. According to these interviews, the Federal University of Minas Gerais performs only some technical services, such as tests, for firms in its labs. Such weak relationships are more surprising since most entrepreneurs are graduates of the university.

These findings also confirm Silva's research of 2004. Her main research question was whether or not one could call the agglomeration of biotech firms in Belo Horizonte a system of innovation. Her conclusions also indicated that in 2004 the relationship both between firms and between firms and the S&T local infrastructure almost disappeared. Silva interviewed not only firms but also local policy agents. Silva concluded that one could observe an almost consensual idea that actors of the local system are organized in an anarchical way, being totally disconnected. According to one person interviewed, what exists in Belo Horizonte 'is not an organized network of firms, but rather an *ajuntado*, a local slang that roughly means that these firms are only geographically concentrated'. In the words of another interviewee, one does not find in the region a social capital set-up, only a reasonable number of producing firms and some

*Table 2.11 Utilization by firms of government innovation programmes, 2008**

| | Yes | No |
|---|---|---|
| Fiscal incentives to R&D | | 11 |
| R&D projects jointly with universities | 4 | 7 |
| Scholarships for innovation | 4 | 6 |
| Risk capital | 2 | 8 |

*Note:* * result for 11 interviewed firms; one firm did not answer some questions.

*Source:* RedeSist's surveys

policy programmes; what did not exist was a network of relationships. Some cases found in Silva's research refer to local firms performing clinical tests for pharmaceuticals MNCs, while some of the interviewed firms mentioned that it was approached by another MNC that was interested in acquiring all its production but wanted to commercialize under its own label.

There was also a consensual idea that local actors could not work collectively because of the absence of an anchor institution capable of fulfilling the leading role, mapping and connecting the available resources. This discussion about the role of such leadership is of course central to this study.

Finally, in the 2008 survey there was an attempt to get an indication of the impact of government policies. The Belo Horizonte local biotech system has been the subject of a series of programmes by all levels of government – local, state and federal. In that survey it was specifically asked if the firm used any of the government's programmes or mechanisms to support its innovation activities. Table 2.11 shows the responses from the 11 interviewed firms. None of the firms used fiscal incentives to R&D, which has been the main policy instrument for innovation during the last 15 years. They also hardly used the other main instrument, which is support for risk capital. The other two instruments – R&D projects jointly with universities, and scholarships (which subsidize the payment of total wages) to foster the hiring of qualified human resources – were used by fewer than 50 per cent of the firms.

Again, this finding also confirms earlier research that addressed this issue. For example, the study of Lemos (2000) indicated the inappropriateness of government programmes and incentives to the specific needs of local biotech. His conclusion was that the local system presented structural deficiencies and policies that only created a set of unconnected firms exploiting different market niches.

This section should be concluded with two connected discussions. On the one hand, one might argue that the causality between the acquisition of Biobrás by Novo Nordisk and the downgrading, of the local system is not demonstrated by the evidence shown in this chapter. It is interesting, however, that *all* empirical studies show the same picture. It was not the acquisition *per se* that caused the downgrading, but the lack of a coherent and progressive industrial and innovation policy for this (as well as other) geographical area(s). The same could be said for the whole industrial complex and not only for the biotech sector which, as pointed out earlier, only exists as a myth.

Finally we come to a second discussion, which refers to the failures of policy mechanisms that for more than 20 years have not strengthened the area beyond academic vigour. In fact all policies implemented have been 'supply side' policies that implicitly perceive innovation as a natural outcome of scientific developments, a vision that was left behind more than 30 years ago. The basic policy suggestion is to incorporate a more modern vision of innovation as a systemic process, and not only to tackle the university–industry relationships, but to address the overall configuration of production chains, such as the pharmaceutical innovation system, that use biotechnologies.

## CONCLUDING REMARKS

Biotechnology scientific capability and firm creation have been two of the cornerstones of the Brazilian S&T policy for the last 25 years. In so far as scientific development is concerned, Brazil's achievements have been remarkable. Building on earlier achievements in biological and health research, a significant number of research groups have been established and have matured. Positive results are to be found in the number and quality of scientific publications, and in participation in breakthrough international networks in biotechnology research. Brazil now presents a considerable degree of development in different areas of genetics science and technology.

However, the results in terms of innovative capability are more debatable. This chapter attempted to discuss this issue by analysing in detail the evolution for 30 years of the health biotechnology local innovation system of Belo Horizonte, in the State of Minas Gerais. As stated in the introduction, this system has been considered the most successful of its kind in Latin America. It has also been one of the leading receivers of government support for science, technology and innovation, at all levels – municipal, state and federal – in Brazil.

Based on two surveys conducted by RedeSist's researchers – one in 2004 and another in 2008 – and on a series of academic studies that addressed its technological and innovative capabilities, this chapter concludes with a bleak picture. Given the amount of government stimulus for more than 30 years, the number of surviving firms and the level of innovative capabilities are quite small.

Universities, in fact, continue to be merely responsible for providing qualified human resources (the novel 'entrepreneurs') and some services (laboratory infrastructure), particularly during incubation. As firms grow they gradually disconnect from the infrastructure.

When and if a firm acquires a certain level of achievement in biotech it does not produce any spillover locally, and does not strengthen other links in its production chain, neither suppliers nor clients. Its achievements connect it with complex networks which are coordinated in other parts of the country or more correctly outside the country. This type of relationship invariably leads to a subordinate inclusion in such networks and the promising local firm eventually is downgraded to performing routinized tasks of lesser intensiveness, technologically speaking. The anchor firm from outside normally subcontracts the valuable local human resources to execute clinical tests, quality control, produce commodities, and so on. In some cases it acquires the control of the local assets, but innovativeness is lost in the process.

The disappointing results confirm the conclusion of several earlier studies (for example Lemos, 2000) that mere proximity to high-level researchers and supply-side policies are not sufficient to foster a virtuous internalization of services and intermediary products, a setting up of a really local innovation system.

The case of Minas Gerais is particularly illustrative as it presented the only single case of a Brazilian pharmaceutical firm – Biobrás – that grew successfully for over 20 years and that internalized a significant biotech capability. During its existence as an independent firm it served successfully as an anchor of the local innovation system. As all the above studies mentioned, even though a totally virtuous cycle of innovativeness was never completed, important achievements occurred as a series of smaller biotech firms gravitated around Biobrás. After its acquisition by a foreign firm – and subsequent technological downgrading – it lost its anchor role. As the data shown and analysed in this chapter suggest, connections diminished and firms came to pursue independent paths.

In fact, the findings of this chapter side with the propositions and conclusions of Hopkins et al. (2007) that claim that there is a great part of a myth in what some call a 'biotech revolution'. Biotechnology is in fact following a well-established incremental pattern of technical change, and the market structure around it has been totally controlled by large MNCs, particularly in pharmaceuticals. In the biotech innovation system of Minas Gerais a sometimes chaotic, but nevertheless heroic attempt to break the barriers developing countries have for entering into such a big game, was tried with the setting up and growth of Biobrás. A lack of coherent general economic policies led to the failure of this attempt, as Biobrás was taken over by a foreign firm. Future policies should ruminate more on the reasons for the failure and not only naively concentrate on fostering university–industry relationships where there are no firms on which to build.

## NOTES

1. RedeSist is the Research Network on Local Production and Innovation Systems. It comprises research groups at 23 Brazilian universities. One of its main activities is empirical research on the dynamics and evolution of local production systems in different parts of Brazil and on agricultural, manufacturing and service activities. From its setting up in the late 1990s to date, RedeSist has analysed approximately 100 local systems (see www.redesist.ie.ufrj.br for the methodology of analysis).
2. The information used for Table 2.1 is found at the directory of research groups of CNPq (the National S&T Development Council). There may be some overlapping as research groups can inform two or more areas of scientific specialization.
3. Proteomics, pharmacogenomics, pharmacogenetics, biopharmaceuticals, vaccines, molecular diagnostics, immunology, biomaterials, toxinology, recombinant vaccines and stem cells.
4. The same areas as in 2002, but without pharmacogenomics.
5. Genomics, proteomics, pharmacogenomics, pharmacogenetics, vaccines, molecular diagnostics, immunology, biomaterials, toxicology, recombinant vaccines and stem cells.
6. Genomics, pharmaceuticals, vaccines, molecular diagnostics, biomaterials, toxicology, recombinant vaccines and stem cells.
7. FAPESP is the state research foundation of the state of São Paulo.
8. *Xylella fastidiosa* is a fastidious xylem-limited bacterium that causes a range of economically important plant diseases, citrus variegated chlorosis (Simpson et al., 2004).
9. In the first year, the selected firm receives BRL120 000 (USD52 200 thousand) as a subvention (not reimbursable), and in the second year it can take the same amount of credit without paying interest.
10. PhRMA members are largely, though not exclusively, US-headquartered companies.
11. Superintendence for the Development of the Northeast of Brazil. This organization was liquidated in the 1990s.
12. A digestive enzyme obtained from cattle pancreas.
13. HALA introduced a liquid product instead a powder one, which is used for long-distance imports. Since Biobrás produced a copy of the imported product; it had to be transformed into a liquid state for industrial use, which required a high-energy consuming drying plant and, thus, a higher unit cost.
14. The other being Novo Nordisk.
15. Biobrás and Eli Lilly got, respectively, 54 per cent and 46 per cent of capital equity.
16. At that time it was in Eli Lilly's interest to import insulin crystals and concentrate its industrial production on making insulin formulates.

# REFERENCES

Achilladelis, B. and N. Antonakis (2001), 'The dynamics of technological innovation: the case of the pharmaceutical industry', *Research Policy*, no. 30, pp. 535–88.

Angell, M. (2004), *The Truth about the Drug Companies: How they Deceive us and what to do about it*, New York: Random House.

Audretsch, D.B. (2001), 'The role of small firms in the US biotechnology clusters', *Small Business Economics*, **17** (1/2), 3–16.

Batista, R. (2002), 'Briga de actionistas da Biobrás pode parar no cade', *Valor Econômico*, 9 September.

Biominas (2001), *Parque Nacional de Empresas de Biotecnologia*, Belo Horizonte: Fundaçáo Biominas, December.

Biominas (2007), 'Estudo Biominas 2007: um instrumento facilitador do desenvolvimento de bionegócios no Brasil, 2007', available at www.biominas.org.br

Bond, E.C. and S. Glynn (1995), 'Recent trends in support for biomedical research and development', in N. Rosenberg, A. Gelijns and H. Dawkins, *Sources of Medical Technology: Universities and Industry: Medical Innovation at the Crossroads*, vol. 5, Washington: National Academy.

Casper, S. and A. Karamanos (2003), 'Commercializing science in Europe: the Cambridge biotechnology cluster', *European Planning Studies*, **11**(7), 805–22.

Cockburn, I. (2008), 'Pharmaceuticals', in J. Macher and D. Mowery, *Innovation in Global Industries: US Firms Competing in a New World,* Washington, DC: National Research Council.

Cockburn, I. and R. Henderson (1996), 'Public–private interaction in pharmaceutical research', *Proceedings of 'Science, Technology and the Economy'*, National Academy Sciences, Washington: 20–22 October, 1995.

Cooke, P. (2001), 'Biotechnology clusters in the UK: Lessons from localisation in the commercialisation of science', *Small Business Economics*, **17**(1/ 2), pp. 43–60.

Cooke, P. (2003), 'The evolution of biotechnology in three continents: Schumpeterian or Penrosian?', *European Planning Studies*, **11**(7), 757–63.

Cooke, P. (2004a), 'The accelerating evolution of biotechnology clusters', *European Planning Studies*, **12**(7), October, Special Issue: Globalization of biotechnology, pp. 915–20.

Cooke, P. (2004b), 'Regional knowledge capabilities, embeddedness of firms and industry organization: bioscience megacentres and economic geography', *European Planning Studies*, **12**(5), 625–41.

Costa, E. (2006), 'Política de medicamentos: tecnologia e produção no País', *Abifina Informa*, n. 216, March. Available at <http:/www.abifina.org.br>.

Costa, E. (2008), 'Estrutura pública de produção de medicamentos como potencializadora da indústria farmoquímica no Brasil', paper presented at the Seminar on the Health Industrial Complex, BNDES, Rio de Janeiro, 19–21 May.

Cruvinel, T. (2004), 'O caso Biobras', *O Globo*, 7 June.

De Vol, R., P. Wong, J. Ki, A. Bedroussian and R. Koepp (2004), 'America's Biotech and Life Science Clusters: San Diego's Position and Economic Contributions', *Technical Report*, Santa Monica, CA: Milken Institute.

*Economist, The* (2005), 'Prescription for change', 16 June.

Fajnzylber, P. (2002), *Fatores de Competitividade e Barreiras ao Crescimento*, Belo Horizonte: UFMG/Cedeplar.

Feldeman, M. and J.L. Francis (2003), 'Fortune favours the prepared region: the case of entrepreneurship and the Capitol region biotechnology cluster', *European Planning Studies*, **11**(7), 765–88.

Ferrer, M. et al. (2004), 'The scientific muscle of Brazil's health biotechnology', *Nature Biotechnology*, **22**, Suppl, pp. DC8–DC12.

Fonseca, M.G.D. (2009), 'Perspectivas do investimento em biotecnologia no Brasil', project PIB, Economics Institute, Federal University of Rio de Janeiro, Rio de Janeiro.

Gadelha, C.A.G. (2002), *Estudo da Competitividade por Cadeias Integradas no Brasil: Impactos das Zonas de Livre Comércio: Cadeia Complexo da Saúde*, Universidade Estadual de Campinas, Instituto de Economia, Campinas, December.

Gadelha, C.A.G. (2003), 'O complexo industrial da saúde e a necessidade de um enfoque dinâmico na economia da saúde'. *Ciência e Saúde Coletiva*, **8**(2).

Gambardella, A. (1995), *Science and Innovation: The US Pharmaceutical Industry During the 1980s*, Cambridge: Cambridge University Press.

Geraque, E. (2003), 'Colaboração do Brasil ao mundo', *Revista Pesquisa FAPESP* Special issue 'Dupla Hélice 50 Anos', April, p. 30–33, available at: http://revistapesquisa.fapesp.br/.

Hopkins, M., P. Martin, P. Nightingale and A. Kraft (2007), 'The myth of the biotech revolution: An assessment of technological, clinical and organisational change', *Research Policy*, **36**, 566–89.

Judice, V.M.M. and A.M.C. Baêta (2005), 'Modelo empresarial, gestão de inovação e investimentos de venture capital em empresas de biotecnologia no Brasil', *Revista de Administração Contemporânea*, **9**(1).

Judice, V.M.M. and E.E. Soares (2004), *Diagnóstico biotecnologia em Minas Gerais 2004: Desafios rumo a 2010: Inovação, Internacionalização e Governança do Setor*, Belo Horizonte: IEL-MG.

Judice, V.M.M. and M.C.R.L. Vasconcelos (2006), 'Condições ambientais, culturais e empreendedorismo tecnológico no Brasil: estudo de caso de uma empresa de biomateriais avançados', paper presented at *XXIV Simpósio de Gestão da Inovação Tecnológica*, ANPAD, 2006, 17-20 October, Gramado RS.

Judice, V. and C. Vedovello, (2008), 'Biotechnology innovation system in Brazil: an exploratory study', *Research Paper* 13/07, the BRICS Project, IE/UFRJ and Globelics.

Kaiser, R.T. (2003), 'Multi-level science policy and regional innovation: the case of the Munich cluster for pharmaceutical biotechnology', *European Planning Studies*, **11**(7), 757–63.

Kaufmann, D., D. Schwartz, A. Frenkel and D. Shefer (2003), 'The role of location and regional networks for biotechnology firms in Israel', *European Planning Studies*, **11**(7), 823–40.

Klevorick, A., R. Levin, R. Nelson and S. Winter (1987), 'Appropriating the returns from industrial research and development', *Brookings Papers on Economic Activity*, **3**, 783–820.

Lemos, M.B. (1992), 'The agro-food system in semi-industrialized countries: the Brazilian case, PhD Thesis, London: University of London.

Lemos, M.B. (2000), 'Local system of innovation under peripheral conditions: a case study of biotechnology in the metropolitan area of Belo Horizonte, Brazil', Technical Note of the project: Arranjos Produtivos Locais e as Novas Políticas de Desenvolvimento Industrial e Tecnológico, RedeSist/IE/UFRJ.

Love, J. (2001), 'Paying for health care R&D: carrots and sticks. MSF/DND', *Working Group*. Geneva: MSF/DND, pp. 1–11, January.

Matraves, C. (1999), 'Market structure, R&D, and advertising in the pharmaceutical industry', *The Journal of Industrial Economics*, XLVIII(2), 169–94.

McKelvey, M. and Luigi Orsenigo (2001), 'Pharmaceuticals as a sectoral innovation system', paper prepared for the *ESSY Project* (European Sectoral Systems of Innovation) and within the Epris Project, November.

Milan, R.D.S. (2006), 'Propriedade intelectual e inovaçáo na UFMG', *Diversa*, Revista da UFMG, **5**(10), 16–17.

Mytelka, L. (1999), 'New trends in biotechnology networking', *International Journal of Biotechnology*, **1**(1), pp. 30–41.

Nelsen, L.L. (2005), 'The role of research institutions in the formation of the biotech cluster in Massachusetts: the MIT experience', *Journal of Commercial Biotechnology*, **11**(4), 330–36.

Niosi, J. and T.G. Bas (2003), 'Biotechnology megacentres: Montreal and Toronto regional systems of innovation', *European Planning Studies*, **11**(7), p. 789–804.

*O Estado de São Paulo* (2003), Biobrás desenvolveu insulina humana com financiamento do órgão', 26 Augest.

Powell, W., K. Koput and L. Smith-Doerr, (2006), *Interorganizational Collaboration and the Locus of Innovation: Networks of Learning in Biotechnology*, Cheltenham, UK and Northampton, MA, USA: Edward Elgar.

Prevezer, M. (1997), 'The dynamics of industrial clustering in biotechnology', *Small Business Economics*, **9**, 255–71.

Silva, E. (2004), 'A rede de inovação no cluster de biotecnologia de Belo Horizonte – um estudo sobre interações sociais, ambiente local e a capacitação tecnológica das empresas', MSc Dissertation, Minas Gerais: PUC.

Simpson, A.J.G., A.A. Camargo, J.A. Ferro, J. Parra and A.T. Vasconcelos (2004), 'Coordinated network-based research as a strategic component of science in Brazil', *Genetics and Molecular Research*, **3**(1), 18–25.

Souza S.G.A. (2001), 'Potencialidades da biotecnologia em Minas Gerais: estudos sobre empresas e suas relações com universidades', MSc Dissertation, Federal University of Minas Gerais, Belo Horizonte.

Souza S.G.A. (2004), 'Transferência de conhecimento nas universidades do estado de Minas Gerais: uma contribuição ao desenvolvimento regional', paper presented at the 'Seminário sobre Economia Mineira', Diamantina: 24–27 August.

Sutton, J. (1998), *Technology and Market Structure: Theory and History*, Cambridge, MA: MIT Press.

Sutz, J. (2005), 'La innovación como problema de la política en América Latina', *Analogía*, **19**(1).

Teixeira, M. (2008), 'Sistemas produtivos baseados na ciência indústria de biotecnologia', paper prepared for the PIB project, IE/UFRJ, Rio de Janeiro.

Thorsteinsdóttir, H., U. Quach, A. Daar and P. Singer (2004) 'Conclusions: promoting biotechnology innovation in developing countries', *Nature Biotechnology*, **22**.

# 3. Linkages between bio-innovation, knowledge production and policy in Uruguay

**Isabel Bortagaray, Isarelis Pérez Ones and Judith Sutz**

## INTRODUCTION

Biotechnology has been a sort of buzzword for the Uruguayan science and technology policy for more than 20 years. During this time, biotechnology never failed to be nominated as a high priority for the country. This was due for bad as well as for good reasons. The bad reason was mainly a blind follow-up of the new technologies world fashion. The good reasons included the agro-industrial bias of the Uruguayan productive structure, which made biotechnology a good cognitive bet, and the relatively strong research capabilities in the whole field of life sciences accumulated in the country. Such prioritization was expressed mainly at the level of competitive funds for S&T administrated by the government. In the late 1980s and the 1990s such funds originated in S&T loans from the Inter American Development Bank; since the middle of the first decade of the twenty-first century budgetary resources were greatly increased. But little was done in terms of calling on biotechnology from the demand side: industrial, environmental or health policies have not thoroughly explored biotechnology as a means to achieve some of their goals. Moreover, Uruguay did not push to be included in the early Argentinean–Brazilian institutional common efforts in biotechnology as part of the MERCOSUR agreement in 1985, another signal that 'high priority' was a label with restricted practical consequences. Twenty years later, in 2005, this imbalance began to be redressed: Uruguay entered the BIOTECSUR Platform, financed with funds from the European Union, with high participation from the academic side, especially the University of the Republic.

Time has brought a noticeable enlargement of research capabilities in the field. Diverse branches of biotechnology have been developed in recent years, both at the level of research and at the level of postgraduate studies.

Researchers working on diverse biotechnology problems have developed strong ties with more fundamental researchers in disciplines such as cellular biology, microbiology and chemistry. The capacity of biotechnology research to solve practical problems of great economic importance has been proven time after time. The productivity of Uruguayan rice, for instance, is in debt to local biotechnology research; the same is true for milk production. These are examples related to public institutions including the public university, where research results were delivered directly to producers. The National Institute for Agricultural Research, in particular, has strong biotechnology divisions that work around cultivations, mainly rice and winter cereals, animal production, fruit and greens production, and forestry, in close contact with related producers. University research has been transferred to a few enterprises in the realm of environmental biotechnology, among other applications.

The scope of public institutions to deliver concrete biotechnology solutions – be they products or services – is limited, though, at least in a market economy. The case of vaccines is a good example of this assertion. Given the huge economic importance of foot and mouth disease it was once intended to build a specific public endeavour to produce animal vaccines against it, but such a goal was not achieved; human vaccines are researched at the university but it is not possible to produce them there.

Accumulated research capabilities around biotechnology need to be catalyzed by firms to fully deploy economic and social impact. This assertion is equally valid for market economies and non-market economies. It simply states that producing some biotechnology goods and delivering them to satisfy the demand of public bodies or other firms is an activity that requires capacities and know-how that are outside the academic world. We stress the obvious fact that Uruguay is a market economy because as part of the research on biotechnology made in Uruguay several interviews were conducted in Cuba around questions similar to those posed in Uruguay. Not surprisingly, the institutional logic of actions in both countries was markedly different; the results, obviously, were also different.

Four actors are needed to make biotechnology work: those that need biotechnology (and know that they need it); those able to perform the research that is needed to find biotech solutions; those that produce biotech solutions; and those that provide for the needed stimulus so the former three actors can usefully interact. The issue of these interactions is at the heart of the systemic approach to innovation; this chapter attempts to study in detail the role those actors have in the innovation decision-making of biotechnological firms in Uruguay. This is a point of entry of particular importance from a developmental point of view because, as said before, the ability of agriculture, of industry and of the health system to solve problems through biotechnology depends crucially on the capacities and willingness to innovate of biotechnology-based firms.

We shall analyse the innovation decisions processes in three Uruguayan biotechnology firms that have turned themselves into niche exporters besides their well-established presence in the local market. One of these firms is devoted to human health, the second to animal health and the third to environmentally friendly waste treatment. The aim of this analysis is to better understand the decision-making processes leading to strategic investments, which include the mobilization of financial resources and the establishment of the right set of incentives that can induce cooperation to implement an innovative strategy. This is what Lazonik and O'Sullivan term 'social conditions of the innovative enterprise', that are both internal and external to the firm (Lazonick and O'Sullivan, 2000; Lazonick, 2005). This understanding is necessary to design innovation policies able to foster effective innovation in biotechnology by enhancing the social conditions for cooperation. Precisely one of the reasons why one size does not fit all in innovation policy is that the social conditions of the innovative enterprise vary greatly in different contexts, including national, regional and sectoral settings.

Innovation is an interactive process (Lundvall, 1988) in which different types of knowledge are combined. In the case of modern biotechnology, the combinations of knowledge almost always involve people in the firm and people outside it, given the strong scientific nature of the expertise involved. So, the analysis of the innovation decision at firm level must include the state of the relationships with the knowledge providers: Are they available? Are they willing to interact in ways that are appropriate for the firm? How does the search process of such knowledgeable partners occur? On the other hand, the firm's clients, the final users of the innovation, can have a key role in steering the innovation decisions. This is specifically the case in Uruguay, where the interaction with old and new clients is, as we will show, a main source of innovation. So, to better understand the setting of the firms' innovation agendas the analysis encompasses the firms incorporating the innovations, the academic or knowledge actors that provide the specialized answers for the problems to be solved and, to a lesser extent, the bio-innovation users.

This approach is actor-based, with the bio-innovator related to a specific bio-project as the entry point, and from there both knowledge producers and knowledge users have been traced back and involved in the study (see Figure 3.1). Policies have been approached from the perspective of the bio-innovator, as policy users, a point of view 'from the demand side' able to usefully inform policy design.

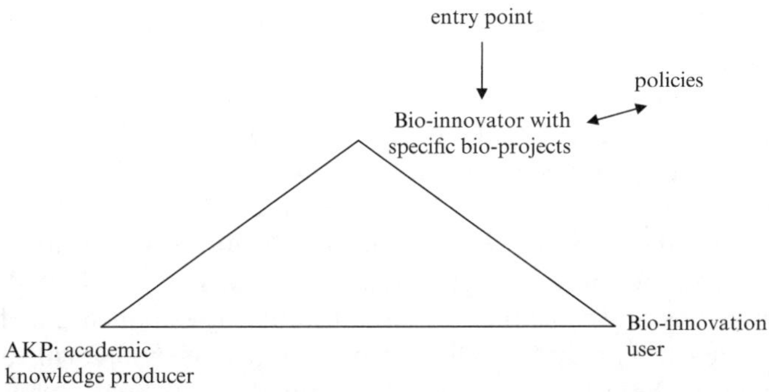


*Figure 3.1 Sketch of the approach*

The study was conducted in Uruguay, following the approach depicted in Figure 3.1 in the three cases, i.e. animal health, human health and environmental issues. Interviews were based on a semi-structured protocol, with slight variations depending on the type of actor. Interviews conducted in Cuba aimed at complementing and contrasting the analysis of the factors affecting innovation decisions by introducing a different national context.

The chapter is organized in five sections. After this introduction, it briefly presents some distinctive features of biotechnology that make it an area of opportunity for developing countries. The next section outlines the analytical framework and sketches the discussion on innovation agendas and decisions. The fourth section connects the analytical framework with the specific cases studied in Uruguay. The fifth section briefly revises innovation agendas and decision-making processes in biotechnology research centres in Cuba, with the aim of establishing a few meaningful comparisons. The last section attempts to wrap up the main issues discussed in the chapter.

## DYNAMIC OF THE TECHNOLOGY: BIOTECHNOLOGY IN MOTION

Biotechnology entails an area of opportunities for developing countries for several reasons. It encompasses a wide range of technologies with varying levels of complexity, costs involved and applicability. Biotechnology applied to agriculture has been one of the areas in which less-developed countries have seen great scope for innovation opportunities. Agro-biotechnology combines knowledge from different scientific fields with knowledge related to species' adaptation to local environmental conditions, both types of knowledge rooted in national research. Another functional specificity of agrobiotech for developing countries is that investment levels can be manoeuvred depending on the budget available, while other types of development, such as chemical synthesis, require larger investments. The role of local knowledge and know-how is of utmost importance as biological systems imply mutation, variation, adaptation and interaction with contextual conditions. Thus, knowledge on local conditions and variation might substantially enhance the efficiency of the product and might be well known and mastered by a local company. This has been the case of vaccines for animal health, which has indeed been an area of opportunity for some local companies in Uruguay.

The case of environmental biotechnology is not so different from agrobiotechnology. The use of micro-organisms for bio-fermentation, for instance, shares the same attributes of agro-biotech: the adaptation and specificity of local flora to degrade waste is a very valuable asset within the whole process. Human health oriented biotechnology is different: The rules of the game are largely permeated by global dynamics, costs are higher, regulations and IPRs impose severe obstacles for developing markets where the ability to pay royalties is very limited, as is the ability to enforce IPRs or actually to file a case against those breaking IPRs. However, as the term '90/10 gap' coined by the Global Forum on Health Research expresses, human health oriented biotechnology is related to local specificities at the level of the research agenda. With 90 per cent of all the world's financial resources on health research devoted to the illnesses of 10 per cent of the world population, the issue of orienting the research agenda can establish a strong bond between human health biotechnology and local conditions.

## THE ANALYTICAL FRAMEWORK AND INNOVATION AGENDAS AND DECISIONS

### The Analytical Framework

To organize the empirical exploration of the innovation decisions at business firms' level in Uruguay, we propose to analyse the aspects as shown in Figure 3.2.

The innovation base of the firm is key to determining the screening scope for innovative opportunities. It depends, first of all, on the personal innovation drive of entrepreneurs, not too different from what happens in academic research where curiosity plays an important role. The statement

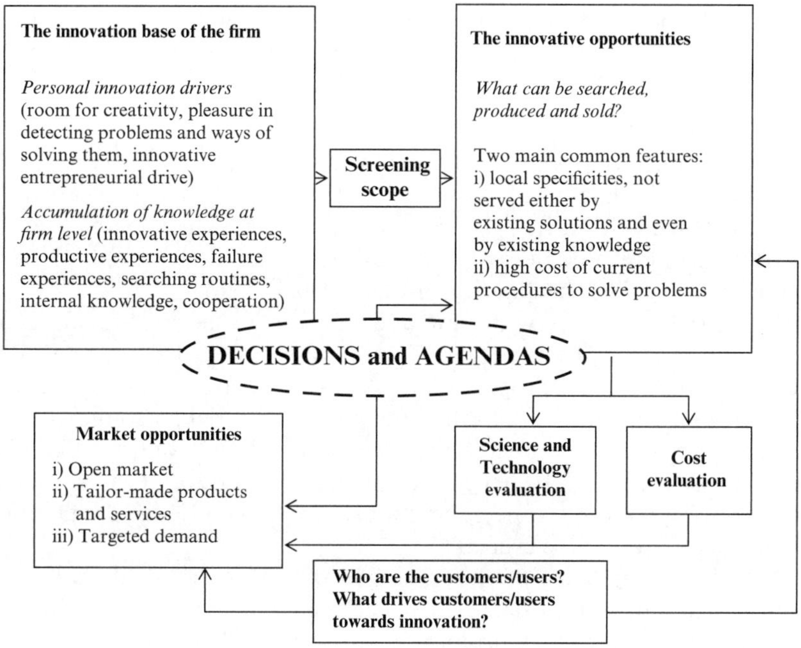


*Figure 3.2 Components of innovation decisions at firm level*

'I invent because developing new things is one of my biggest joys – as satisfaction of my curiosity', was asserted by one of the co-founders of Sony Corp. (Dasgupta, 1996, p. 26), but it could be applied to the managers of the interviewed firms.

The urge to explore new research avenues and ideas entails some doses of irrationality, intuition and guts which shed light on a specific course of action over others. As one of the interviewees put it: 'if I have an idea and like it, I of course analyze pros and cons, I counter-assess it against my experience, but then if I like it . . . I just jump on it'.

The screening direction is greatly influenced by the accumulation of knowledge at the firm. Such knowledge includes scientific and technical components, as well as commercial components, including customers' needs, preferences and prejudices. Such accumulation or 'knowledge trajectory' has been a central part of the appreciative theory of innovation. As Nelson and Winter point out (1982):

> In many technological histories the new is not just better than the old; in some sense the new evolves out of the old. One explanation for this is that the output of today's searches is not merely a new technology, but also enhances

> knowledge and forms the basis of new building blocks to be used tomorrow [. . .] The result of today's searches is both a successful new technology and a natural starting place for the searches of tomorrow (Nelson and Winter, 1982, pp. 255–7).

From another viewpoint, the screening scope relates to the capacity of the firm to assimilate new knowledge, or the firm's 'absorptive capacity', strongly dependent on its prior related knowledge, which enables actors 'to recognize the value of new information, assimilate it, and apply it to commercial ends' (Cohen and Levinthal, 1990, p. 128).

Filling the skill gaps with new and knowledgeable individuals is an important part of firms' strategies. The relevance of 'skilled and new hired individuals' has been clearly recognized in the literature and in empirical work around innovation,[1] and the role of recently hired personnel regarding the propensity to innovate has been explored and confirmed empirically (Nielsen, 2007). The experience of these Uruguayan firms indicates the same trend and relevance; the importance of knowledge cooperation circuits has proven to be crucial too.

One of the outcomes of the screening efforts is the identification of innovative opportunities. The range of emerging opportunities is wide and diverse, and includes some promising paths that will become part of their innovation agendas, as well as the identification of 'non-opportunities', that is, alternatives that remain outside the scope of the firm. Such is the case of pharmaceutical innovations that are out of the range of what is possible for them, as expressed by the owner of the Uruguayan health oriented biotechnology firm:

> The difference between the biological and the pharmacological worlds is enormous. In the former we can start from zero and reach a final product with high value added, because in general it is about knowledge and working with microorganisms in a culture milieu, and after several processes, though complex ones, we get the final product. In the latter however, developing a molecule requires a huge investment. We could purchase drugs from India or China, those [whose] patents are about to expire, and produce something like a commodity. But in that case the recognition is to the original drug and to the original company. There is room for differentiation by combining existing drugs and by making them efficient for specific conditions. We have done that, because there are still vacuums, and special conditions that require special drugs not available as such. But we emphasize the biological aspects.

In developing countries in general and in Uruguay in particular, two types of innovative opportunities, that is, opportunities to solve problems that are of importance to productive actors, are present in the most diverse settings: the need to cope with specificities, and the need to find ways of solving a given problem in a cheap enough way to be producible or accessible to potential customers.

Specificities are particularly strong in the life-related sectors. Let's take as an example the vaccine against Leptospira, a pathogen agent that is a main cause of miscarriage in bovine cattle.[2] There are several varieties of Leptospira, and there is no one-size-fits-all vaccine. Ineffective vaccines are detected by epidemiological vigilance carried out by the public authorities when a higher than normal abortion rate occurs. In these cases, it is possible to detect the variety of the pathogen that is present; a specific bacteriological vaccine against this variety is needed to immunize the cattle. The doses involved are not too high, around some tenths of thousands, and no foreign firm will undertake the process of research and development needed to produce the vaccine: this opens one type of innovative opportunity stemming from specificity.

The need to substitute imported solutions for reasons of cost is a permanent source of innovative opportunities in developing countries. In fact, it can be argued that this need has led to a sort of idiosyncratic heuristic problem-solving strategy, 'the capacity to innovate in scarcity conditions' that can be seen as a variety in its own terms within the different types of innovation (Srinivas and Sutz, 2008). Two quite different examples of this kind of opportunity can be given from the Uruguayan biotechnological firms' experiences. In the first example the situation is straightforward, as described by the environmental firm's owner: 'You have two possibilities: either you have half a million dollars and you buy the fuel oil furnace, or you have the knowledge of these three young engineers and one and half hundred thousand dollars, plus the metal-mechanical capabilities (to do an alternative furnace).'

The problem arose when, as a result of environmental regulations, the blood from the slaughter industry could no longer be thrown into water pools in the countryside. What to do with the blood became a difficult problem, given the volume to be disposed of. Even if the response to the problem was not foreseen, the opportunity was. The target was then to cope successfully with finding a good solution to a new problem: gather the blood from the slaughterhouse and transform it into blood flour, rich in proteins and able to be used in the animal-feed industry. The knowledge resources were mainly drawn from previous experience and a reservoir of routines. These included going out to firms and universities in the region, looking for similar problems and the available portfolio of solutions, a significant amount of intuition and vision, technical knowledge not exactly on the same problem but complementary, and a pool of professionals able to help with the process of adjusting the identified general solutions to the specific contextual requirements. Thus, the proposed methodology to transform liquid blood into blood flour was adapted from the dairy industry and from the technique used to produce milk powder. A main bottleneck was how to supply the needed energy. The production costs were first estimated considering the use of conventional furnaces fed with fuel oil, but the economic equation did not fit, given the very high costs of fuel oil. The alternative was to use wood instead, which is abundant and relatively cheap in Uruguay, but no furnace based on wood and with the needed characteristics was available (the production volume and the type of blood, among others). As a result the firm manager decided to become involved in the design and manufacturing of such a furnace, which then became patented in Uruguay and Brazil. The process went even a step further as they decided to apply for carbon bonuses through the Kyoto Protocol, which is now under way, because the furnace substituted fossil carburant with a renewable fuel. In sum, the innovative manager started the process by searching for similar situations, studying the process of transforming liquid into powder and adapting if from milk to blood. He then searched for engineers who could design an alternative furnace and draw on the previous knowledge they had from a metal-mechanic workshop, so that the experimentation stage for manufacturing the furnace would be assured. The furnace was subsequently improved several times, until it was able to support the drying process of 1200 litres of blood per day, from an initial level of 400 litres per day. As stated above, the innovative opportunity arose from the need for a good solution in terms of cost to an environmental and production problem. This is also an example of a more general feature: innovation in biotechnology, particularly in developing countries, must often be coupled with innovation in other areas in order to deliver solutions (e.g. in mechanical engineering, electronics, software).

The second example related to innovation opportunities arising from cost considerations belongs to the area of human health. It relates to developing, through scientific and technical capabilities, an inexpensive way of obtaining useful diagnostic information for a particular genetic disease, 'X fragile', which affects the ability to learn, but which has no other manifestations during early childhood. A diagnosis protocol to detect the existence of the X fragile disease exists, but it is extremely costly. For this reason, when a child with persistent learning problems has a medical consultation, the last test he/she is given is related to the X fragile diagnosis. The scientific leader of the Uruguayan biotechnological firm thought differently. The question should be inverted: instead of having a diagnosis based on a residual process where a long checklist has to be fulfilled first, deciding whether or not the child has the X fragile disease should be the first thing to do. The triggering question was what should be done to assure that a child does not have the X fragile disease. The outcome is a diagnostic kit that costs USD20 and hence can be used at the beginning of the diagnosis protocol. The test does not confirm that the child has the X fragile disease; it is only able to assure, if it is the case, that the child does not have the disease, rationalizing the search for alternative explanations of the observed symptoms.

The evaluation of innovative opportunities involves a twofold process: first, an assessment of the real scientific and technological possibilities of solving the problem in a reasonable time; and second the cost of finding and implementing the solution. This is a substantial stage in the innovation decision process. In the case of the biotechnological firm devoted to animal health, the hiring of a group of young biologists and biochemists had the effect of multiplying the scope of opportunities and alternative paths to follow that were never before considered by the firm. This explosion of innovative ideas leads to a re-configuration of the R&D department to enable the business prospects of the new ideas and proposals to be properly analysed.

Market opportunities for these biotechnological firms could be schematized into three types: open markets, tailor-made solutions and targeted demand. We shall briefly exemplify them as they appear in the case of the animal health biotechnology firm, but they also capture the market opportunities for the other firms; moreover, these patterns are part of other knowledge-based sectors in Uruguay, such as the case of professional electronics (Snoeck et al., 1992).

By 'open market' we mean serving a market without any particular producer in mind. The examples in this case are vaccines against common diseases of cattle that the firm can provide at a better cost or with better biological characteristics, based on their frequent and close interactions with farmers which has enabled them to know better what is needed and take into account the specificities of the Uruguayan production system.

By 'tailor-made solutions' we refer to a situation where a specific client asks for a solution to a specific problem. The example in this case is a vaccine to protect sturgeon, an incipient and encouraging new export industry in Uruguay. In this case, the pathogen was isolated by the sturgeon fish farm and the vaccine was developed by the biotechnological firm; it was possible because of the accumulated experience of the latter. It meant more than just making the antigen: they had to adapt it and develop an immunization methodology totally different from their previous experience on cattle immunization.

By 'targeted demand' we refer to a situation where a specific demand is presented by a specific actor, but not just for solving the single problem of a given client but for solving a more general problem. The experience in vaccine production of the biotechnology firm dedicated to animal health drew on the production of vaccines against very different diseases, from which one was cattle anthrax. An entrepreneur interested in the development of a vaccine against human anthrax contacted the firm with this specific demand. The antigen was successfully developed and tested within the firm's premises, but the process of encapsulation had to be done in a specially built lab in a free trade zone.

Finally, it is important to explore the reasons why a customer becomes innovative. Specificities and high costs have already been mentioned as innovation drivers; in the biotechnological realm, particularly in everything related to environment, regulations play a major role. A case in point, which combines specificities and regulations, is that of the disposal of ruminal fluid (the contents of the stomach of cattle) from the slaughter industry. Before the current environmental regulations were adopted, ruminal fluid was just disposed of by throwing it away without further treatment: this is no longer allowed. This situation moved a slaughter firm to search for innovation. The owner of the environmental biotechnology firm contacted by the slaughter enterprise to find a solution, first went to Spain for advice, but then he realized that understanding and knowing about bio-digestion does not guarantee the ability to design a bio-digester for ruminal contents. This is so because only countries where cattle are fed mainly on grass on a year-round basis, which leads to a specific digestion cycle, face the problem of huge rumen content production and therefore have the problem of how to dispose of it. As an interesting additional feature, rumen content has the potential to be transformed into a high quality organic fertilizer. But to do that, it must go through a specific bio-digestion process which requires specialized equipment. The biotechnological firm, in alliance with researchers of the Chemistry Institute at the Faculty of Engineers, developed a specific bio-digester of rumen content that produces a very effective organic fertilizer on the one hand, and biogas on the other, turning the problem of rumen content disposal, the trigger of the whole process, into an opportunity to develop useful sub-products.

## Innovation Agendas: Moving from Ideas to Projects

Innovation agendas are made up of both ideas and projects. They coexist along a continuum, varying in terms of their maturity, feasibility and concreteness. Innovations were ideas at some point. Ideas are key resources for innovation, as triggers of innovation processes. As such, mechanisms for hunting new ideas are institutionalized in some organizations through 'offices of innovation' which are aimed at nurturing the emergence of new ideas (Ilori and Irefin, 1997), or through innovation managers, innovation midwives (Vincent, 2005), and so on. But even if ideas are at the birth of materialized innovations, there is not a linear sequence between them. There are circularities, gaps, conscious efforts as well as chance and casual circumstances that confabulate to make a brilliant idea become an innovative project and then an innovation per se, while others do not get to follow such a path or might follow a different one.

The three cases under analysis share a common driving force: their strong reliance on innovativeness for expanding their core strengths and finding new niches of production. Innovativeness is embedded in their routines as a mechanism for their development. More concretely, their innovativeness translates into an attitude characterized by a continuous quest for finding new problems not yet solved at the local level.

Firms' innovation agendas tend to articulate two types of strategies, in spite of variations across the mechanisms implemented in the innovation processes. On the one hand they rely on their intramural capabilities to carry on strategies aligned with their core strengths, which allows them to build on and accumulate established trajectories. Within the set of internal efforts, there are differences in the extent to which the focus remains on envisioning and screening new opportunities or on advancing their technological frontier, including more technological vigilance. It is a dual strategy in which firms take the advantage of skills and cumulated strengths and explore new initiatives. Exploration, though, is not random or totally blind, but focused and adjusted by a series of factors and successive steps. Core strengths set the broad boundaries within which screening and scanning processes take place, as firms tend to build on previously accumulated strengths, shaping a path-dependent process, where more or less directly, incremental efforts lead to the expansion of that focus.

On the other hand, innovation is triggered from the outside, based on a strategy oriented to searching and responding to problems arising in other sectors. The three firms, through different trajectories, have developed an antenna system for identifying and interacting with potential users to better understand both the problems they have and the ways they can be solved. This stream of exploration is complementary to the exploitation of their core strengths in the areas around which accumulated capabilities tend to concentrate, and constitutes an important source of continuous innovation while nurturing their innovativeness.

Often existing competences drive the emergence of new ideas, shaping a loop, as firms tend to search for solutions within the neighbourhood of their practice, but known terrains embody new questions and problems. That process leads to new ideas even though sometimes a third party is required to support that process and fill in the cognitive gap that might exist internally. New areas of application drive new search efforts, shaping an iterative process. New ideas do not emerge in a vacuum; they build up in a more or less articulated dialogue with existing competencies, which in turn set the boundaries from where to depart in the search for the new.

As mentioned earlier, doses of irrationality, intuition and guts are important forces underlying the three cases. If guts and intuition define part of the boundaries, the level of acceptable risk sets the ceiling, one that cannot be too high and that is defined by plausibility in this type of firm; the unimaginable remains outside the scope of considered ideas. Other triggers of new ideas have to do with synergies and cross-fertilization between different actors, such as users, or junior researchers, which certainly constitutes the case of the animal health biotechnology firm, as mentioned earlier.

The path from new ideas to new projects could be rough, depending not only on those internal decisions but also on contextual factors. The following paragraphs attempt to disentangle some of the knots that mediate between new ideas and new projects.

## Innovation Decision Processes

Decision-making is often tackled through phases: the entire process is divided among a series of discrete and sequential phases moving from the identification of the problem/opportunity, evaluation of alternatives, choice of the 'best' one, followed by its implementation. This type of approach is utilized to study decision processes in other areas such as policy-making (Birkland, 2001).

Overall decisions stand on a combination of positive, explicit and pursued factors with others that are brought into the process as residuals. Decisions are of different types and concern different themes, even when focusing at the level of innovation decisions. They shape circuits that are sometimes interconnected while others run in parallel, which might move along one dimension and then suddenly shift into another one. Decisions are not linear, nor is it always possible to establish a temporal or continuous line between them.

One of the first steps in the process of scanning the alternatives for growth and development of a small firm from a small country entails a demarcation between the possible and strategic, and the non-plausible. Thus the first assertion in an attempt to conquer new areas and/or markets has to do with being aware of what is not going to be part of the strategy. In the words of a small biotech firm's manager: 'we cannot compete in large markets, so we either must look for niches or segments, or innovate in terms of markets that have not yet been targeted'.

This initial orientation and screening of new niches is in tandem with the strengthening of firms' capacities. At a certain time there is a compromise between expanding and reinforcing capacities and capabilities on the one hand, and pushing for accessing new markets on the other hand. These two strategies constitute a bi-directional movement in which the process of pushing aside for expanding core strengths nurtures and complements the process of pushing forward to access new markets.

Within these boundaries, varying alternatives and combinations remain. Some are more incremental decisions, which refer to a succession of events that build on previously accumulated strengths, shaping an interwoven decision process. Others are more radical as they refer to something that did not exist earlier. But between these extremes is where most decisions stand, based on a combination of incrementalism and radicalism.

After that first innovative idea pops up, the immediate step is almost certainly to assess how feasible it is, at least in two different senses: in terms of the knowledge required and in economic terms. These two dimensions are very distinctive in the series of decisions they involve, the type and loci of advice they call for, and so on. Deciding whether to commit to a new development implies much more than the consideration of its pros and cons as it interacts with other considerations such as the existing portfolio vis-à-vis dimensions such as the technological, economic, markets and human capital. In a small country it also implies assessing the difficulty of getting the required inputs for production. In the pharmaceutical industry for instance, inputs for drug production are an important bottleneck as most of them are imported, thus the possibility of substituting them with locally produced inputs is crucial. So, the decision to enter into the production of a new drug implies a series of assessments, including, among others, the chance to get the required inputs produced locally, that is, the identification of a potential producer, the feasibility in terms of costs and production process and how successful the marketing strategy would be for this new product.

## THE CASES IN THE ANALYTICAL FRAMEWORK

In the next paragraphs, the three biotechnological firms are briefly put into the analytical framework just described above.

### The Firm's Innovation Base

#### Personal innovation drivers

*Biotechnology human health* The owner, a university researcher, incubated the firm in the Faculty of Sciences; after four years it was acquired by a national pharmaceutical firm, though retaining its distinctive molecular biology knowledge base. One of the strategies of the firm was to hire young biologists and biochemists attracted by the firm's challenging intellectual work. The firm was clearly oriented by the owner's scientific interests and resembled his former full-time academic job. In fact, more than 80 per cent of the firm activity is R&D.

*Biotechnology animal health* The owners are veterinarians with a large amount of entrepreneurial experience. They almost lost the firm when in 1994 a new sanitary policy threw them out of their star product's market, an innovative vaccine against foot and mouth disease, around which they made an important investment. After that, and after downsizing the firm, they tried different products and finally tried out a new strategy. They invested again heavily in state of the art facilities to develop a new series of biological cattle vaccines, mainly reproductive vaccines. In parallel they introduced organizational changes to better steer and monitor the firm's innovation portfolio, which for them is a key asset for the market position of the firm.

*Environmental biotechnology* The owner has trained himself, moved by his concerns and interests in environmental issues. Part of that knowledge was driven by curiosity and the interest in how other people solved problems, as well as through participating in seminars, congresses and being in touch with researchers (Spanish or Portuguese-speaking researchers) related to environmental issues, more particularly to the environmentally safe disposal of different types of waste. He started providing services to the Municipality in the countryside based on collecting and disposing of urban waste, evolving towards innovative as well as productive ways of coping with new environmental regulations.

**Accumulation of knowledge at firm level**

*Biotechnology human health* The members of the firm involved in R&D, particularly the original founder, are dedicated to actively searching for knowledge related to the firm's core area in molecular biology and medical applications, complemented by investments in laboratory equipment needed to cope with state of the art experimental work. Strong relations with academic researchers in the biomedical domain as well as open doors for young biologists and biochemists, some hired by the firm and some hosted for temporary projects, contribute to turn this firm into an intellectually strong and dynamic environment.

*Biotechnology animal health* The firm always had a 'technical periphery', including university researchers in Uruguay as well as Uruguayans abroad, mainly veterinarians. They also developed strong linkages with international bodies regulating and monitoring animal health and providing standards for best practices. Most recently they hired young biologists and biochemists on a permanent basis, while strengthening their relations with academic researchers in domains such as microbiology, virology, animal physiology and molecular biology.

*Environmental biotechnology* Relationships revolving around knowledge exchange are a core component of this firm strategy. The owner established cognitive relations with different kind of actors, from engineers to academic researchers driven by the need to find answers to difficult questions. A graduate of a new university field, food technology, was also given a permanent position to take charge of the technical demands of the blood flour plant and to contribute to the prospective plan to develop new value-added products from the same raw material. He has proven successful in establishing fruitful knowledge relations with very different kinds of contacts, probably through his ability to pose problems from the vantage point of actual practice.

### Screening Scope

For the three firms, the combination of personal drive, accumulated experience and reliance on external cooperation has substantively widened their screening scope. Innovative projects proposed to the firms by external actors have been an important factor through different modalities (direct contacts, bids). Thus, the firms' innovation agendas are fuelled by internal proposals and complemented by ideas and projects coming from outside. Even though the core strength in the three firms is the original one, particularly in the case of the oldest and best-established firm, the animal health firm, external demands have substantially widened its activities' scope.

#### Innovative opportunities

*Biotechnology human health* A main generic source of innovative opportunities for this firm stems from the capacity to 'navigate through the genome' and find new ways to use the related knowledge to determine genetic diseases. Biosciences in general and in this case in particular, entail the advantage of specificity and importance of local contexts, with important policy implications. For instance, the discussion in the Uruguayan Ministry of Public Health around whether or not to supply a vaccine for the human papilloma virus (HVP) which causes cervical cancer revolved around the efficacy of the imported vaccine: the hypothesis was that the

Uruguayan strain of the virus was different and the vaccine could be ineffective. So, this firm was hired to determine the characteristics of the local strains of this disease, an expertise that could be applied to other diseases, such as Hepatitis B. For this firm, the transformation of innovative opportunities into innovation projects is heavily dependent on the technological public procurement of the Ministry of Public Health and, more broadly, on the national health policy.

*Biotechnology animal health* As already mentioned, the specificities of the local milieu are a main source of innovative opportunities for this firm, given that bacterial strains are diverse and changing over time, so new specific vaccines need to be developed. Moreover, the diversity of animal diseases in need of immunizations is a permanent source of innovative opportunities, at least as candidates to enter its innovation agenda. At present, the transformation of innovative opportunities into innovation projects is mainly a result of an internal process of evaluation and decision-making.

*Environmental biotechnology* The innovative opportunities for this firm are associated with the capacity to provide new types of environmental services by transforming environmentally damaging waste into saleable products. The need to provide such environmental services at low cost calls for innovation in a diverse range of issues, from the kind of energy used to ways of exploiting every possible facet of the disposal to be treated. The innovative opportunities related to environmental activities fundamentally depend on regulations and the ability to enforce them, both at national and international levels.

**Market Opportunities**

The three firms share, to different extents, market opportunities. Part of their business is done through the 'open market', that is, to markets with no previously identified clients. This type of market can change due to external circumstances, mainly through public policy, particularly in businesses associated with human health and the environment, which are affected by such policies.

The tailor-made market opportunities have a different level of importance for these firms. While for the environmental firm it represents a main part of its business (the privileged relationship with a big slaughter industry), for the animal health firm it represents the way to explore new avenues outside their main area of experience. Targeted demand is a particularly important driver of market opportunities for the animal health and human health firms. In the latter case, targeted demand is associated with its main clients, medical doctors. Through different procedures such professionals become acquainted with state of the art diagnostic practices and push to have them available in their working environments, leading to a type of targeted demand particularly suitable to the firm expertise.

What kind of public policies to support the development of biotechnology firms' innovations can be derived from the analyses just made? Interesting indeed, more than innovation policies per se, what seems to be needed are pro-innovation measures within specific policies: agriculture policies, health policies and environmental policies. If one of the marketing features of Uruguayan meat exports is the exceptional animal health in the country, the market for animal vaccines will grow as a consequence of a public policy that will enforce the need to have healthy animals. If this is combined with a policy to develop local providers of vaccines, the market growth will be a business as well as an innovation opportunity for the animal health firm. A similar reasoning is valid for the other two cases. Public procurement from the Ministry of Health for cheap innovative solutions to different types of diagnostic kits can open up a market that stabilizes the business activities of small health-related biotechnology firms and incentives innovation by lowering the risk of innovation investments. Carefully designed environmental regulations, financial instruments to help polluting firms to cope with such regulations and a robust system of enforcement, would open up a market for innovative solutions to several environmental problems, widening the business opportunities for biotechnology-based environmental innovative firms.

## SIMILARITIES AND DIFFERENCES IN DECISION-MAKING IN BIOTECHNOLOGY IN URUGUAY AND CUBA

Biotechnology-related research in Cuba is concentrated mainly in centres dealing with specific aspects, including agriculture, animal science and human health. Three distinctive features are worth mentioning in relation to these centres; they shape the unique landscape in which the extremely successful Cuban biotechnology has developed. The first feature is the high scientific status of those in charge of the centres: the centre's administrative chiefs are first-rate scientists themselves. The second feature is perhaps the most 'Cuban' in the sense of being related to the non-market economy of the country: even if they are research centres as any other research centre in the world, they are involved as well and sometimes command the whole process up to production. That is, they establish links along the chain, closing the full circle, or what they call '*proceso a ciclo completo*', and so their decision-making process regarding research agendas and innovation agendas has points in common with those that would be followed by a firm. A third distinctive feature, also fundamental in the trajectory of Cuban biotechnology development, has been the continuous influx of support in the form of a compulsory and explicit strategy for public health procurement. This feature is not coherently present in Uruguay, but it has been present in several developed countries, acting as 'market builders' for firms faced with weak internal private markets. It is important to stress that in any innovative endeavour related to human health the quality of public health policy is paramount: each vaccine, each pharmacological product, must undergo a very complex and internationally regulated procedure of clinical trials to be validated, and thus be able to be patented and commercialized. The Cuban health policy system has provided an excellent companion to biotechnology health innovations by allowing efficient and reliable validation.

The Cuban biotechnology business model has another distinctive feature: the engagement with international partners for distribution and commercialization, and sometimes for research itself. The control of patents, however, when they belong to national centres, is maintained. They might let the partner participate in the exploitation of the patent but never lose their command in that process, and always keep the right to manufacturing. This has been the strategy of the Center of Molecular Immunology at the West Scientific Pole, which today has more than 400 patents around the world. They sell the product and then receive a royalty from the sales of their partners, keeping control both at the beginning and at the end of the process.

To learn something meaningful for Uruguay from the comparison with the Cuban situation, the differences in the logics of decision-making at firm level induced by economic and political differences must be taken into account. Uruguayan firms face bankruptcy if their decisions turn out to be consistently wrong or misplaced; even a single mistake can have such an effect: market rules are merciless. The Cuban biotechnology centres '*de ciclo completo*' can better afford eventual mistakes, first because they have many more initiatives as an entrepreneurial unity than any small biotech firm in Uruguay and they can spread the damage among several projects, and secondly because they are strongly backed by the state budget. Cuban biotech innovative projects have a clear strategic goal: to sustain and further develop the scientific, technological and health policies trajectories that put the country among those in the world with better health indicators. Patriotism can provide a great innovative framework (history gives us many examples); the same goes for solidarity. The key point that

Cuba incarnates in terms of innovation decision-making is the 'solidarity-through-innovation' strategy: the demonstration that this strategy is possible, efficient and can additionally have a positive commercial side, is formidable.

Uruguay is a country long distinguished by having lower inequality among the market economies of Latin America; moreover, solidarity and substantial concern for social inclusion have been explicitly present in public policy strategies since a leftist government took office in 2005. However, unlike the Cuban experience, public policy related with the well-being of marginalized people has not been addressed thoroughly through science, technology and innovation; STI have been mainly conceived as a driver of competitiveness and concomitantly of economic growth. But the examples we have given from the three Uruguayan biotechnology firms show that if a sound governmental 'biotechnology procurement' strategy is put in place by the Ministries of Health, Environment Agriculture and Industry, there are local entrepreneurial capacities to respond. In fact, despite the difficulties faced by innovative firms in a market economy that does not have a clear innovation strategy, the firms we have examined have developed working solutions for problems that enhance the quality of life of different layers of the population. Our assumption is that much more can be accomplished, but for that to happen an aggressive, consistent and long-term oriented hands-on public policy is needed. This is also what the owners of the firms themselves have said.

## CONCLUDING REMARKS

This chapter analyses the chain of processes at the micro level through which innovation decisions and agendas are shaped in three biotechnology firms in Uruguay. These underlying processes are sketched in an analytical framework that structures the narratives around the cases. Innovation decisions and agendas are embedded in a complex setting with interacting dimensions. Some of these are: the firm's innovative base, innovative and market opportunities, intertwined with the evaluation of what science and technology have to offer, costs, and consideration of the type of customers and their driving forces towards innovation.

Decisions revolve around a complex set of trade-offs between the accomplishable and the desirable, between existing resources and capacities, and foreseen opportunities, or between complementary strategies that bridge the present with the future, and existing constraints with targeted goals. Decisions are then sieved through the world of the possible, but not totally constrained by it. They stand on processes that combine doses of bounded rationality with craftsmanship in the art of finding opportunities around market niches, problems in search of a solution and pushing for the knowledge frontiers looking for a gap to fulfil, given severe constraints at the firm and national levels. In the same vein, agendas involve both ideas and projects with varying levels of plausibility, concreteness and materialization. The factors influencing the transition from ideas to projects are diverse, some interconnected and even sequential, while others run along independent dimensions. An important factor pulling ideas and projects into the agenda is based on a user–producer relationship. External demands are important triggers in the three cases. Internal strategies implemented to search and screen new ideas vary, depending on the intramural cognitive strengths available. Still in all of them, the role of internal triggers is very important and strategic. The difference lies more in the extent to which the whole process can be completed within the firm or whether it calls for external actors to get involved in the process.

The Cuban case confirms the paramount importance that policies play in the orientation of innovation agendas and decisions. Their scope and continuity contributed to the building of a learning path paved by the accumulation of existing and new capabilities, while nurturing a supportive environment for biotechnology research and a specific mode of entrepreneurship. The Uruguayan firms also built a learning path based on the accumulation of capabilities, both in terms of nurturing new knowledge and mastering upgraded know-how. Their business environment is fragile, though, and this leads to a more conservative strategy in terms of innovation decision-making. Having room to explore and inevitably commit mistakes without being ruled out of business is part of what any sound innovative strategy should guarantee. A strong demand from the biotechnology-user policies – health, environment, agriculture, industry – rooted not only in productive concerns but very heavily in social concerns will be able to provide for this, enhancing the Uruguayan capacity to respond to developmental needs through innovation.

## NOTES

1. See, for instance, the Innobarometer (European Commission, 2004).
2. The economic importance of this problem is great, particularly nowadays, given the price of meat in international markets (and given the cost of high quality semen). In the case of milk cattle it is even more evident, because cows need to be out of milk production for around six months to be able to get pregnant, so abortion implies a severe economic loss.

## REFERENCES

Birkland, T.A. (2001), *An Introduction to the Policy Process: Theories, Concepts and Models of the Policy Process*, New York: M.E. Sharpe.
Bortagaray, I. (2007), 'The building of agro-biotechnological capabilities in small countries: The cases of Costa Rica, New Zealand and Uruguay', PhD dissertation in Public Policy, Atlanta, Georgia Institute of Technology.
Cohen, W.M. and D.A. Levinthal (1990), 'Absorptive capacity: A new perspective on learning and innovation', *Administrative Science Quarterly*, **35**(1), 128–52.
Dasgupta, S. (1996), *Technology and Creativity*, New York: Oxford University Press.
European Commission (2004), *Innobarometer 2004: Flash Eurobarometer 164*, Brussels: European Commission.
Ilori, M.O. and I.A. Irefin (1997), 'Technology decision making in organisations', *Technovation*, **17**(3), 153–60.
Kingdon, J.W. (1984), *Agendas, Alternatives, and Public Policies*, Glenview, IL: Scott, Foresman and Company.
Lazonick, W. (2005), 'The innovative firm', in J. Fagerberg, D.C. Mowery and R.R. Nelson (eds), *The Oxford Handbook of Innovation*, Oxford: Oxford University Press.
Lazonick, W. and M. O'Sullivan (2000), 'Perspectives on corporate governance, innovation, and economic performance', *Targeted Socio-Economic Research Programme*, Brussels: European Commission.
Lundvall, B.-A. (1988), 'Innovation as an interactive process: from user–producer interaction to the national system of innovation', in G. Dosi, C. Freeman, R. Nelson, G. Silverberg and L. Soete (eds), *Technical Change and Economic Theory*, London: Pinter.
Lundvall, B.-A. (1985), *Product Innovation and User–Producer Interaction*, Industrial Development Research Series, **31** Aalborg: University Press.
Nelson, R. and S.G. Winter (1982), *An Evolutionary Theory of Economic Change*, Cambridge, MA: The Belknap Press of Harvard University Press.
Nielsen, R. (2007), 'Human resources in innovation systems, with focus on introduction of highly educated labour in small Danish firms', PhD dissertation, Department of Business Studies, Aalborg: Aalborg University.
Snoeck, M., J. Sutz et al. (1992), *Tecnología y Transformación: La Industria Electrónica Uruguaya como Punto de Apoyo*, Montevideo: Trilce.
Srinivas, S. and J. Sutz (2008), 'Developing countries and innovation: Searching for a new analytical approach', *Technology in Society*, **30**, 129–40.
Vincent, L. (2005), 'Innovation midwives: Sustaining innovation streams in established companies', *Research Technology Management*, **48**(1), 41–9.

# 4. Biotechnology, university and scientific and technological policy in Cuba: a look at progress and challenges

**Jorge Núñez Jover, Isarelis Pérez Ones and Luis Félix Montalvo Arriete**

## INTRODUCTION

Biotechnology is an important and strategic area in Cuba's scientific and technological policy. The primary locus of the Cuban biotech industry lies in the West Scientific Pole [cluster] of Havana (here referred to as the 'Scientific Pole'). The highest priority has been given to the application of biotechnology to the medical-pharmaceutical industry. Moreover, also underway are important projects for agriculture, food production, animal husbandry and environmental protection. The results of biotechnology benefit the Cuban health care system and they represent an increasing share in national exports.

The development of biotechnology, the establishment of the Scientific Pole and its breakthroughs, underscore the priority that the Cuban state has given to scientific and technological development and the training of researchers during five decades. Different studies have been made on this topic (including, Lage, 1994; Kaiser, 1998; Majoli, 2002; López et al., 2006; López et al., 2007).

This chapter emphasizes the following issues:

1. It establishes the relationship between the development of biotechnology in Cuba and the S&T policy and the decision-making mechanisms in this field.
2. It analyses the role of the university in biotech development, building on previous work (Núñez and Pérez, 2007; Núñez et al., 2008; Pérez and Núñez, 2009).
3. It illustrates, through three success stories, some of the progress made and challenges faced both by biotech industry and the S&T policy.

We will begin by commenting on the institutional framework created by the Cuban state, regarding the country's S&T policy decision-making. Then, we will show how the decisions were adopted to drive the development of the biotech industry. These decisions, like many others in the field of science and technology, were adopted at the highest level of the Cuban political system. The biotech industry has had the personal initiative and support of Cuban President Fidel Castro. This is a very important element in the comprehension of the decision-making process in science and technology in Cuba.

Next, we will assess the involvement of Cuban universities in the field of biotechnology, through the success stories of the Molecular Immunology Center, the Biomolecular Chemistry Center and the Institute of Plant Biotechnology; we will provide examples to illustrate decision-making in this area and the cooperation dynamics between the centres of the Scientific Pole and the universities. We will conclude with a discussion on some issues regarding the national S&T policy and its links with biotechnology.

## THE NATIONAL SCIENTIFIC AND TECHNOLOGICAL POLICY: AN INTEREST IN THE SOCIAL USE OF KNOWLEDGE

Since the 1960s the Cuban state has made political decisions that have led to major breakthroughs in science and technology. Unlike other countries where other factors, such as the market, have played a decisive role in the scientific and technological development, probably the most important element to explain such a development in Cuba has been the political priority given to it. Since the 1960s, the highest leadership of the state has backed science and education to become driving forces of development. Also since the early 1960s, the country devoted great efforts to the health care sector. The case of biotechnology reflects that priority very well. The fruits of the previous investment in education, science and health care have been reaped in this field.

The scientific community that led the major scientific transformation in the country emerged in parallel with the consolidation of the revolution that took place in January 1959. National science, with a background in medicine, for example, is basically the outcome of the revolution in power. Generally speaking, this scientific community took as its own the purposes of the new political project and has since then supported and helped shape the scientific strategies of the country.

It is known that Cuba's economic situation worsened considerably[1] in the second half of the 1980s, but the interest in science and technology, particularly biotechnology, remained unchanged. For example, in December 1991 Cuba's President proclaimed that the survival of the revolution, socialism and the preservation of national independence relied on science and technology (Castro, 1990; Castro, 1991a). It is very unlikely that you will find this kind of political perception on the role of science and technology in other Latin American countries in the second half of the twentieth century.

The starting point of the national S&T policy has always been the priorities for the economic and social development of the country. As a result, special attention has been given to the use of scientific and technological knowledge in furthering the goals of inclusion and social equity. Health care, science and education have always been considered key elements to achieving these goals. This is an important aside, in order to understand what will subsequently be presented regarding the development of the biotech industry.

We will show how the decision-making process has been formulated to be able to cater to the priorities of economic and social development of the country. Our focus is on the actors and the organization of the Science and Technology Innovation System. We do not seek to make a historical analysis of the evolution of the S&T policy (see for example García Capote, 1996), although we have considered some historical references in order to grasp how the bases of national science were shaped as well as the role played by the university.

Of great importance to understand the evolution of the Cuban S&T policy are the transformations that took place in the field of higher education in the 1960s. The University Reform (National Council of Universities, 1962), implemented in the 1960s, played a very important role. Before 1959 there was barely any scientific research going on in Cuban universities (there were at the time three public and some private universities). In the early years of the decade, a number of professors and many professionals left the university and the country. In the case of medicine, it is estimated that nearly 3000 professionals, around 50 per cent of the total, left for other countries. The structure of the university enrolment, very much centred on liberal professions and humanities, did not favour scientific education either.

The education in Cuba and abroad in science and engineering programmes were among the main priorities of the university reform. Research became part of the obligations of university professors and the students' curriculum.

The curriculum was drawn up, again emphasizing research and the involvement of students in the social praxis. Quite often, the locus in the training of students stopped being the university campus and expanded into research centres and the productive sectors; the link with social praxis became the axis of the educational model.

The scholarship plans were very important as they allowed for many young students from diverse social backgrounds to begin training as scientists and researchers. These changes generated increased social mobility and transformed the social composition of the faculty and student bodies.

As pointed out above, the years following the university reform saw important processes taking place regarding the S&T policy:

1. Incorporation of research and high-level training into university life.
2. Establishment of an institutional basis for science. Research institutions and groups were established or added to the universities within a decade, and tens of research groups and centres were created inside and outside the universities.
3. Important progress was made in the organization of scientific research. It included the setting up of scientific agendas based on the country's needs, as well as instructions for the practical application of results.
4. Emphasis on high-level training of human resources (master's, specialization and doctoral programmes) with the support of international exchanges.

The aforementioned point can best be illustrated with the founding of the National Center for Scientific Research (CNIC) in 1965. CNIC was founded on 1 July 1965, with Fidel's personal involvement, and it became part of the University of Havana in 1966. CNIC played a fundamental role in the training of researchers and the use of international cooperation. In the history of Cuban science, CNIC became a breeding ground for research institutions and high-level researchers. Many of the centres that are currently part of the Scientific Pole can trace their origin back to CNIC.

Dr Wilfredo Torres, general manager of CNIC from 1966 to 1976 (Torres, 2006) says CNIC is the expression of Fidel Castro's policies as he had visualized the future of the country as a future of men of science since the early 1960s. That policy required the establishment of teams of researchers. CNIC was established with a mission to conduct research and development and training of human resources at the highest level in areas of national interest.

Torres has said that in order to undertake these ambitious plans they relied on a core of Cuban professionals and a significant number of foreign researchers. Their selection and hiring was the result of a tour that took Torres to some 40 research centres in several countries (Torres, 2006).

In the early 1970s CNIC was entrusted with the training of professors of basic sciences for the universities, in particular for the different specialities of the medical sciences (CNIC, 1975). In the second half of the 1970s changes in the political system of the country transformed and institutionalized the mechanisms through which S&T policy decisions were made.

The strategy for the economic and social development of the country is defined at the congresses of the Central Committee of the Communist Party, and the decisions adopted by the state's highest body, the National Assembly of the People's Power (ANPP), which holds two sessions every year. Every year the assembly approves the national economic plan, the state budget, and sets the country's priorities. From these priorities stem the priorities of the Science and Technological Innovation sector in the country. For example, the priorities for 2008–15 include: food production, sustainable energy development, health care and the people's quality of life, ICT development, environment and climate change, social sciences and basic sciences (CITMA, 2008).

The Council of State of the Republic of Cuba, elected by the ANPP, occupies a very important place in the scientific development of the country. Under this body, for instance, fall two of the main technological undertakings in the country: the Scientific Pole, main actor in Cuba's development in biotechnology, and the University of Information Sciences (UCI), key actor for the computerization of Cuban society and a source of important revenues that the country earns from exports of ICT products and services. Considerable resources have been channelled to these objectives.

The Ministry of Science, Technology and the Environment (CITMA) is the ministry in charge of 'proposing and assessing the strategy and the scientific and technological policies consistent with the economic and social development of the country, establishing the relevant objectives, priorities, working lines and programs and overseeing its implementation'. CITMA recommends the approval of the budget for R&D and, in tandem with other bodies of the Central State Management Agencies (OACE) and research institutions, structures general guidelines for research. Once approved by the Council of Ministers, such guidelines are translated into R&D plans that define the objectives, stages and required resources.

Since 1994 the national S&T policy has set out to attain a better link between science and technology activities and the productive sector, emphasizing innovations, trying to integrate the main innovation actors with a system of Science and Technological Innovation (STI).

The main actors of the STI system include the OACE, universities and centres for research, centres for scientific and technological services and units for scientific and technological development, a total of 211 institutions. Also part of the STI system are the financial institutions and a number of social movements (Forum on Science and Technology, National Association of Innovators and Downsizers and the Technical Youth Brigades) and organizations that seek to strengthen the connections between these actors and the links between innovation and the needs for social and economic development (Núñez Jover and López Cerezo, 2008). These organizations include the Productive Scientific Poles, whether regional or sectoral. The latter include the West Scientific Pole of Havana.

Biotechnology ranks very highly in the Cuban S&T policy. Biotech projects are included within the national priorities with regard to the production of food, health care and energy and the environment (CITMA, 2011).

There are three National Science and Technology programmes[2] to respond to these priorities. The National S&T programmes are instruments of the S&T policy that intend to achieve the greatest coherence possible between the development strategy of the country and the development objectives of science, technology and innovation.

Directly linked to the biotech industry is the National S&T programme of the agricultural biotech sector, including plant and animal biotechnology; the National S&T programme for the Development of Biotech, Pharmaceutical and Green Medicine Products; the National S&T programme for Human and Veterinary Vaccines.

However, the mechanisms through which science and technology agendas are defined, as well as their relation to the economic and social strategy of the country, are not limited to the above-mentioned schemes. Exchange takes place at the highest levels of government, ministries, companies, universities and research centres; the political leadership of the country meets with the scientific community of Cuba and elsewhere (as in the case of biotechnology, as will be seen later on), contacts that generate initiatives that influence the production and social use of knowledge.

## ORIGIN AND DEVELOPMENT OF THE BIOTECH INDUSTRY

The interest of the Cuban government to develop a biotech industry in the country was first displayed in the early 1980s. Around that time, Professor Randall Lee Clark, President of the MD Anderson Cancer Center in Houston, Texas, visited Cuba and in a meeting with Fidel Castro he advised him to initiate production of interferon in the country[3] (Herrera, 2008). In January 1981 the decision was made that a small group of Cuban researchers were going to start working on the project. Two of them were sent to the United States to acquire training under Professor Clark; another six went to Finland to the laboratory of Professor Kari Cantell in Helsinki, the world's first interferon-producing laboratory (Limonta, 2002). Another researcher went to France in order to learn the recombinant form of production of interferon (Herrera, 2008). This group, comprising researchers from CNIC and the Clinic of the Ministry of the Interior, was established and subsequently directly supervised by the President (Herrera, 2008).

Back in Cuba, with interferon production technology in place, the group began working in a government facility turned laboratory. In May 1981, the group presented the outcome of their work to President Castro. This first Cuba-produced interferon was taken to Finland to be tested at Professor Cantell's laboratories, where it was found to meet all required quality standards and very soon was introduced into the Cuban health care system to fight an epidemic of haemorrhagic dengue (Limonta, 2002).

Later the same year, the Center for Biological Research (CIB)[4] was created in the facilities where interferon had been obtained. The founding of CIB was seen as 'an early and decisive support to the vision of rapid development of medicine advocated by the Cuban government' (Limonta, 2002, p. 4).

In June 1981, the government made an additional important decision: founding the 'Biological front' whose objective was geared to strengthening and coordinating the work and research of various institutions and scientific groups in the field of biology and biotechnology in Cuba (Majoli, 2002; Limonta, 2002). The objective was to promote interactions and synergies between scientific and productive institutions and the government.

But the interest of the Cuban government went beyond that: 'at the highest level of government there was a prospect of development geared to achieving the highest levels of biotechnological science in the world' (Limonta, 2002, p. 5). In December 1982, UNESCO/UNIDO called for the creation of an International Center of Genetic Engineering and Biotechnology (CIGB). Cuba was hoping to be selected as the host country for the institution. Instead, Italy and India were chosen as venues in December 1983, and two facilities were built in Trieste and New Delhi. However, the Cuban Government decided to build a centre with its own resources[5] (Limonta, 2002; Herrera, 2008). The universities contributed professors and researchers to the establishment and development of the new institution.

On 1 July 1986, the Center for Genetic Engineering and Biotechnology (CIGB) opened with cutting-edge facilities and, above all, a group of highly trained and motivated young scientists (Majoli, 2002).

The institutionalization of biotechnology in Cuba (Limonta, 2002) was based on some key tenets. The objective was the establishment of an institution for research and production, which entailed, for instance, making all the necessary research to obtain interferon and producing the required amounts to meet the demand of the country. Objectives and phases to achieve this were clearly defined. Communication and interaction of researchers allowed for the full familiarization with the project as a whole, even when the tasks of each of the participants were very well defined. The continued training of professors and researchers became a priority, as well as the search for advanced methods of quality control.

These conceptual foundations have hitherto characterized biotech research in Cuba, at least in the centres of the Scientific Pole. Lage (1999) states that many of them can be identified as features of what he calls 'the experiment' of Cuban biotechnology.

In sum, the biotechnology efforts were powered by two fundamental objectives: meeting the social needs, particularly the progress of the Cuban health care system; searching for sources of revenue from foreign markets for the national economy. This explains the formidable take-off of biotech research, essentially in association with the medical-pharmaceutical sector. The interest in biosciences was also influenced by the need to deal with biological attacks against Cuba[6].

As mentioned before, the interest in biotechnology did not decline in spite of the strong economic crisis of the 1990s and the tightening of the US blockade. One of the most outstanding examples of the progress made by the biotech industry is the development and obtainment between 1980 and 1989 of the antimeningococcal vaccine VA-MENGOC-BC®[7], led by the Finlay Institute[8]. This is the first and only vaccine in the world that is effective against meningococcus type B. Unlike the production of interferon, VA-MENGOC-BC® is an absolute innovation. Not only was the vaccine new to Cuba but also to the world[9]. In 2010 the leading Cuban biotech product was, for the third consecutive year, Heberprot-P10, with sales exceeding 40 million CUC. In March 2011 this vaccine received the Gold Medal from the World Intellectual Property Organization for its remarkable contribution to the treatment of more than 18000 patients in Cuba and other countries (Díaz, 2011, p. 2).

The Scientific Pole was formed rapidly and now comprises more than 52 institutions, over 12000 staff workers, 7000 scientists and more than 900 patents submitted (Lage, 2006; López et al., 2006). Broadly speaking, the Scientific Pole is not only limited to the institutions that are geographically located to the west of the Cuban capital. There is a whole group of institutions in the country that, while subordinated to different OACEs and located in various territories, share working objectives that contribute to the development of the research going on in Scientific Pole. Fidel Castro's concept is reflected here when he said that the Scientific Pole is an 'instrument of cooperation and mutual support between research centers whose objective is to get all centers to cooperate, help each other in teams, in personnel, in experience, in knowledge, which multiplies their possibilities' (Castro, 1991b).

Within this broad concept, some of the centres that are the core foundation of the Scientific Pole include:

1. Center for Genetic Engineering and Biotechnology (CIGB)
2. Molecular Immunology Center (CIM)
3. National Center for Scientific Research (CNIC)
4. Center for Immuno Assays (CIE)
5. National Center of Laboratory Animals (CENPALAB)
6. National Center for Biopreparations (BIOCEN)
7. Center for Neurosciences of Cuba (CNC)
8. Institute of Tropical Medicine Pedro Kouri (IPK)
9. Finlay Institute
10. National Center for Plant Protection and Animal Health (CENSA)
11. International Center for Neurological Restoration (CIREN)
12. Biomolecular Chemistry Center (CQB)

In addition to the research facilities, auxiliary institutions have been created in order to further the commercial aspects of the whole biotech venture. The marketing of the biotech production began in late 1983. In connection with this, the trading company Heber Biotec SA was incorporated. It currently distributes its main products in over 45 countries and owns over 400 registrations for pharmaceutical products. In 2010 sales were expected to reach over USD9 million. Vacunas Finlay SA is the company that has exclusive representation for the negotiation and marketing of products, technical and consulting services for the Finlay Institute. CIMAB SA is the exclusive representative for the marketing of products and services of the Molecular Immunology Center. Consequently, nearly all centres of the Scientific Pole have their own marketing companies, thus closing the research–production–marketing cycle. Currently, these companies export to more than 60 countries, making their parent institutions at present the second exporter of goods in the country with a sales turnover in the vicinity of USD400 million.

In recent years, the Scientific Pole and their business system have strengthened their work together with foreign companies from China, Vietnam, India and Iran, where similar centres to those of the Scientific Pole have been established. There has also been cooperation with companies from Canada and the United States. There are technology transfer agreements or negotiations under way with around 20 countries.[10] The negotiations include both products and intangible assets (patents, technology, tacit knowledge, scientific capability).

The development model of Cuban biotechnology has been successful from scientific, social and economic points of view and is characterized by the following (Lage, 2006):

1. Taking place in a developing country with limited material resources and little industrial development; it has basically been conducted by the state and practically without any direct foreign investment, although cooperation through joint ventures is making progress. The financing of the investments come from revenues resulting from sales.
2. Integration of research, production and marketing activities. Very similar to what is claimed by Gibbons et al. (1994), regarding 'Mode 2' production of knowledge, in the biotech sector there is work on the 'context of implementation'. Interaction and networking are encouraged, appropriate quality requirements are used (the production of papers[11] or the obtainment of academic acknowledgement is less appreciated than the solution of a health condition or the obtainment of economic success through sales abroad) and cross-disciplinary work is fostered. As can be seen, from the point of view of the production of knowledge, the work takes place, up to a certain point, within a model that is similar to the international biotech industry, although in the Cuban experience, unlike that in other countries, the research centres engage in a 'full cycle' scheme, from the production to the use of knowledge.
3. Differences in the social ownership of the results: the Cuban population – and quite often persons from other countries who receive the benefits of the Cuban medical cooperation – acquire ownership of these breakthroughs free of charge or at very low prices. As these are state-owned companies, the benefits belong to the people. These benefits are used to help maintain Cuba's health care system.
4. The ability to work simultaneously on issues at the scientific and technological cutting edge and on more traditional products, generating economic opportunities for the country. The latter allow for major learning that is reflected in the consolidation of the industry and, together with medical services, becomes an important factor in terms of trade with several countries. Some of the selected technological trajectories are consistent with epidemiological profiles different from the one that characterizes Cuba. This is the case, for instance, for the vaccine against cholera, which does not pose a health care issue in Cuba but that nevertheless is a chief priority for other countries. The

objectives of Cuban biotechnology are consistent with the interests of countries of both the South and the North.

5. The excellent communication with the state and the government. Relations with the Council of State are continuous and permanent. Some tens of researchers are members of the Cuban Parliament and the general manager of the CIGB is a member of the Council of State, which affords this sector of the scientific community a high political representation.
6. The scientific and technological community of the Cuban biotech industry is characterized by values such as devotion to work and public-service orientation. Although there are different ways to provide the members of this community with economic incentives, the benefits obtained are much less than the services rendered and the wealth created.
7. The cooperation of the Scientific Pole centres with other Cuban institutions, such as the Cuban health care system and its network of hospitals and family doctors, as well as the system of higher education.

The progress has been based on: priority of biotechnology within the S&T policy; strength of the health care system; implementation of various management modalities suitable to the purposes of the biotech industry (this management has always been led by ambitious indicators and the permanent search for the new); closed cooperation between actors; expansion of international markets; use of international cooperation; relatively rapid learning in terms of quality management and handling of intellectual property rights; appropriate social perception on the contribution of biotechnology to development, based on the trust of the population in Cuban medicine.

## HIGHER EDUCATION AND BIOTECHNOLOGY

The Higher Education System in Cuba comprises a total of 65 institutions (IES)[12] All higher education institutions carry out continuous education, postgraduate training and research, with various degrees of intensity and depending on their academic profiles. Some ten of these IES carry the main burden of research and doctoral training. Most of them are connected to the Ministry of Higher Education.

Higher education contributes chiefly to the development of biotechnology through the training of graduates in the programmes of interest for the Scientific Pole as well as postgraduate training in specialities, master's and doctoral programmes. For instance, around 30 per cent of the doctors trained in the University of Havana (UH) every year (some 100 in total) are connected to areas of interest for the development of biotechnology. An example of such a connection is the cooperation between UH-CIGB in doctoral studies in vegetable biology, where there exists a joint academic committee with exchange of professors. This collaboration led to ten doctoral dissertations by researchers at the Scientific Pole.

Hundreds of researchers and technologists of the Scientific Pole have been trained in master's programmes at the UH, and the same is true for programmes at other universities. Within the higher education system training programmes have been established, based on the needs of the industry in areas like bio-informatics and biomedical engineering. It has also created teaching facilities in the Scientific Pole to train professionals in programmes of interest. When the students of programmes such as chemistry, biology, biochemistry and pharmacy, among others, and related engineering programmes complete their university studies, one priority workplace offered to them is a job at the centres of the Scientific Pole.

Several universities carry out activities in the fields of bioscience and biotechnology in three main areas: human, plant and animal biotechnology. The main areas of study by universities and research centres in the field of human biotechnology include the development of vaccines, medicines and diagnostic kits. Vaccine development, in particular against cholera, *haemophilus influenzae* type B and classic swine fever, as well as the introduction of natural products through synthetic means, are areas in which several research centres are working.

The UH is developing a series of industrial applications including the microbial treatment of waste water polluted with organic and inorganic compounds, the selection and physiological manipulation of microbial strains with enhanced capacity in metal removal, and molecular microbiology and biology that is oriented to the biodegradation of textile effluents, colouring and others.

In the field of animal biotechnology, one of the developing areas is animal health, in particular the development and application of advanced technologies for diagnosis of animal diseases, characterization of pathogens, making new generation vaccines as well as genetic and animal breeding.

The field of plant biotechnology is enjoying more development in the universities and higher education research. The Institute for Plant Biotechnology (IBP) under the Central University of Las Villas (UCLV), the Bioplant Center of the University of Ciego de Avila (UNICA) and CENSA are leaders in this field, primarily in lines related to genetic transformation and breeding of plants in order to increase resistance to biotic and abiotic factors; the development and transfer of new methodologies for efficient and cost-effective micropropagation of plant species and the development of effective methods for the production of biomass to obtain secondary metabolites. Genetic transformation of plants is of utmost importance. Recently Cuba stated that it is making progress in the development of genetically modified crops[13] in order to contribute to the country's food production. Some university teams are also working on this objective.

Other areas of development in the field of plant biotechnology include the production of biological products, such as bioplaguicides, biofertilizers and biostimulants and growth regulators, and the use of micro-organisms as agents for biological control.

Most of this research and production lines were established in the 1960s and 1970s. There have been small biotechnology-related research teams working in many universities and research centres ever since. For example, there was a biotech team working on fruit trees at UNICA, based at the Higher Agricultural Institute back then, where they made the first *in vitro* cultures of citrus trees.[14] This group led to the establishment of the Bioplant Center in that university in 1990.

## S&T POLICY DECISIONS IN THE FIELD OF BIOTECHNOLOGY THROUGH SOME SUCCESS STORIES

The following section will examine the experiences of three institutions that work in the field of biotechnology. These are interesting examples that reveal the progress and challenges of biotechnology in Cuba and also the priorities of the national S&T policy. These cases also illustrate the involvement of higher education in the field of biotechnology.

### Molecular Immunology Center (CIM)

The Molecular Immunology Center (CIM) was founded in December 1994 and its main objective is to obtain and produce new bio-pharmaceuticals geared to the treatment of cancer and other chronic, non-communicable diseases. The workforce of the centre hovers at around 600 workers, most of them scientists and engineers from various speciality backgrounds.

CIM is one of the most important centres within the Scientific Pole and is focused on two main areas: the culture of higher cells, and cancer research.

Work in this centre is based on some basic ideas (Pérez, 2008b). First of all the articulation of science and economy: ensuring that its products generate sustained economic results. In this connection, the clear identification of objectives and rapid results become indispensable characteristics. The economic goal is an important driving force of research. Secondly: competitiveness based on innovation. Competitiveness ranges from technological capability to scientific capacity and the added value provided by new knowledge. The novelty of products resulting from research is what allows market positions to be taken, which is called differentiation strategy. Thirdly: the advance of the institution through cooperation networking. Networks encourage connectivity between institutions, which becomes a factor of competitiveness and integration. Fourthly: motivation as a fundamental element of human capital. Upgrading, master's and doctoral programmes, recruitment and selection of new staff as well as obtaining academic ranks in the field of research and teaching are essential.

CIM undertakes different types of projects. Depending on the characteristics of the projects, the universities have a greater or lesser level of involvement in project development. An important feature that was discussed in general terms in the biotech sector can also be noted at CIM: the ability to work simultaneously on issues at the scientific and technological cutting edge and on more traditional products, thereby generating economic opportunities for the country.

Consequently, problems leading to innovations and inventions can be identified as well as projects that can bring about patents. If there is no possibility of generating the expected patent in a period of three to four years, the project is abandoned. If the project continues, animal tests are run to ensure technological feasibility, followed by 'concept tests'[15] (with patients) to obtain preclinical data. These tests significantly reduce the risk of failure and increase bargaining power, even when the product is not finished.

In sum, the logic behind this type of research is as follows: discovery–concept test–patents–negotiations–financing. Obviously, cooperation with universities is more important in the discovery phase.

The objective of these research projects is market penetration in industrialized countries, through innovations originating in the South but of interest to the North. The idea that the South can be innovative and solve problems for which the North has found no solution is an idea that not only makes economic sense, but also expresses certain political views associated with the purpose of ending technological dependence.

For example, one of the tumoural targets chosen by CIM is related to the receptor of the epidermal growth factor (EGF), a protein in the cell of the tumoural membrane. 'We have a monoclonal antibody that blocks this receptor and is being used in Cuba and developed countries', says Dr Rolando Pérez, deputy director of CIM (2008a, p. 26). This is a very important result and is likely to become the first product of the Cuban biotech industry to be registered in Europe. At present Phase I, II and III clinical trials have begun in Japan and Canada–USA, the latter with the approval of the Treasury Department.

Against this target, they have a vaccine that removes the EGF (through the sequestration of the tumour-stimulating hormone). They are also investigating a type of antigen – gangliosides – that is present in tumours. CIM has targeted a ganglioside whose expression is altered in tumours and not in normal tissue. Based on this finding, they are developing antibodies and vaccines to attack tumours. CIM has a big chance with this type of ganglioside because there is no other group working on it. There are other examples of results that are very novel worldwide.

The projects underway are classified in closed-cycle projects, basic and development structured projects and projects called 'the scientific question of the year'. 'The scientific question of the year' projects deal with exploratory research, where scientists pose interesting questions. The goal is to promote the asking of questions and the identification of novel scientific issues that increase the likelihood of achieving relevant results from the scientific point of view and social use.

CIM also develops a 'bridge strategy' consisting of obtaining low-risk products, mainly 'me too' types of product. These products let you have a positive cash flow and they let you gain experience in industrial scaling. The most important is the production of erythropoietin, with which the markets of Cuba and Brazil are completely covered.

CIM has established a legally independent trading company (CIMAB) registered with the Chamber of Commerce of the Republic of Cuba. The company ended 2008 with a sales turnover of more than USD50 million and exports made to 26 countries (Lage, 2008). During 2010 sales reached USD60 million with exports to 29 countries. There were 682 patent applications, of which 262 are abroad, and 462 are being processed abroad. The centre has developed two models for product marketing. In the countries of the South they send representatives who register and distribute the products. In the countries of the North, the strategy is to set up a joint venture. They also promote project licensing, and getting advanced payments prior to the arrival of the products in the market. CIM keeps the manufacturing rights in Cuba, except for China and India, where they have production plants. CIM's main marketing strategy is product and market diversification, in a way that no product or market amounts to more than 20 per cent of the total revenue of the centre, thus ensuring a robust export system.

Currently, as part of its cooperation with universities, CIM is developing several theme areas including the development of a breast cancer vaccine, adjuvants for therapeutic vaccines, study of complex systems, and simulation of biological systems and fermentation of higher cells. It cooperates with the University of Havana, the Jose Antonio Echeverria Higher Polytechnic Institute, the University of Information Sciences, among other higher education centres. Particularly outstanding is the cooperation with the Centro de Estudio de Proteínas[16] (CEP), at the University of Havana, where researchers are working on the design of inmunotoxins and novel systems for delivery of macromolecules to the cellular cytosol and are also engaged in research projects linked to the calculation of free energy used together with enzyme-inhibitor complex and antigen antibodies (Chávez et al., 2010, pp. 14–15). CIM also has a 'teaching unit', where university students do their on-the-job training. Some of them can later become researchers at CIM.

CIM exhibits some of the distinctive features of the Scientific Pole:

1. Integration of research, production and marketing activities.
2. Ability to work simultaneously on issues at the scientific and technological cutting edge and on more traditional products, generating economic opportunities for the country.
3. Close collaboration with the state and the government.
4. A scientific and technological community characterized by values of hard work and serving the public.
5. Close cooperation with other Cuban institutions, such as the Cuban health care system and its network of hospitals and family doctors, as well as the system of higher education.

**Biomolecular Chemistry Center (CQB)**

As of the year 2008 Cuba has a new research centre that is part of the Scientific Pole: the Biomolecular Chemistry Center (CQB) is a result of the merger of the Laboratory of Synthetic Antigens (LAGS) of the University of Havana with the Center for Pharmaceutical Chemistry (CQF) of the Ministry of Public Health.

LAGS was the research centre that created the Quimi-Hib, the world's first synthetic vaccine for human use. This vaccine attacks the *haemophilus influenzae* type B (Hib) bacteria that causes meningitis, pneumonia and otitis, among other diseases in children under five, and kills half a million children worldwide every year.[17] Cuba was spending USD2.5 million every year purchasing the conjugate vaccine (Majoli, 2002) and went off to develop the vaccine by synthetic means.[18]

The vaccine is considered 'the first major product of the Cuban biotech industry with origin in university laboratories' (Vérez, 2006). At least ten institutions and over 300 people were involved in obtaining the vaccine. During the process, close cooperation was established with several centres of the Scientific Pole, including the National Center of Biopreparations, the Finlay Institute, the Center for Genetic Engineering and Biotechnology (CIGB), the Pedro Kouri Institute of Tropical Medicine (IPK) and institutions from the Ministry of Public Health of the province of Camaguey. The state played a decisive role in its collaboration.

By 1999, the Ministry of Public Health (MINSAP) and the Council of State identified the vaccine as the number one priority of the Cuban biotech industry. The Council of State itself made the decision that the CIGB place itself at the disposal of the development of the Hib vaccine. After two years of clinical trials, in 2003 the vaccine proved to work with infants and induced a very high level of protection. The Center for State Control of Medicine Quality (CECMED)[19] issued a manufacturing licence and the registration of the vaccine. A new plant was set up at CIGB for its production.

As a result of these efforts over 15 years, with the cooperation of several institutions, led by a small lab in the University of Havana, the study was completed and showed that the Quimi-Hib vaccine, developed from a totally synthetic antigen, fulfils all safety and efficiency requirements.

The vaccine is currently mass produced and administered to all Cuban children. It has been patented in several countries, and export agreements have been signed. It is part of the world's only pentavalent vaccine against diphtheria, tetanus, whooping cough, hepatitis B and *haemophilus influenzae* type B.

The obtainment of Quimi-Hib is also a great success because the method of obtainment can be applied in the development of new vaccines for the treatment of other diseases, such as cancer and AIDS.

This success story illustrates some aspects of the Cuban S&T policy:

1. The results obtained are the outcome of a set of policies: the policy giving a high priority to public health care, combining advanced services, own technology, with free services; the policy that promotes the biotech industry emphasizing the health care sector; the policy that has favoured the training of human capital, both inside and outside the university; the policy favoured by higher education of promoting centres of research oriented to innovation (Núñez and Pérez, 2007). These policies provide a framework conducive to obtaining scientific and technological results, like the Hib vaccine.
2. Although Cuba has an organized system of science where the most relevant decisions are made top-down, with the state playing a very important role, this case shows the existence of important 'bottom-up'

initiatives. Originally LAGS used to consist of a small team of researchers devoted to the chemistry of carbohydrates who decided to take up the country's offer in the late 1980s of conducting research that would bring about benefits for the country, with emphasis on public health care and in support of the development of biotechnology. This group understood that novel strategies could be developed for the production of vaccines based in chemistry. As the group made progress, it drew increasing support from the state, but it was the actual group of researchers that charted out this techno-scientific path. It can be seen that the needs identified by the state, the availability of the universities to back up those efforts and the initiative and commitment of researchers were combined to obtain a relevant breakthrough that has sparked new opportunities.

3. The commitment by researchers must be underscored. There is no doubt that there is a scientific community in Cuba working with motivation in the social and economic projects developed by the country. It is an ethical and political dimension, very visible in the biotech industry but not limited to it, which favours the advance of policies.

The results obtained by LAGS and the new commitments that ensued called for more resources and space to grow. The decision was made to approve an important investment in the university campus but it soon became apparent that the pace of the investment was slower than the demand by the projects. In this context, decisions were made in order to generate new institutional arrangements. After a negotiation process between the University of Havana, the Ministry of Higher Education and the Council of State, the LAGS was merged with the Center for Pharmaceutical Chemistry, a scientific institution under the Ministry of Public Health and devoted to pharmaceutical chemistry research and the development of natural products. The merger of the two centres allowed for the combination of the scientific potential of both institutions and a more efficient use of the existing infrastructure. As a result of this process, the Biomolecular Chemistry Center (CQB) was established in late 2008.

The CQB can be considered a university spin-off and its main line of general research is the synthesis of antigens for vaccine development. The centre currently conducts a mega project for the obtainment of a vaccine against pneumococcus.[20] One hundred and four children under one year of age died in Cuba from pneumococcus in 2005. The country spends considerable resources on the purchase of antibiotics to combat the disease. Consequently, a vaccine against it has become a priority for Cuba's health care system.

In addition to pneumococcus, the CQB is working on other vaccines, including the NGM3 vaccine (ganglioside N-glycolyl-GM3) for the treatment of breast tumours and melanoma and in improving the meningococcal vaccine.

The CQB's objectives also include encouraging strategic links between the country's higher learning institutions and centres from the Scientific Pole, as it is considered the first scientific institution that continues under university tutelage, while at the same time it is part of the Scientific Pole and responds directly to the priorities determined by it. It is expected that this institutional arrangement will pave the way for a new stage in the relations between higher education and the Scientific Pole. The linking strategy is based on the joint development of undergraduate and postgraduate training and scientific research.

The strategy includes participation of students from the Chemistry and Biochemistry programmes at the UH conducting internships at CQB as part of their training. It also covers university studies at CQB for workers and technical staff of the Scientific Pole, who need to continue their training to better respond to the tasks entrusted to them by the Scientific Pole. The faculties of the UH and the CQB will provide teaching staff and other resources.

The postgraduate strategy is designed for the UH to provide all the master's and doctoral training required by the CQB, including the tutorship of theses by professors of the UH. Joint research projects are projected to complete the integration of both institutions. The Scientific Pole–university integration through the CQB poses an important challenge. If achieved, not only will the Cuban biotech industry come out stronger, but the university will too.

For the moment, the establishment of the CQB illustrates the opportunities offered by cooperation between the universities, the state, the government and the centres of research. The integration of players sharing objectives can become a very important driver of the scientific and technological development, generating, at the same time, scientific breakthroughs, benefits to the health care system and economically important results.

## The Institute of Plant Biotechnology (IBP)

The IBP is a centre of research that also came into being against the backdrop of innovation-oriented university research policies. It was part of the Central University of Las Villas (UCLV), the third university founded in Cuba, which opened in 1948.

One of UCLV's more experienced scientific groups and with greater links to the productive sector is the Institute of Plant Biotechnology (IBP).

The IBP[21] was founded on 19 November 1992, as a response to the priorities defined in the national S&T policy. IBP has defined three fields of work: research, production and technical services. The IBP is one of the three institutions devoted to plant biotechnology in Cuba. The scientific activity of the centre is organized in cross-disciplinary research projects, with the participation of several universities and scientific institutions of the country. All research projects are backed by economic contracts and they include precise provisions for all aspects relating to the environment, as most of these projects deal with genetic breeding, searching for resistance to diseases and the reforestation of the country. There are currently a total of eight national projects underway, and they are all part of the National S&T programme and respond directly to the country's priorities. Moreover, the IBP participates in several international projects.[22] Finally, the institute publishes the *Plant Biotechnology* journal, which is indexed in 14 international data bases.

The establishment of this centre can trace its background to the results obtained by a group of young agronomists at the UCLV who since 1981 have focused their research on two main lines: plant breeding and propagation. This group worked under two renowned agronomists from the Center of Agricultural Research (CIA) at the UCLV's Faculty of Agriculture, who had received training in plant biotechnology at CNIC since 1980.

In the second half of the 1980s a group on plant biotechnology was operating within the Center for Agricultural Research. The changes in the S&T policy sent new signals to the university and the group. On the one hand, it was influenced by the government's decision to promote the country's development in biotechnology and the establishment of the relevant institutions. On the other hand, back then the Ministry of Higher Education had ordered the establishment of research groups focusing on solving economic and productivity problems, which included the strategy to work 'full-cycle', that is, from basic research to the obtainment and use of scientific results in production.

At the time there was a demand in the country for the production of *in vitro* plants that could not be met with the technology existing in the world for the mass propagation of plants, the reason being that its design was too sophisticated and costly for Cuban conditions at the time; consequently, Cuba could not afford it. The alternative was to look for a project suitable for the country's conditions. These circumstances favoured the emergence of biofactories in the country. The first one opened on 24 September 1987, together with the Provincial Delegation of the Ministry of Agriculture in the province of Villa Clara.

The biofactory is a Cuban technology designed by IBP whose role is the mass multiplication of plants. This technology consists of a facility with culture chambers that includes an aseptic area with filtered air for a micro-organism-free setting. It can produce healthy plants that, once in the field, have greater strength, improved development and initial yield compared to those obtained in natural conditions. It is a flexible technology that allows other technologies and innovations to be added, including automated systems for the mass propagation of *in vitro* plants. The biofactory stems from a cross-disciplinary and integration setting developed by the UCLV and a university policy geared to supporting national priorities.

Biofactories are designed with a managerial approach emphasizing the organization, efficiency and effectiveness of production. Biofactories respond to the need to make *in vitro* propagation efficiently and under full control. With the invention of biofactories, Cuba has made an innovation in the field of micropropagation[23] of plant species.

Biofactories grew rapidly in Cuba between 1987 and 1990. A total of ten were set up in several provinces and this allowed for the production of *in vitro* plants increasing from 1 to 3 million. The onset in 1990 of black Sigatoka, a disease that attacks the banana population, became a new challenge, but also an opportunity for the biofactories. All traditional varieties of banana in the country proved to be sensitive to this pest. An American agronomist from the Honduran Federation of Agricultural Research (FHIA) who had developed varieties of clones resistant to black Sigatoka (FHIA varieties) and learned of the existence of the network of biofactories, offered them to Cuba. The use of the biofactories for the micropropagation of the FHIA clones made it possible for all banana plantations sensitive to the disease to be replaced in just four years. It also let Cuba discontinue the application of fungicide, with savings under this heading of 72 million pesos.

In 1992 the National Network had already formed. It comprised four generations of Cuban biofactories, all of which had been developed in just five years. In total, the network is composed of 16 biofactories, which represents a potential production of 50 million *in vitro* plants per year – with a field survival rate more than 95 per cent – higher than the entire potential of Latin America as a whole (Suárez, 2007).

The financing provided by the Council of State played a fundamental role in the consolidation of this network. By 1995, and pursuant to an agreement between the Ministry of Higher Education and the Council of State, the IBP was authorized to market their products. Biofactories began to be exported as a technological package under sales contracts including assembly of technology, and the quality, organizational and incentive system. The IBP experts design, assemble, start and provide technical advice on the operation of biofactories.

The technology of biofactories has been transferred to other Latin American countries. There are currently several technological parks in the region, including the biofactories in the Misiones Technological Park, Posadas, Argentina; the Antioquia Technological Park, Medellin, Colombia, and the FENORTE Technological Park, Rio de Janeiro, Brazil. Projects with Brazil, Venezuela, Colombia and Argentina are being evaluated at present. The technological package includes training programmes which range all the way from courses and training to master's programmes.

This new technology transfer activity has reaped benefits for the IBP, as it allowed for the development of a fifth generation of biofactories that are modern, multi-use and flexible. For instance, fifth-generation biofactories feature the appropriate inside temperature, regardless of the outside temperature, so that they can conform to the new conditions resulting from climate change.

Furthermore, Cuban researchers believe that biofactories constitute a good policy approach for Latin America. Biofactories reduce production costs, thereby allowing for the socialization of quality seeds that are usually in the hands of mid-size and large producers. As a result of this technology, IBP has already received several awards from the Cuban Academy of Sciences in the years 2000 and 2002. In 2006 it received the National Award for Technological Innovation, together with other institutions.

The biofactories installed in Cuba now require new investment. The biofactories' productive potential of plants is not being utilized efficiently. From a production potential of 50 million, only 5 million plants were produced in 2007, that is, 10 per cent of the installed productive capacity.[24] 'Today we have around seven or eight, out of sixteen, facilities producing – not at full strength though, but just to meet the country's needs. There is a great potential that we are not using for the lack of resources' (Agramonte, 2007).

## FINAL COMMENTS

As we have seen, biotechnology ranks very highly in the Cuban S&T policy. Under very difficult and unique conditions the country was still able to build a dynamic biotech industry based on its scientific capacity and showing economic and social relevance. It ensures important revenues for the country and supports the Cuban health care system. Its breakthroughs also reach out to other countries through Cuba's wide international cooperation.

The state and the scientific community have made the greatest effort through the process of unique processes of articulation and consensus. Decision-making includes both top-down and bottom-up processes.

The idea that biotechnology must be maintained as an important priority is well consolidated within the S&T policy imagery. The growing interest for the production of food determines that the areas of plant and animal biotechnology will become increasingly important. Applications in the field of human health have made the greatest progress so far.

Over the last 15 years, Cuba has been able to build a system of sectoral innovation in connection with the medical pharmaceutical industry where biotechnology plays a leading role. This is the result of the priorities defined in the S&T policy. However, for this successful process to continue several challenges must be addressed, including the following:

1. In spite of the progress in the sales of products in the foreign markets, there are limitations due to several factors, including the difficulty of accessing markets dominated by large transnational corporations and the exaggerated inflation of the 'regulatory context', which increases technical barriers and puts potential companies out of the competition through increasingly restrictive intellectual property systems (Lage, 2007). This is all heightened by the US blockade on Cuba. Notwithstanding, as seen earlier, from time to time the US has authorized negotiations in cases of their own interest. Other economic sectors of the country and their innovation capabilities do not show the same dynamics as the biotech sector, which, in turn, can make it difficult for the biotech industry to receive the resources required. This is true, for example, in the case of the chemical industry. The chemical industry has been decreasing since the beginning of the 1990s, as a result of the crisis in the productive sector in the country. According to Marquetti (2009), the recovery of the productive sector is progressing but it is still only in its initial stage. As a consequence of this crisis, the chemical industry still does not have the capacity to respond to the needs of the biotech sector. The industry sector and the scientific sector do not function as a system in this case.
2. There are areas of great importance to biotechnology, like nanotechnologies, bioinformatics, basic sciences, and so on, and they will have to make great efforts to be able to respond to the needs of biotechnology. Work is currently underway in all these areas, but in a context of serious economic constraints.
3. A special mention goes to higher education. In spite of good examples in terms of cooperation for research projects and training programmes between the Scientific Pole and the universities, the picture,

as a whole, is not entirely satisfactory. The research infrastructure of the universities was severely damaged during the crisis of the 1990s. Additionally, there is a lack of financial mechanisms through which the Scientific Pole can support university research in areas of their potential interest. This is compounded by the classic problem of differences in pace between university research and industry needs. This lack of connection impacts universities and probably the Scientific Pole too, as it limits the conduction of basic research of potential interest. Apparently we are lacking S&T policy decisions and other initiatives to increase synergies between the Scientific Pole and the universities. Another potential problem is the university's limited capacity to supply the Scientific Pole with the graduates needed in the coming years. Mainly due to material constraints, the science and engineering programmes, except for Information Sciences, are producing a relatively small number of graduates. This may lead to the conclusion that the national S&T policy has succeeded in creating innovation capacities in a very important sector for the development of the country, but that this sector nevertheless runs the risk of not finding the answers and incentives needed by other sectors, particularly higher education. It will be necessary to continue working on encouraging systemic interaction between innovation players with a more inter-sectoral view. This more inter-sectoral view could benefit from more participatory decision-making processes.

4. Innovation in Cuba has not been as successful in the case of the sector of agriculture. The innovation system of the biotech and medical pharmaceutical industry, in which two of the examined cases fall, works better than the organization for innovation found in the sector of agriculture. There are asymmetries in the priority and consistency of the policies applied in both sectors. The case of the biofactories can illustrate this. The issue has been discussed publicly and important transformations are under way that will benefit the IBP and other institutions of the agricultural sector. Food production is one of the country's biggest priorities today.[25]

## NOTES

1. In the early 1990s the GDP dropped by 35 per cent, exports by 85 per cent and fuel supply by more than 75 per cent.
2. One of the main components of the STI system is the System of Programs and Projects, whose objective is to arrange the process of organization, financing and control of programmes and projects comprised under the STI system and to promote full-cycle research (CITMA, 2003).

3. In the early 1980s, interferon was thought to be a potential cure in the fight against cancer (Limonta, 2002).
4. CIB had two main purposes: increasing the production of leukocyte interferon by four times the existing production in the original lab and introducing recombinant DNA technology in order to produce interferon, initially, and gradually other recombinant medicines and vaccines (Limonta, 2002).
5. This decision was made in consultation with the members of the CIB and other Cuban institutions, as well as prestigious international personalities of biological sciences such as Professor Albert Sasson (Limonta, 2002).
6. For example, African swine fever in the 1970s, dengue haemorrhagic fever in 1981, and *Thrips palmi* pest which was apparently introduced in Cuba intentionally in late 1996, among others.
7. In the late 1970s and early 1980s, the country was hit by an epidemic of meningococcal disease with some 2000 cases per year. This became the country's most important epidemiological problem.
8. The Finlay Institute is one of the centres devoted to vaccine research and development in the West Scientific Pole.
9. In July 2006 the World Health Organization (WHO) asked Cuba for help in the production of millions of doses of the meningitis vaccine in face of an emergency in the so-called 'meningitis belt' in Africa. The crisis was produced when the transnational corporations that supply the vaccine stopped production because its sales were not cost-effective. In order to respond to the production volume requested by the WHO, a new plant was built with a production capability of up to 100 million doses per year (see http://www.bvv.sld.cu).
10. These include India, China, Brazil, Egypt, Malaysia, Iran, Russia, South Africa, Great Britain, Venezuela, Mexico, Tunisia, Algeria and Belgium, and negotiations are underway with The Netherlands, Spain, Germany and the United States. Brazil buys from Cuba 100000 doses of hepatitis B vaccine and 1 million doses of meningitis B vaccine.
11. Notwithstanding, some centres such as CIGB have published 680 peer-reviewed papers in scientific journals, from 1986 to 2006. It is also worth noting that CIGB's papers have been cited in more than 3000 papers (López et al., 2006).
12. The process of 'Universalization of Higher Education', effective since the year 2000, has led many IES to establish university venues in all 169 municipalities of the country. They are not currently engaged in scientific research.
13. During the 2008 Agro Biotechnology Congress, Carlos Borroto, assistant manager of the Center for Genetic Engineering and Biotechnology (CIGB) said that Cuba is making progress in the development of genetically modified crops with favourable results in several crops like corn, GM sweet potato and tomato, in order to develop resistance against viral pests and diseases (see http://exthome.cigb.edu.cu).
14. Cuba was the third country (after Spain and Israel) that achieved *in vitro* flowering of citrus trees in 1990.
15. In this phase, the participation of the oncological community is critical; the community has been consolidating a network for years that includes the founders of CIM, as many of them came from the National Institute of Oncology.
16. This is an academic institution where scientific research is oriented towards the production and characterization of proteins, as well as other new aspects that constitute advanced technologies in biochemistry and protein biotechnology.
17. Since the late 1980s, conjugate vaccines have been successfully used against Hib. With different configurations, these vaccines are very efficient, highly safe and have limited adverse effects. However, only 2 per cent of the children in the world who are at risk of catching the disease are protected. For developing countries prices are relatively high and the Hib kills half a million children every year with pneumonia.
18. In 1987 Dutch scientists theoretically proved the possibility of obtaining the vaccine by synthetic means and LAGS set out to make efficient the process of chemical synthesis for the reproduction of capsular polysaccharide.

19. The drug regulatory body of the Republic of Cuba, CECMED performs the basic roles of access control to laboratories, registration of medicines and diagnostic kits, clinical trials, post-sales surveillance, good practice inspections, lot releases and issuance of licences to establishments.
20. According to the Pan American Health Organization (PAHO), the Johns Hopkins University and the US Center for Disease Control and Prevention, pneumococcus kills two children every hour in Latin America, with a toll of 18 000 deaths every year.
21. It comprises 26 researchers, 38 technicians and 16 workers, and it gathers specialists in microbiology, radiochemistry, biology, veterinary medicine, pharmaceutical sciences, industrial engineering and some 50 per cent of agricultural engineers. It includes 11 doctors and masters in sciences.
22. The main foreign institutions cooperating with the IBP include Belgium's Council of Flemish Universities (VLIR), the International Network for the Improvement of Banana and Plantains (INIBAP), the Swiss Development Cooperation (COSUDE) and the Union of Latin American Universities (UDUAL).
23. This is a technology applied to plant species in order to obtain a larger population as soon as possible.
24. In the last ten years only 160 million *in vitro* plants have been produced, out of a capacity of 500 million for this period. The high competitiveness of this technology can be seen by the low levels of production losses (under 1 per cent) and the spending structure, where 5 per cent of the total cost goes to energy and 1 per cent of the total cost is for reagents.
25. An example of this is that the 2008 Havana Biotech Congress, held between 30 November and 5 December 2008 in Havana, was called 'Agro Biotechnology: new approaches before great challenges'. The congress focused on the impact that biotechnology has had and will have in order to help meet the food and health care needs of the growing population in the world.

# REFERENCES

Agramonte, D. (2007), Interviews given to the authors.

Castro, F. (1990), Closing speech at 'Pedagogia 90', Karl Marx Theater, 9 February, available at http://www.cuba.cu/gobierno/discursos/1990/.

Castro, F. (1991a). Opening speech at the 4th Forum on Spare Parts, Equipment and Advanced Technology, available at http://www.cuba.cu/gobierno/discursos/1991/.

Castro, F. (1991b), Closing speech at the 7th Congress of the National Trade Union of Workers of Education, Science and Culture, available at http://www.cuba.cu/gobierno/discursos/1991/.

Chávez, M., J. Díaz, M. Lanio and C. Álvarez (2010), 'El Centro de Estudios de Proteínas: contribución científica y compromiso social', *Havana University Review*, no. 271, pp. 56–70.

CITMA (2003), 'Reglamento sobre el sistema de programas y proyectos de ciencia e innovación tecnológica', Resolution 85/2003 by CITMA, available at http://www.aenta.cu/archivo/aenta/pol%C3%ADtica%20-documentos.htm.

CITMA (2008), 'Science and technology: strategic planning: research priorities, 2008– 2015', draft copy.

CITMA (2011), 'Advice for the reorganization of the system of programs and projects according to the development strategy of the country', draft copy.

CNIC X Aniversario 1965–1975 (1975), 'Informe de los 10 años de actividades del Centro Nacional de Investigaciones Científicas', CNIC annual report.
Consejo Superior de Universidades (1962), 'La reforma de la enseñanza superior en Cuba', collection of documents from University of Havana.
Díaz, O. (2011), 'Heberprot-P. Medalla de Oro para biofármaco cubano', *Granma*, 24 March, p. 2.
García Capote, E. (1996), 'Surgimiento y evolución de la política de ciencia y tecnología en Cuba (1959–1995)', in E. García and R. Faloh (eds), *Ibero-American Seminar Workshop on Updating of Technological Management*, Havana: GECYT.
Gibbons, M., C. Limoges, H. Nowotny, S. Schartzman and M. Trow (1994), *The New Production of Knowledge: The Dynamics of Science and Research in Contemporary Societies*, London: Sage.
Herrera, L. (2008), 'El camino hacia el éxito está lleno de fracasos', interview given to Enrique Ubieta for *La Calle del Medio*, Volume 5, November.
Kaiser, J. (1998), 'Cuba's billion-dollar biotech gamble', *Science,* **282**(5394), 1626–8.
Lage, A. (1994), 'Biotechnology: Offering new opportunities for developing countries', *Biopharm*, **7**(5), 18–19.
Lage, A. (1999), 'Las biotecnologías y la nueva economía: crear y valorizar los bienes intangibles', available at: http://www.nodo50.org/cubasigloXXI/economia/lage1_310502.htm.
Lage, A. (2006), 'La economía del conocimiento y el socialismo: ¿hay una oportunidad para el desarrollo?', available at http://www.cubasocialista.cu/cs41/cseconomia.html.
Lage, A. (2007), 'Propiedad y expropiación en la economía del conocimiento', in J. Núñez and M. Macías (compilers), *Reflexiones sobre Ciencia, Tecnología y Sociedad*, Havana: Editorial Ciencias Médicas.
Lage, A. (2008), Personal communication given to the authors.
Limonta, M. (2002), 'Historia exitosa de una visión de futuro: la biotecnología médica en Cuba', available at: http://www.catedradh.unesco.unam.mx/catedradh2007/SeguridadHumana/prospectiva%206/revista.
López, E. et al. (2006), 'Biotechnology in Cuba: 20 years of scientific, social and economic progress', *Journal of Commercial Biotechnology*, **13**(1), 1–11.
López, E. et al. (2007), 'Taking stock of Cuban biotech', *Nature Biotechnology*, **25**(11).
Majoli, M. (2002), 'Ciencia y desarrollo en Cuba: aspectos del desarrollo científico y tecnológico cubano', FLACSO, Havana: University of Havana.
Marquetti Nodarse, Hiram (2009), 'Cuba: una nueva etapa del desarrollo industrial', Boletín Cuatrimestral, 'Economía y Gerencia en Cuba: Avances de Investigación', April, CEEC, Havana, available at: http://www.ceec.uh.cu/Sitio800/paginas/Publicaciones/Boletin/.
Núñez, J. and I. Pérez (2007), 'La construcción de capacidades de investigación e innovación en las universidades: el caso de la Universidad de La Habana', *Educacion Superior y Sociedad*, **12**(1), 146–73.
Núñez Jover, J. and J.A. López Cerezo (2008), 'Technological innovation as social innovation', Science, Technology, and the Rise of STS Studies in Cuba', *Science, Technology, & Human Values*, **33**(6), 707–729.
Núñez, J., L.F. Montalvo, I. Pérez Ones, J.L. García Cuevas and A. Fernández (2008), 'Universidad, innovación y sociedad: la universidad cubana en el sistema

nacional de innovación', in various authors, *Cátedras de Integración Andrés Bello: la Universidad y los Procesos de Integracion*, no. 4, Bogotá: Convenio Andrés Bello, pp. 123–84.
Partido Comunista de Cuba (PCC) (2010), 'Proyecto de lineamientos de la política económica y social', VI Congreso del PCC, November.
Perez, I. and J. Núñez (2009), 'Higher education and socio-economic development in Cuba: high rewards of a risky high-tech strategy', *Science and Public Policy*, **36**(2), 97–101.
Pérez, R. (2008a), Interviews given to the authors.
Pérez, R. (2008b), 'Inmunoterapia. Artillería anticáncer', Interview given to *Bohemia*, Year 100, vol. 26, December.
Pérez, R. (2011), 'Alianza estratégica entre universidad y biotecnología. La experiencia de colaboración entre el CEP y el CIM', paper presented in the Biology Faculty of the University of Havana, February.
Polo Científico (2011), 'Biotechnology and pharmaceutical industry', Annual Report, 2010.
Suárez, M. (2007), Interviews given to the authors.
Torres, W. (2006), 'Taller Científico: "Política Científica Nacional"', University of Havana, April.
Vérez, V. (2006), Interviews given to the authors.
Vérez, V. et al. (2004), 'A synthetic conjugate polysaccharide vaccine against *Haemophilus Influenzae* type B', *Science*, **305**(5683), 522–5.

# PART II

# Africa

# 5. The role of product development partnership for the appropriation of knowledge and innovation in biotechnology in Tanzania

**Emmarold Mneney, Bitrina D. Diyamett and Burton L.M. Mwamila**

## BACKGROUND

Innovations in science and technology in Africa and other developing countries are essential means of alleviating poverty and stimulating economic growth and development. Certain types of social, economic and technological advances can only be achieved by private firms, public sector organizations, universities, NGOs or civil society working together. A rapid rise in the incidence of research partnership (public–private, private–civil society, public–private–civil society and many other forms) has stimulated a policy debate regarding their effectiveness and whether these relationships enhance overall economic performance in the developing countries. Research partnerships have become an increasingly important means of creating and diffusing scientific and technical knowledge. This is of great importance, particularly in biotechnology-related activities. Application of biotechnology provides new and promising opportunities for achieving greater food security, reducing poverty and improving the quality and standard of living of marginalized people in the developing world. The number and scope of biotechnology applications are expanding, not only in the developing countries, but also in African countries.

The level of biotechnology research, development and utilization in Tanzania is still in its infancy. There are several public, private and non-governmental institutions that are involved in biotechnology research at different levels in Tanzania. Although progress in the adoption and utilization of this technology in Tanzania has remained rather slow, the country is picking up quickly, with the agricultural sector being the most active. It is being envisioned that public–private partnership could speed up the commercialization of research in biotechnology. The major aim of this chapter is therefore to examine how the role of partnerships in biotechnology can be enhanced and sustained in order to improve efficiency and effectiveness of research and technological innovation for the poor. Information is scarce, particularly regarding the role of universities, and what incentives drive successful partnership in product development, technology transfer and in product delivery.

## OBJECTIVES OF THE STUDY

The primary goal of the research project has been to study and recommend functional policies to enhance partnerships in biotechnology product development and delivery in Tanzania. Specific objectives were to:

- Analyse baseline status of biotechnology R&D in Tanzania and examine the range and characteristics of R&D partnerships in biotechnology.
- Review the role of universities in biotechnology R&D in Tanzania, including examining the status of research capacity in terms of number of researchers/laboratories and discussion on the specific problems university biotechnology research is facing.
- Discuss who determines private and public biotechnology research agendas in Tanzania and identify the key actors and their rationale of pursuing the research agenda.
- Examine how Product-Development-Partnerships (PDP) are managed, analyse the influence of these partnership in terms of enhancing scientific and technological capacities, stimulating collaboration and promoting intellectual property management.
- Assess the extent to which partnerships engage the participation and interest of a range of stakeholders, including resource-poor farmers and other technology user groups.
- Draw up policy options to create sustainable partnerships, resource mobilization and linkages for biotechnology product development and dissemination.

## METHODOLOGY

To achieve the objectives of the study, questionnaires (Appendix A5.1) were developed and sent to 23 different institutions (Appendix A5.2). The study also included a literature review on biotechnology innovation, and product development partnership and secondary data available from previous studies. In addition, the team conducted face-to-face interviews with scientists and other staff involved in biotechnology research in academia and at National Research Institutes (NRIs) .

## FINDINGS

The findings presented below are based on the information received through the questionnaires, discussions held and on previous relevant reports. The finding represents a summary of information gathered from 4 universities and 16 National Research Institutes (NRIs).

### Baseline Status of Biotechnology Research and Development in Tanzania

Although Tanzania considers biotechnology a tool that may provide new opportunities for achieving productivity gains in both agriculture and food, the progress in the adoption and utilization of this technology in Tanzania has remained rather slow. The level of biotechnology research, development and utilization is still in its infancy but the country is picking up quickly, with the agricultural sector being the most active. Most biotechnology laboratories in Tanzania are for teaching and/or research purposes. None is operating at a large scale or at a commercial level.

The application of biotechnology in Tanzania must be considered in the context of the country's need for more food and a sustainable development for its people. However, besides agriculture, biotechnology has also been applied, albeit to a small extent, in the areas of medical and public health, industry and environment.

There are several public and private R&D institutions that are involved in biotechnology research at different levels in Tanzania. Tables 5.1 and 5.2 give a list of these laboratories, number of researchers and area of specialization in the National Research Institutes (NRIs) and universities, respectively. Most of these laboratories are either involved in or have been designed to deal with agricultural biotechnology research. The following are some of the main biotechnology applications in Tanzania.

#### Tissue culture and micropropagation

Of the biotechnology techniques, tissue culture and micropropagation are the most widely applied in Tanzania. The application of *in vitro* culture techniques to address constraints of availability to farmers of adequate disease-free planting materials and rapid improvement in crop production is now routinely applied in several African countries. Plant tissue culture

*Table 5.1 Biotechnology R&D at universities*

| Research area | University name | Qualification | No. | Discipline |
|---|---|---|---|---|
| Agricultural biotechnology | SUA | PhD | 10 | Immunology, Molecular epidemiology, Molecular genetics, Molecular pathology, Tissue culture, Molecular biology |
| | | Msc | 3 | Molecular genetics, Tissue culture |
| | UDSM –Botany Department | PhD | 2 | Molecular physiologist, Molecular taxonomist |
| | | Msc | 1 | Molecular physiologist |
| Medical biotechnology | MUCHS | PhD | 4 | Microbiology, Molecular pathology |
| | | Msc | 2 | Microbiology, Molecular pathology |
| | Tumaini University (KCMC) | PhD | 4 | Microbiology, Molecular pathology |
| | | Msc | 2 | Microbiology, Molecular pathology |
| Industrial and environmental biotechnology | UDSM–MBB Department | PhD | 4 | Molecular biology, Microbiology, Environmental biotechnology, Industrial biotechnology, Biosafety |
| | | Msc | 5 | Microbiology, Environmental biotechnology, Industrial biotechnology, Molecular taxonomy |
| | UDSM – Chemical and Processing Engineering Department | PhD | 4 | Biofuel, Food technology, Environmental biotechnology |
| | | Msc | 1 | Biofuel |

is of particular interest to Tanzania because it is not very demanding at the technological level and it also belongs to the category of comparatively low-cost technology compatible with the country's economic status.

In Tanzania, tissue culture techniques have been employed in

*Table 5.2 Biotechnology research capacities at the National Research Institutes*

| Research area | Name of institute | Qualification | No. | Discipline |
|---|---|---|---|---|
| Agricultural biotechnology | Mikocheni Agricultural Research Institute | PhD | 6 | Molecular genetics, Molecular pathology, Molecular virology, Molecular breeding, Tissue culture |
| | | Msc | 2 | Molecular pathology, Genetic engineering |
| | ARI-Ukiriguru | PhD | 2 | Molecular breeding |
| | | Msc | 2 | Molecular breeding, Molecular virologist |
| | ARI-Tumbi | Msc | 1 | Molecular breeding |
| | ARI-Mlingano | MSc | 2 | Tissue culture |
| | ARI-Uyole | MSc | 1 | Tissue culture |
| | Horti Tengeru | MSc | 2 | Tissue culture |
| | Central Veterinary Lab (CVL) | PhD | 2 | Molecular virologist |
| | | MSc | 1 | Molecular breeding |
| | NPGRC | PhD | 2 | Tissue culture, Biosafety, Molecular biology |
| | | Msc | 1 | Molecular ecology |
| | Kizimbani | Msc | 1 | Tissue culture |
| Medical biotechnology | NIMR (Tanga) | PhD | 4 | Immunology, microbiology and biochemistry |
| | | Msc | – | |
| | AMRC | PhD | 2 | Molecular pathology, Molecular entomology |
| | | Msc | 1 | Molecular pathology, Molecular entomology |
| | IHI | PhD | 2 | Immunology, Drug resistance |
| | | MSc | 2 | Molecular pathology, Immunology |
| | | BSc | 2 | Molecular pathology, Drug resistance, Molecular entomology |
| Industrial biotechnology | TFNC | PhD | 2 | Food science |
| | | MSc | 1 | Microbiology |

a number of laboratories. These include Sokoine University of Agriculture (SUA), Kizimbani Agricultural Research Centre in Zanzibar, Tanzania Coffee Research Institute, National Plant Genetic Resources and four Agricultural Research Institutes (ARI) under the DRT, namely ARI Mikocheni, ARI Uyole, Horti Tengeru, and ARI Mlingano.

Mikocheni Agricultural Research Institute (MARI) has been collaborating with other international laboratories in the development of a protocol for coconut embryo culture, to facilitate safe exchange of germplasm within and between countries and for micropropagation of various crops such as cashew, banana, pineapple, cassava and sweet potato. MARI is also currently implementing a project for developing capacity in the production of disease-free cassava and sweet potato varieties through tissue culture, in order to facilitate safe exchange of germplasm within and across borders. Another tissue culture laboratory is based at ARI Mlingano in Tanga. The laboratory was established in collaboration with Katani Ltd for multiplication of sisal (Mneney, 2000), as well as for mass propagation of any other crop, provided that the capacity is extended to accommodate them.

At Horti-Tengeru (Arusha) and Kizimbani Agriculture Research Station (Zanzibar), basic tissue culture facilities are available. Although the Horti-Tengeru laboratory is not fully operational, the institute is currently working on mass propagation of disease-free planting materials for horticultural crops, specifically banana and sweet potato, in collaboration with the Kenya Agricultural Research Institute (KARI). On the other hand Kizimbani Agricultural Research Centre (KARC), the only biotechnology laboratory in Zanzibar, is currently propagating banana planting material *in vitro* with technical assistance from the International Network for the Improvement of Banana and Plantains (INIBAP) laboratories and other horticultural crops.

The Sokoine University of Agriculture (SUA), based in Morogoro, and National Plant Genetic Resources Centre (NPGRC) in Arusha, have both established tissue culture laboratories for micropropagation of horticultural crops and for long term *in vitro* conservation of plant genetic resources, respectively.

Finally, Tanzania Coffee Research Institute (TaCRI), a private institute that was established in 2000 to oversee all coffee production and promotion activities in Tanzania, will soon have a tissue culture facility with an annual capacity of producing about 250000 plantlets through somatic embryogenesis, with technical support from Centre de Coopération Internationale en Recherche Agronomique pour le Développement (CIRAD) and with financial support from the EU.

**Application of DNA marker technology**

Marker assisted breeding is now routinely used in agricultural research throughout the world to improve the accuracy and the efficiency of exploiting and incorporating single and polygenic traits in plants and also for livestock breeding. In Tanzania, molecular genetic markers such as AFLP, RFLP, RAPD, ISTR, SSR (Microsatellites) and SNP[1] have been used to detect genetic polymorphisms between and within different plant species, including coconuts (Rohde et al., 1999; Mneney et al., 2006), cashew (Mneney et al., 2001; Croxford and Mneney, 2007), cassava (Fregene et al., 2003; Masumba et al., 2006) and coffee (Masumbuko et al., 2002).

Currently MARI is implementing five projects on the application of molecular marker assisted selection (MAS). The first is on 'Genetic enhancement to increase productivity in rice through breeding for resistance to rice yellow mottle virus (RYMV) disease in Tanzania' with assistance from The Rockefeller Foundation; the second is on 'Molecular marker-assisted and farmer participatory improvement of cassava germplasm for farmer/market preferred traits in Tanzania' (Kullaya et al., 2005) funded by Rockefeller. The third is on 'Development of Marker Assisted Selection (MAS) technologies to support breeding for tolerance to Al toxicity and improved P uptake in sorghum' (funded by BIOEARN and involving collaboration with institutes from Kenya Uganda and Ethiopia). The fourth is on the 'Application of molecular tools for breeding of tomatoes for drought and temperature tolerance' funded by GTZ, and the fifth is 'Delivering new sorghum and finger millet innovations for food security and improving livelihoods in Eastern Africa'. The objective of these projects is to use new tools for molecular marker-assisted selection (MAS) and farmer participatory breeding to develop elite varieties of food crops that many rural communities in Tanzania depend on for their food security and livelihood.

With respect to animal biotechnology research, the Faculty of Veterinary Medicine of the SUA and the Central Veterinary Lab (CVL) have made a significant contribution to the understanding of genetic diversity in livestock and wildlife species using microsatellite, RFLP and other molecular markers (Msoffe et al., 2006; Stephen, 2006).

**Molecular disease diagnostics**

Plant and animal disease diagnostic kits, based on the products of biotechnology such as monoclonal antibodies and recombinant antigens, are of critical importance to Tanzania. The molecular techniques are quicker and more accurate in identification, and hence of practical application in disease forecasting, as well as in monitoring the spread and control of various diseases.

The MARI has been collaborating with other institutions in the development of molecular techniques for diagnosing different diseases, such as the lethal coconut disease caused by phytoplasma (Rohde et al., 1993), and sweet potato viral disease complex (Tairo et al., 2004; 2005).

With respect to animal disease, scientists in the faculty of Veterinary Medicine, SUA and the Animal Disease Research Institute (ADRI) are applying a number of DNA-based techniques for diagnosis of various animal diseases including CBPP, FMD and surveillance of rinderpest.

**Genetic engineering**

Genetic transformation activities were initiated in Tanzania in early 2011 after the establishment of Biosafety level II laboratory at the Mikocheni Agricultural Research Institute. Ongoing work includes development of cassava varieties tolerant to Cassava Mosaic Disease (CMD) and Cassava Brown Streak Disease (CBSD). Efforts are also underway to initiate a Confined Field Trial (CFT) for maize tolerant to drought at Makutopora Dodoma through a collaborative project entitled Water Efficient Maize for Africa (WEMA).

**Treatment and use of waste production of agricultural bioinputs and bioenergy**

Waste treatment has become one of the most important areas of application of biotechnology today. The biotechnological treatment of waste relies on the utilization of a vast diversity of micro-organisms and their versatile catabolic processes to degrade or transform waste either into useful biomass or into compounds less damaging to the environment. Apart from its positive effect on public health and environmental hygiene, the application of biotechnology to waste treatment can also contribute towards production of energy. With the increasing awareness of the general public and mounting international pressure for environmental protection, the biotechnological treatment of waste has globally become the method of choice for the control of pollution, since traditional strategies, based on dilution, burial or incineration, have proven their ineffectiveness.

At the University of Dar es Salaam, the Department of Molecular Biology and Biotechnology (DMBB) is engaged in developing various biotechnology techniques such as use of micro-organisms for bioconversion of waste and other organic residues for production of biofertilizers. In addition, the DMBB is also involved in bioprospecting for micro-organisms with potential applications in crop protection as biopesticides, such as the use of microbial-based insecticides (*Bacillus thuringiensis*).

**Production of useful substances through industrial biotechnology**

The use of microbes and plant cell cultures has an increasing role in the agro-industrial sector today, in the production of food and other useful substances. The utilization of micro-organisms or plant cells as bioreactors for producing proteins and other metabolites in bulk quantities at low cost provides enormous scope for the industrial transformation of primary agricultural products into food and fine chemicals. The production of enzymes, additives, flavourings, colourings and other components have transformed the food processing industries just as the large-scale production of secondary metabolites such as antibiotics and antitumuor drugs have transformed the pharmaceutical industries.

Application of biotechnology in Tanzanian agro-industries is still very limited. Most of the fermentation and brewing processes used locally form part of traditional food processing and do not receive any significant input from modern biotechnology. Some limited research on the possible use of microbes and plant cell cultures in production of novel products is being carried out at the University of Dar Es Salaam and the National Food and Nutrition Centre (TFNC). The TFNC has three professional staff (2 PhD and 1 MSc) involved in food processing biotechnology-related activities (Table 5.2). The main areas of research collaboration include: germination and fermentation in collaboration with Chalmers University of Technology, Sweden; food processing and fermentation in collaboration with Natural Resources Institute (NRI) in the UK; and analysis of β-carotene/carotenoids in orange fleshed sweet potato under the Programme for Vitamin A for Africa (VITAA).

**Medical biotechnology**

Medical biotechnology is practised in Tanzania but on a very limited scale. The leading institutions are Ifakara Health Institute (IHI), National Institute for Medical Research – Tanga, Medical Research Centre, Amani Medical Research Centre (AMRC) and the Muhimbili University College of Health Services (MUCHS).

The IHI is a non-profit foundation that operates as a trust under the leadership of the Ministry of Health. The mission of IHI is to develop and sustain a rural district-based health research and resource centre, capable of generating new knowledge and relevant information, regarding priority problems in health systems at the district, national and international level through research, training and services aiming at better health and community development. To achieve its mission, the centre is conducting research projects within the fields of molecular epidemiology, diagnostic immunology and parasitology. The centre has 20 professional staff (7 PhD and 13 MSc). Six of these researchers are actively involved in biotechnology R&D (Table 5.2). The laboratory for molecular epidemiology is well equipped for research in genetics, molecular biology cell biology and molecular modelling.

The biomedical department is involved in testing and applying current biomedical tools to understand the epidemiology of endemic diseases and to evaluate their potential in monitoring control programmes and intervention studies. Ongoing research projects include: studies on markers for drug resistance; *Falciparum* genetic diversity studies; gene expression studies; studies on HIV; and molecular entomology.

Another institute involved in medical biotechnology is National Institute for Medical Research – Tanga. This is a government institution under the National Institute for Medical Research (NIMR), located within the Tanga Regional Hospital compound in Tanga. It has 13 research scientists (4 PhD and 9 MSc), four of whom are involved in medical biotechnology research (Table 5.2). Currently, the centre runs a modern high technology biomedical research laboratory established in 2004, with a capacity to carry out immuno-molecular studies. The main focus of the centre includes lymphatic filariasis, malaria and tuberculosis, HIV/AIDS, enteric diseases and non-infectious diseases. Biotechnology research areas include immunology, disease diagnostics, bio-informatics, vaccine development, microbiology (protozoology and helminthology) and biochemistry.

Currently the centre is collaborating with different institutions in malaria research (drug resistance and vaccine development), HIV and traditional medicine. At the national level the centre is collaborating with other NIMR centres, MUCHS and KCMC, while at the regional level it is collaborating in malaria research with Kenya Medical Research Institute (KEMRI), Malaria Centre in Kilifi, Mombasa, as well as institutions in Zambia, Mozambique, Zimbabwe and Ghana. The centre also has international collaborative linkages with the following organizations: World Health Organization (WHO); Danish International Development Agency (DANIDA) and China on malaria research, and drug and vaccine development; Bill & Melinda Gates Foundation, USA and DANIDA on HIV research; and Japan International Cooperation Agency (JICA) on malaria research, and on human and infrastructure capacity building.

Amani Medical Research Centre (AMRC) is another government institution under NIMR dedicated to medical research. It has 12 research scientists (2 PhD, 4 MSc and 6 BSc), three of whom are involved in biotechnology-related research activities (Table 5.2). The scientific team at the centre has the technical and scientific capacity to use molecular tools in the identification of disease vectors, detection of malaria and filariasis parasites in mosquito vectors, and detection of insecticide resistance genes in mosquitoes. The centre has conducted molecular epidemiology studies of *Plasmodium falciparum* infections in Tanzanian children, and the ongoing studies related to malaria include studies on mosquito resistance, efficacy of pyrethroids, mosquito biology and ecology and bed net testing.

In terms of research collaboration, the centre is collaborating with UDSM and Ifakara Health Institute in malaria resistance studies; the Department of Veterinary Sciences at SUA in vector ecology; and with MUCHS in malaria research. At the regional level the centre is collaborating with the International Centre for Insect Physiology and Ecology (ICIPE) in vector studies, while at the international level it has collaborative linkages with the Danish Bilharzia Laboratory, London School of Hygiene and Tropical Medicine, Liverpool School of Hygiene and Tropical Medicine, and American Biophysics Cooperation.

## The Role and Status of Biotechnology at Tanzanian Universities

The level of biotechnology R&D in Tanzanian universities is very low. Only three universities out of ten are presently practising biotechnology. These are: Sokoine University of Agriculture (Agricultural Biotechnology), University of Dar es Salaam (Industrial and Environmental Biotechnology) and the Muhimbili University College of Health Services (Medical Biotechnology). The main mandate of these universities is teaching, research and provision of services.

Two universities have now established formal training programmes in Biotechnology. These are the University of Dar es Salaam, Department of Molecular Biology and Biotechnology, which is offering a BSc in Molecular Biology and Biotechnology, and the Sokoine University of Agriculture offering BSc in Biotechnology and Laboratory Studies. There is no explicit biotechnology training for MSc and PhD, but training in biological sciences and agriculture has a biotechnology component. There is no formal training for technicians or other technical personnel. This is creating a serious imbalance between professionals and technicians.

Much as there are three university laboratories that are involved in biotechnology research, most activities are limited in scope and have not gone beyond the laboratory. A review of institutional capacity and research activities of the three universities is summarized below.

### Agricultural biotechnology

*Sokoine University of Agriculture, Morogoro (SUA)* The SUA is a public university established on the 1 July 1984 by Parliamentary Act No. 6. It has three core mandates: teaching, research and consultancy services, in agriculture, veterinary medicine, forestry, natural resources and related fields including food science and biotechnology. The university has laboratory facilities for tissue culture, immunology and molecular biology and offers undergraduate biotechnology training, leading to the award of degrees in BSc (Biotechnology and Laboratory Studies). SUA has about 170 professional staff (100 PhD and 70 MSc), of whom only nine scientists are engaged in biotechnology research activities (Table 5.1). The university has conducted the following biotech activities/projects:

- Assessment of genetic diversity in indigenous animal populations using RAPD, microsatellite and Mitochondrial DNA polymorphism, with financial support from ENRECA-DANIDA and collaboration on similar projects at ILRI, Kenya.
- Molecular studies to assess immunodulative effects of *Theileria parva* parasite expression of cytokine genes in cattle using RT PCR.
- Assessment of molecular genetic diversity of rural chicken ecotypes in Tanzania (IHEPRUCA project) with collaborators in Kenya, Uganda, Malawi and Demark (Msoffe et al., 2006).
- Application of molecular techniques to detect *Theileria parva* and other disease causing parasites in livestock.
- *In vitro* multiplication of disease-free banana cultivars, with financial assistance from International Atomic Energy Agency and PANTIL-NORAD.
- Anther culture of rice – funded by International Atomic Energy Agency.

The SUA has also recently established an ultra modern seed pathology laboratory and the Genome Sciences Centre for research in functional genomics and bioinformatics. The Genome Sciences Centre accommodates a wet lab, facilities for printing, hybridizing and scanning microarray systems, and a computer infrastructure necessary for data storage, manipulation and analysis.

*University of Dar es Salaam* The Department of Botany of the University of Dar es Salaam has a total of 17 scientists (8 PhD & 9 MSc), of whom only four are involved in biotechnology activities (Table 5.1). The department has a small tissue culture laboratory and a molecular biology laboratory, both of which are ill-equipped. Currently the university is implementing two collaborative projects on biotechnology. The first project is entitled 'Genetic physiological and molecular approaches to improve heat and drought tolerance of tropical tomato'. It is funded by GTZ and involves four collaborative institutions: AVRDC – The World Vegetable Centre, Taiwan; Mikocheni Agricultural Research Institute (MARI), Tanzania; Leibniz University of Hanover, Germany; ICRISAT and Academia Sinica, Taiwan. The second biotech project entitled BiosafeTrain started in 2004 and is funded by DANIDA under the programme on Enhancement of Research Capacity (ENRECA). The long-term vision of this project is to build capacity to cope with the challenges of introducing genetically modified crops in East Africa by developing a platform for capacity-building on biosafety impact assessment of transgenic plants.

**Medical biotechnology**

*Muhimbili University College of Health Services (MUCHS)* MUCHS is a public medical university located in Dar es Salaam. Apart from training, it is also involved in research and consultancy services. MUCHS has eight research scientists (4 PhDs and 2 MSc) who are involved in different biotechnology research activities (Table 5.1). The areas of research include immunology, biosafety, diagnostics, virology and biochemistry.

Currently MUCHS is conducting research on HIV/AIDS, tuberculosis and malaria. Its collaborative partners at national level include NIMRI and the Ministry of Health, while at international level it is collaborating with Harvard University in the USA and Sida/SAREC.

*Kilimanjaro Christian Medical Centre (KCMC)* The Kilimanjaro Christian Medical Centre is a referral hospital for over 11 million people located in the foothills of Mount Kilimanjaro. It was opened in March 1971 by the Good Samaritan Foundation. The KCMC also has established a well-equipped biotechnology laboratory. The facility is manned by three professional staff (1 PhD, 2 MSc, Table 5.1).

In terms of research, the KCMC Biotech Lab is collaborating with different national, regional and international institutions in malaria, HIV/AIDS, tuberculosis research, parasitology and non-parasitic diseases. The collaborating institutions include: Kigali Hospital, Kenyatta Hospital on HIV/AIDS research; London School of Hygiene and Tropical Medicine; Duke and Harvard Universities in USA on tuberculosis and malaria; Copenhagen University on malaria and parasitology; Gates Malaria Partnership; Bill & Melinda Gates Foundation on tuberculosis; Dutch agencies (WOTRO, NACCAP) on non-parasitic diseases; and other organizations such as EDCTP, IDRC, WHO and Welcome Foundation on tuberculosis and HIV/AIDS research.

**Industrial and environmental biotechnology**

The application of industrial and environmental biotechnology in Tanzania is limited to a few institutions. Very few cases of biotechnological treatment of agricultural waste are known in Tanzania. A few projects on waste treatment are still at the research stage at the University of Dar es Salaam.

*Department of Molecular Biology and Biotechnology (DMBB), UDSM* The Department of Molecular Biology and Biotechnology (DMBB) of the University of Dar es Salaam is involved in research and training in industrial and environmental biotechnology application. The department produces about 20 BSc graduates in molecular biology and biotechnology annually. Some of the research activities being conducted at DMBB include the following:

- Screening of genetically modified maize, with financial support from Sida/SAREC;
- risk assessment of BT cotton, funded by DANIDA;
- capacity-building in biosafety and environmental impact assessment of transgenic plants of East Africa, with financial support from the Danish Institute of Agricultural Sciences;
- molecular taxonomy of mushrooms, with financial support from Sida/SAREC;
- studies on lignocellulolytic enzymes from mushrooms, funded by Sida/SAREC;
- studies on bioactivity of traditional plants used in the control of fungal infections associated with HIV/AIDS in Lake Victoria, with financial support from Sida/SAREC;
- utilization of industrial and agricultural waste for bio-energy and value-added products, with financial support from Sida/SAREC;
- biosafety research, funded by DANIDA;
- mycorestoration of degraded forests, with financial support from McArthur Foundation.

*Department of Chemical and Process Engineering, UDSM* The Faculty of Chemical and Process Engineering of the UDSM is involved in research and training in chemical and process engineering. It has four researchers (PhD) who are directly involved in industrial/environmental biotechnology research activities. The department has a small biotechnology laboratory, two medical waste incinerators and a food processing laboratory.

The department is currently conducting the following projects on:

- Production of biodiesel from Jatropha, funded by Sida/SAREC;
- production of biofuels and biochemicals from sisal waste, funded by Sida/SAREC;

- determination of food micronutrients in collaboration with the Tanzania Food and Nutrition Centre (TFNC), with financial support from NORAD;
- improvement of indigenous banana juice production technology in collaboration with Makerere University under Lake Victoria Research (VicRes) programme;
- wetland waste water treatment in collaboration with University of Nairobi under VicRes programme.

**Constraints of biotechnology R&D at Tanzanian universities**

Biotechnology R&D in the universities is limited by several technical, administrative and financial constraints. The efforts by the universities in promoting biotechnology will produce an impact only if the following shortcomings are addressed.

*Insufficient financial and related support mechanisms* In general, the current level of infrastructure and human resource development in the universities and other research institutes is not adequate to conduct reasonable research and development in the application of biotechnology (Kullaya et al., 2001). Consequently, very few, if any, biotechnology products have gone outside the doors of the laboratory. In countries where biotechnology has taken off, it is observed that high government investment has been at the base of the building of national and international enterprises. One of the fundamental reasons for the early emergence of biotechnology in USA, China, India, Brazil and in Singapore has been the strong background of national research and development, and the creation of a dedicated budget to promote the growth of commercial biotechnology.

Due to high investment costs, private–public sector collaborations as well as new and innovative sources of funding need to be sought. Except for ARI-Naliendele and ARI Mlingano (Mneney, 2000) none of the other available institutions are being supported by the private sector in terms of biotechnology R&D. All the surveyed universities were to a large extent dependent on government and donors' funds. In this study about 90 per cent of the reported projects were funded from external sources while 80 per cent of the surveyed institutes indicated funding as the main reason for collaborating with other institutes.

*Inadequate infrastructure and related support services* Most of the university laboratories surveyed were found to lack the necessary infrastructure that will enable them to conduct biotechnology on a competitive level. In several universities basic infrastructure and facilities even for the simplest biotechnology methods, such as micropropagation/tissue culture, are not available. Modern facilities for networking such as access to email and the Internet are also limited in many universities of Tanzania, which seriously hampers the acquisition of relevant knowledge and the application of biotechnology. Furthermore, unreliable water and power supply is a serious constraint. Application of biotechnology in universities and other R&D institutions is also limited by unavailability of chemicals and lack of capacity to supply, service and repair scientific equipment. With the exception of the Genome Sciences Centre at the Sokoine University of Agriculture, none of the other universities are supported by an adequate infrastructure required to keep up to date in biotechnology. Universities have traditionally been regarded as a source of new technologies and innovation, and one would have expected such institutions to be equipped with state of the art facilities for biotechnology.

It is notable that some national research institutes were better equipped than the universities. The survey has revealed there are a number of institutions with modern items of equipment, such as automatic DNA sequencer at the Government Chemist Laboratory Agency, micro-array facilities at IHI and Real Time PCR to be installed at CVL and MARI. This finding came as no surprise because both NRIS and universities are competing for the same donors/funding agencies.

In parallel with the scientific infrastructure, there is a need to build up an infrastructure of supporting services for equipment repair and maintenance. The present lack of trained engineers for servicing sophisticated equipment compels frequent recourse to the manufacturing firms abroad for repair.

*Donor dependency* Most biotechnology and biosafety-related research and development activities being carried out in Tanzania are funded through foreign grants from development partners. A study commissioned by the Vice President's Office (Anonymous, 2005) revealed that government funding for biotechnology research was less than 5 per cent. Similar results were recorded in this study as 90 per cent of the projects recorded were receiving funds from donors. This is not a viable situation. In order to implement a home-grown biotechnology agenda that addresses national priorities on a sustainable basis, it would be necessary for the government to allocate more financial resources to support biotechnology research in the different disciplines.

*Lack of critical mass of skilled human resource* The unavailability of qualified scientific, technical and supporting staff considerably handicaps the carrying out of biotechnology research in universities and research institutes of Tanzania. The SUA, for example, has about 170 professional staff (100 PhD and 70 MSc), of whom only nine scientists (5.27 per cent) are engaged in biotechnology research activities. At the University of DSM the situation is the same. Out of about 700 professional staff only 12 scientists (1.71 per cent) are actively engaged in biotechnology research.

The survey also revealed that most of the currently available research scientists in our universities have been trained in 'conventional' biotechnology applications such as tissue culture, genotyping, markers assisted breeding, microbiology molecular pathology, immunology and so on (Table 5.1). This is now an opportunity for the Tanzanian universities to start investing in cutting-edge technologies such as DNA microarrays, proteomics, genetic engineering, bioinformatics, computational biology, genomics, genetherapy and nanobiotechnologies.

Another serious constraint is the loss of skilled personnel who have received training in developed countries, but don't return to their native countrry, and thus add to the brain drain. Conducive working opportunities in Tanzania are often inadequate. Funds to pay salaries and absorb running costs of biotechnology projects are either limited or inadequate. Furthermore, biotechnology training gained abroad is in some cases not attuned to local needs because of the different research and infrastructural environments in many African universities. As a result the demands and opportunities present in the home country are often not met or remain unanswered.

*Inadequate information and public awareness* Lack of public awareness and clear understanding of both the potential promise and perils of biotechnology was noted. This is conducive to the creation of an atmosphere of suspicion among consumers and for decision-makers to take uninformed decisions. The private/public sector needs a better understanding on what innovations/products can offer universities and vice versa.

*Weak linkage between universities and private/business sector* Whereas in most developed countries the private sector engagement in biotechnology has been the driving force, the investment of the private sector in biotechnology in Tanzania is negligible. In the developed world, public–private partnerships for biotechnology are not a new phenomenon; although limited originally to relationships between universities and industries, collaboration between the private industry and the public sector in the industrialized countries has existed for several decades.

There is therefore a need for the government to introduce mechanisms such as provision of incentives, tax write-offs for industry-funded university research and development, research levies and so on in order to promote formation of strategic partnerships between universities and the private sector. The Tanzanian government can also provide 'seed' money to finance research arrangements consisting of consortia of universities and industries.

On the part of universities, there is a need to put in place mechanisms that may enhance linkage and collaboration between universities and the private sector/business community. This can be achieved by putting in place supportive institutional policies and structures, such as IPR and technology transfer offices.

## Setting the Research Agenda for Biotechnology R&D

The government's policy guiding research in Tanzania strongly emphasizes demand-driven research where stakeholders set the research agenda and influence the selection of research projects and resource allocation.

In the agricultural sector, research priority setting starts with researchers receiving specific requests from farmers or other stakeholder groups. These requests are then translated into draft research proposals and presented at the annual Zonal[2] Internal Programme Review (IPR) meetings, which are attended by researchers, farmers, extension agents, NGOs and policy-makers. The IPR reviews the proposed agricultural R&D projects to ensure that they are demand-driven and address farmers' or stakeholders' priorities, while also taking into account total resource availability (zonal, national, public, private and so on). Although stakeholders have been represented in planning and in project review committees, their influence/impact on the overall decision-making has been minimal. This situation is partly caused by research managers selecting stakeholders on the basis of recommendations made by the government's regional and district officials rather than the stakeholders themselves deciding who should represent them. Many of these individuals largely represent their own personal interests and concerns, and contribute little to real downward accountability.

Most biotechnology research projects and partnerships have successfully used participatory methods to incorporate the needs of beneficiaries in their research (Mneney et al., 2002). Regional biotechnology research programmes such as BIOEARN[3] and ASARECA,[4] for example, have involved stakeholders in the priority setting, in planning and in the implementation of the projects. However, given the situation of decreasing research budgets and the high cost of investing in biotechnology, setting priorities and research agendas of some biotechnology projects has involved other players, such as donors and international research organizations. Cases of donors or international institutes dictating research priorities, technology/tools to use, type of stakeholders to engage and so on, were reported in the current work and in a study by Mello and Mneney (2004). Stakeholders such as farmers, consumers and industries have largely been excluded from setting the research agenda of biotechnology projects, especially those receiving external funding. Hence conventional 'top-down' project planning and implementation structures are still the norm in Tanzania's biotechnology sector. In order to implement a national biotechnology agenda that addresses the priorities of Tanzania on a sustainable basis, it would be necessary for the government to find its own resources and develop local technological capabilities to support biotechnology research in the different disciplines.

## Product Development Partnerships (PDP)

Innovations in science and technology in Africa and other developing countries are essential means of alleviating poverty, and stimulating economic growth and development. Certain types of social, economic and technological advances can only be achieved by private firms, public sector organizations, universities, NGOs or civil society working together. A rapid rise in the incidence of research partnership (public–private, private–civil society, public–private–civil society and many other forms of partnership) has stimulated a policy debate regarding their effectiveness and whether these relationships enhance overall economic performance in the developing countries. Research partnerships have become an increasingly important means of creating and diffusing scientific and technical knowledge. This is of great importance particularly in biotechnology-related activities. The partnerships may be complex (combining all activities in the product development chain such as research, product/technology development, marketing dissemination) and may require involvement of multiple players such as individual innovators, research organizations, universities, small and medium-sized private companies, resource-poor farmers and so on.

### Current status of PDP and biotechnology product dissemination in Tanzania

Although biotechnology in Tanzania has a promising future, there is lack of clear roadmaps for getting the biotechnology from the laboratory to farmers or other end users. A problem facing Africa and Tanzania in particular is the lack of a dynamic private sector to deliver technologies to those in need.

Private sector involvement in biotechnology in Tanzania has been very limited. However, a few private institutions have shown an interest in participating in biotechnology through partnership with the public sector, in crops such as coffee, pyrethrum, sisal, cotton, cashew and tobacco. Specific examples include KATANI Ltd (private) collaborating with Mlingano Agricultural Research Institute (public) on micropropagation of sisal planting material through tissue culture for rehabilitation of the sisal industry in the country; and Mbegu Technologies Limited for promotion of quality seed and cassava seedling production through partnership with government institution (Mneney and Diyamett, 2003). The tobacco, tea and sugar cane sectors are also interested in R&D collaboration with the public agricultural research sector in the development of new improved varieties and seed dissemination systems.

**Major issues emerging from the study**

From the analysis of the responses from the interviewees, the following lessons can be drawn:

- Contrary to the private partners in the developed world who are normally very strong, the private partners in Tanzania are very weak and can hardly be expected to finance research in the universities or other R&D institutions.
- There is lack of information and awareness in both the private and public sector on the potential of biotechnology and benefits of public–private partnership (PPP) collaborations.
- Lack of skills to manage PPPs is another serious constraint. The skills required for entering into public–private partnerships are considerable. In developing countries like Tanzania, there is no legal capacity or support for the negotiation of contracts. Seeking a legal opinion can be extremely time-consuming and may cause delays in negotiating with the private sector. Private companies are much more experienced than the public sector in contract management and are much more flexible and less bureaucratic than the public sector.
- Implementation of PPPs requires trust, goodwill and support of people who are prepared to consider new and innovative ways of delivering services and overcoming the overwhelming bureaucratic administrative processes. Universities and other public research institutions are unfamiliar with these working modes and cultures.
- Lack of a strong research programme is another serious constraint to the formation of PPPs. Research collaboration is not likely to take place in countries where either the public or the private sector do not have strong research programmes. Universities and public research institutions need to be able to meet the demands of

industries. Correspondingly, industries need to have the capacity to invest in modern technologies that require innovative solutions from the universities or from public R&D institutions.

## CONCLUSION AND RECOMMENDATIONS

Biotechnology clearly offers tremendous promise for addressing key problems in food and agriculture, the environment, industry, and human and animal health. However, resources for biotechnology R&D are limited in developing countries like Tanzania and, as a consequence, their policy-makers are faced with a series of very difficult choices. How much importance should they give to biotechnology research, how should they allocate the biotechnology research resources with respect to the different sectors (i.e. agriculture, environment, medical) or to different kinds of biotechnologies currently available (i.e. tissue culture, genetic engineering)? The last and most important choice to make, especially for poor countries like Tanzania, is how they should prioritize different kinds of problems (and specifically those affecting resource poor farmers) that might be addressed by the research.

Despite the fact that Tanzania considers biotechnology a tool that may provide opportunities for achieving productivity gains in major sectors of the economy, the progress in both R&D and utilization of this technology in Tanzania has remained rather slow. The level of biotechnology research, development and utilization is still in its infancy, although it is picking up quickly in the agricultural sector. Most biotechnology laboratories in Tanzania are for teaching and/or for research purposes. None is operating at a large scale or at a commercial level, and there is no difference between a university and a public research institute in terms of depth and size of biotechnology research. If anything, some of the research institutions are better placed in terms of research facilities and human resources. It was expected that universities could be performing their teaching functions, and therefore could be a major source of human resources for the public R&D institutions. However, this is not the case; the finding of this study indicates that there is no specialized training for biotechnology at the MSc and PhD levels. This means that most of the human resources for biotechnology in universities and in R&D institutions received their training abroad, and to a large extent this has aggravated the problem of brain drain in the country.

With the existing human resource and infrastructure capacity at the different institutions, it would be possible to conduct more research if financial resources were available. This would therefore increase the level of utilization of the existing capacity. Institutions are therefore urged to increase their efforts in seeking additional financial resources by engaging the private/commercial sectors within and outside the country, in order to fund priority projects that are likely to contribute to the attainment of national development goals.

As noted in this study, most biotechnology and biosafety-related research and development activities being carried out in Tanzania are funded through foreign grants from development partners. This is not a very healthy situation. In order to implement a home-grown biotechnology agenda that addresses national priorities on a sustainable basis, it would be necessary for the government to find its own resources to support biotechnology research in the different disciplines.

Engaging opportunities, partnerships and collaboration with other organizations is the only rational way of getting biotechnology to work. As is the case with developed countries, Tanzanian governments should provide an enabling environment and incentives that can attract the participation of the private industry in biotechnology R&D. Partnerships do not merely happen, they are built. Contracting partnerships among diverse actors is hard work and needs organizational reforms. Partners need to participate in biotechnology development with a clearly articulated goal. A company attempting to introduce a novel biotechnology product must ensure that the target public has a very clear local demand for the product (i.e. demand driven). Creating partnerships is a continual learning process. As PPPs are evolving, it is important to implement regular training programmes for public and private sector scientists, so as to increase their understanding as well as their deployment to undertake negotiations in biotechnology.

## NOTES

1. AFLP = Amplified Fragment Length Polymorphism
   RFLP = Restriction Fragment Length Polymorphism
   RAPD = Random Amplified Polymorphic DNA
   ISTR = Inverse Sequence Tagged Repeats
   SSRs = Simple Sequence Repeats
   SNPs = Single Nucleotide Polymorphisms.
2. The Ministry of Agriculture has seven research zones. These broadly represent seven agro-ecologies: Eastern, Southern, Northern, Western, Southern Highlands, Central and Lake Zone.
3. BIOEARN = East African Regional Programme and Research Network for Biotechnology, Bio-safety and Biotechnology Policy Development.
4. ASARECA = Association for Strengthening Agricultural Research in Eastern and Central Africa.

# REFERENCES

Anonymous (2005), 'Biosafety and biotechnology in Tanzania: a country study', Vice President's Office, United Republic of Tanzania, March 2005.

Cohen, J.I. (1994), 'Biotechnology priorities, planning and policies: A framework for decision making', A biotechnology Research Management Study, ISNAR Research, Report No. 6, The Hague: International Service for National Agricultural Research.

Croxford, A. and E.E. Mneney (2007), 'Review of current and future use of plant biotechnology in the genetic improvement of cashew in Tanzania', paper presented at the Second COSTECH Scientific and Technological Conference, Dar es Salaam, Tanzania, 23–25 May.

Fregene, M.A., M. Suarez, J. Mkumbira, H. Kulembeka, E. Ndedya, A. Kullaya, S. Mitchel, U. Gullberg, H. Rosling, A.G.O. Dixon, R. Dean and S. Kresovich (2003), 'Simple sequence repeat marker diversity in cassava landraces: genetic diversity and differentiation in an asexually propagated crop', *Theoretical and Applied Genetics*, **107**, 1083–93.

Kullaya, A.K., A.A. Mpunami and E.E. Mneney (2001), 'Agricultural biotechnology capacity in Tanzania: national and regional initiatives', in *Proceeding of the First National Biotechnology Stakeholders Workshop*, 8 October, Dar es Salaam, pp. 59–73.

Kullaya, A., K. Mtunda, H. Kulembeka, M. Ferguson, J. Marin, C. Ospina, E. Barrera, A. Jarvis, N. Morante, H. Ceballos, J. Tohme and M. Fregene (2005), 'Molecular marker-assisted and farmer participatory improvement of cassava germplasm for farmer/market preferred traits in Tanzania', in *Proceedings of the Second General Meeting of the Rockefeller Foundation on Breeding, Biotechnology and Seed Systems in Africa*, 24–27, January, Nairobi.

Masumba, E.A., S. Nchimbi-Msolla, A.K. Kullaya, H. Kulembeka, M. Fregene and M. Ferguson (2006), 'The genetic diversity of cassava (Manihot esculenta Crantz) varieties commonly grown in the eastern, southern and Lake Victoria zones, Tanzania', in *Proceedings of the First COSTECH Scientific and Technological Conference*, 24–26 May, Dar es Salaam, Tanzania.

Masumbuko, L.I., T. Bryngelsson, E. Mneney and B. Salomon (2002), 'Genetic diversity in Tanzanian Arabica coffee using Random Amplified Polymorphic DNA (RAPD) markers', in *Proceedings of the Second National Workshop on Plant Genetic Resources and Biotechnology*, 6–10 May, Arusha, Tanzania.

Mello, D. and E.E. Mneney (2004), 'Agricultural biotechnology research partnerships in Tanzania', *UNU-INTECH-Technology Policy Briefs*, **3**(1), 8–10.

Mneney, E.E. (2000), 'Katani Ltd/MARI Partnership in the Meristemic Tissue Culture (MTC) Project. A Tanzanian case of Public–Private Partnership (PPP) in biotechnology', paper presented at the BIOEARN regional workshop on enlarging public–private partnership in biotechnology', 20–25 September, Kenya.

Mneney, E.E. and B.D. Diyamett (2003), 'Public Private Partnership (PPP) for technology development in Tanzania: The case of product development and transfer in biotechnology', paper presented at the PPP stakeholders workshop, 30 May, Dar es Salaam, Tanzania.

Mneney, E.E., S.H. Mantell and B. Mark (2001), 'Use of random amplified polymorphic DNA (RAPD) markers to reveal genetic diversity within and between populations of cashew (Anacardium occidentale L)', *Journal of Horticultural Science and Biotechnology*, **76**(4), 375–83.

Mneney, E.E., A.A. Mpunami, P.S. Gwakisa, B.D. Diyamett, J.S. Kasonta and R. Kingamkono (2002), 'Needs assessment and priority setting in agricultural biotechnology for Tanzania', paper presented at the Regional Workshop on Needs Assessment and Priority Setting in Agricultural Biotechnology, 23–27 September, Nairobi, Kenya.

Mneney, E., R. Sallu, M. Seth, F. Luka, Z. Mchomvu, S. Chalamila and A. Kullaya (2006), 'Molecular marker-based characterization of coconut (Cocos nucifera L) germplasm in Tanzania', in *Proceedings of the First COSTECH Scientific and Technological Conference*, 24–26 May, Dar es Salaam, Tanzania.

Mpunami, A., P. Jones, A. Tymon and M. Dickinson (1996), 'The use of DNA probes and PCR for detection of coconut lethal disease (LD) in Tanzania', in *Proceedings of the Brighton Crop Protection Conference*, 18–21 November, Brighton, UK, pp. 577–82.

Msoffe, P.L.M., M.M.A. Mtambo, U.M. Minga and P.S. Gwakisa (2006), 'Genetic diversity in free range local chicken ecotypes of Tanzania based on microsatellite DNA typing: Implications on poverty reduction efforts', in *Proceedings of the First COSTECH Scientific and Technological Conference*, 24–26 May, Dar es Salaam, Tanzania.

Rafalski, J.A. and S.V. Tingey (1993), 'Genetic diagnostic in plant breeding: RAPDs, microsatellites and machines',*Trends in Genetics*, **9**(8), 275–80.

Rohde, W., A. Kullaya, A. Mpunami and D. Becker (1993), 'Rapid and sensitive diagnosis of mycoplasma-like organisms associated with lethal disease of coconut palm by a specially primed polymerase chain reaction for the amplification of 16S rDNA', *Oleagineux*, **47**, 511–15.

Rohde, W., A. Kullaya, J. Rodriguez and E. Ritter (1995), 'Genome analysis of Cocos nucifera L. by PCR amplification of spacer sequences separating a subset of copia-like EcoRI repetitive elements', *Journal of Genetics & Breeding*, **49**, 179–86.

Rohde, W., J. Dowe, A. Kullaya, A. Santos, J. Rodriguez and E. Ritter (1999), 'Analysis of genetic biodiversity in Palmae by DNA marker technology', *Acta Horticulturae*, **486**, 65–72.

Sariah, J.E. and H.F. Lyimo (2006), 'Simultaneous detection of viruses in planting materials by Enzyme linked Immunosorbent Assay (ELISA) technique', in *Proceedings of the First COSTECH Scientific and Technological Conference*, 24–26 May, Dar es Salaam, Tanzania.

Stephen, J. (2006), 'Characterization of sheep populations in Tanzania using DNA markers', *in Proceedings of the First COSTECH Scientific and Technological Conference*, 24–26 May, Dar es Salaam, Tanzania.

Tairo, F., A. Kullaya and J.P.T. Valkonen (2004), 'Incidence of viruses infecting Sweetpotato in Tanzania', *Journal of Plant Diseases*, **88**(9), 916–20.

Tairo, F., S.B. Mukasa, Roger A.C. Jones, A.K. Kullaya, P.R. Rubaihayo and J.P.T. Valkonen (2005), 'Unravelling the genetic diversity of the three main viruses involved in Sweetpotato Virus Disease (SPVD), and its practical implications', *Journal of Molecular Plant Pathology*, **6**(2), 199–211.

# APPENDIX A5.1: A QUESTIONNAIRE FOR BIOTECHNOLOGY PRODUCT-PARTNERSHIP STUDY

## 1. Introduction: Important Definitions

Research activities in biotechnology should include:

- DNA-based: genomics, pharmacogenetics, gene probes, DNA sequencing/synthesis/amplification, genetic engineering
- Proteins and molecules – the functional blocks: proteins/peptide sequencing/synthesis, lipid protein, glycol-engineering, proteomics, hormones and growth factors, cell receptors/signalling/pheromones
- Cell and tissue culture and engineering: cell/tissue culture, tissue engineering, hybridization, cellular fusion, vaccine/immune stimulants, embryo manipulation; process biotechnologies: bioreactors, fermentation, bioprocessing, bioleaching, biopulping, biobleaching, biosulphurization, bioremediation and biofiltration
- DNA and RNA vectors: gene therapy, viral vectors
- Others: Bio-informatics, nanobiotechnologies , microbiology

## 2. Definition of R&D

R&D comprises creative work undertaken on a systematic basis in order to increase the stock of knowledge. Any activity classified as R&D is characterized by originality. Investigation is a primary objective. R&D consists of basic research, applied research and product development.

## Section A: Basic Information about the Organization

### 1. Type of organization (*Please mark with an X*)

a) Public R&D organization
b) Private R&D organization
c) Public university
d) Private university
e) Policy and regulatory body
f) Private productive sector

**2. Please state number and qualification of employees working *regularly* on biotechnology projects as defined above**

Names of staff involved in biotechnology and biosafety-related activities at . . . . . .

| Area of research/expertise | Number | | |
|---|---|---|---|
| | BSc | Msc | PhD |
| | | | |
| | | | |
| | | | |

**3. Please list all completed and currently ongoing biotechnology projects, and their major objectives and funding sources (***Major categories of funding sources to be included: government, private commercial company, donors, own sources, and any other***)**

| Project | Objective | Funding sources | Status |
|---|---|---|---|
| | | | |
| | | | |
| | | | |

**4. Are you collaborating with any organization? (***Please select from the list below. You can select more than one***)**

a) With a university
b) With R&D organization
c) With a regulatory body
d) With the private sector
e) Other. Please mention.

**5. What is the primary purpose of collaboration? (***Please select from the list below. You can select more than one***)**

a) R&D per se (*please note the definition of R&D above*)
b) Product/process development
c) Clinical field trials
d) Marketing
e) Finance
f) Policy and regulatory issues
g) Other. Please mention.

6. **How is the research agenda set? What is/was the influence of the collaborating institution mentioned in Section 4 above? Identify other key actors and their role in defining the research agenda.**

7. **Were you able to commercialize or put into any economically and socially useful practice, any of your R&D projects? If so, what do you think is the role of the collaborators you mentioned (if any)?**

8. **What do you think are major constraints to commercialization of R&D projects?**

9. **What do you think are the major constraints to useful collaborations?**

# APPENDIX A5.2

*Table A5.1 List of institutions contacted during the study*

| Type of institution | Research area | University name |
|---|---|---|
| University | Agricultural biotechnology | Sokoine University of Agriculture (SUA) |
| | | University of DSM – Botany Department |
| | Medical biotechnology | Muhimbili University College of Health services (MUCHS) |
| | | Tumaini University (KCMC) |
| | Industrial and environmental biotechnology | Molecular Biology and Biotechnology Department, UDSM |
| | | UDSM – Chemical and Processing Engineering, UDSM |
| National Research Institutes (NARIs) | Agricultural biotechnology | Mikocheni Agricultural Research Institute |
| | | Agricultural Research Institute (ARI) – Ukiriguru |
| | | ARI-Tumbi |
| | | ARI-Mlingano |
| | | ARI-Uyole |
| | | Horti Tengeru |
| | | Central Veterinary Laboratory (CVL) |
| | | National Plant Genetic Resources Centre (NPGRC) |
| | | Kizimbani Agricultural Research Centre (KARC) |
| | Medical biotechnology | National Institute for Medical Research (NIMR) (Tanga) |
| | | AMRC |
| | | Ifakara Health Institute (IHI) |
| | Industrial | Tanzania Food and Nutrition Centre (TFNC) |
| Others | Regulatory | Tanzania Food and Drug Authority (TFDA) |
| | | Tropical Pesticides Research Institute (TPRI) |
| | | Government Chemist Laboratory Agency (GCLA) |

# 6. Biotechnology in Mozambique: present situation and future trends

**Luis Neves, Paula Macucule, Carlos Miguel Ribeiro and Ana Maria da Graça Mondjana**

## INTRODUCTION

Modern biotechnology, which is based on cell biology, molecular biology and genetic engineering, is considered one of the most dynamic branches of the biological sciences and has been the basis for an unprecedented technological and industrial expansion. This fact is testified by the exponential increase in the investment in biotechnology, which is proportional to the increase in the number of patents and technological processes presently in use in agriculture, animal production, aquaculture, medicine and engineering.

Biotechnology is generally perceived as a multi-disciplinary complex technological platform that is playing a decisive role in triggering development, notably in presenting a range of solutions to the increase of quantity and quality of food products, in the generation of modern diagnostic tools, therapeutic drugs, and vaccines for combating diseases in humans, animals and plants as well as in presenting alternative processes for industry and environmental management.

Although the intellectual and applicative epicentre of this technology was and still is in many aspects located in developed countries, there are many examples of developing countries that have very quickly capitalized on the potential of biotechnology and are important players in devising products, processes, services and other applications which make biotechnology an important vector in their own socio-economic development. However, the vast majority of developing countries could be considered biotechnology marginalized countries, just because they do not meet the criteria either of technology producer or of technology consumer.

In spite of all the potential benefits of modern biotechnology, this phenomenon, as has happened in other processes of technological revolution that have been experienced by human societies, brings ethical, cultural, ecological and socio-economic problems and polemics which are intrinsically associated with it. This should be considered and widely debated in order to assure that any technological option that would be adopted takes into account the respect to human dignity and the peculiarities of each society.

Mozambique is a developing country located in south-east Africa. The country has as its major socio-economical goal the reduction of absolute poverty. Presently, biotechnology in Mozambique has an extremely limited social expression both in terms of application and research.

Acknowledging the positive and innovative role that this type of technology might play in the socio-economic development of Mozambique, it is vital to adopt a strategy that will promote, in a balanced way, the introduction, adoption and expansion of biotechnology processes, respecting the general lines of socio-economic development established for the country.

This chapter focuses on the evaluation of the present situation regarding biotechnology research and development, as well as the current application of classical biotechnology processes in Mozambique. The range, characteristics and management of partnerships and the role of the university in stimulating them are also examined. Special attention is given to the analysis of different institutional positions and perceptions towards biotechnology.

In order to achieve our objectives, a comprehensive documentation review was performed, and semi-structured interviews were applied to relevant stakeholders, namely, academics and researchers, policy-makers, business representatives, NGOs and cooperation agencies' representatives.

## THE PRESENT SITUATION

'Classical biotechnology' is a concept used to define the application of biological processes for the production of materials. These processes rely on the activity of various fungi and bacteria and, by definition, exclude the use of genetic engineering. Antibiotics, cheese and wine are typical examples of products from classical biotechnology.

The present level of industrial application of classical biotechnology in Mozambique is extremely low. In urban areas, breweries and a few dairy factories are the sole representatives of this type of technology. Usually the technological processes used in those industries are imported from other countries and there is no link with the local scientific community in research or development activities.

Yet in the scope of classical biotechnology, it is worth mentioning the experience of the National Institute of Veterinary Research, which could be considered a pioneer in the production of bacterial and viral vaccines. Unfortunately, the technological processes were not updated and the production lines were shut down due to the combined effect of lack of efficiency and bio-safety problems. Presently, the institute is producing, on an experimental scale, the Newcastle virus vaccine, using an Australian technological process.

In the rural areas of Mozambique the production of fermented beverages is an established tradition, with a wide range of regional variation. However, historically they never had links with formal markets, and the technological processes involved in their production are mastered only by limited communities, which pass this knowledge on solely through oral means.

This is an area that deserves intensive research not only on sociological grounds, where increased knowledge would promote and preserve essential elements of local culture, but also, looking from a technological point of view, the transformation of the indigenous knowledge into defined biotechnological processes could lead to its application in new products, hence contributing to the economic development of the communities.

Modern biotechnology, which implies a strong association with molecular biology and genetic engineering is applied in Mozambique by small groups from universities and research institutes. The Biotechnology Center of the Eduardo Mondlane University (UEM), the Immunology Laboratory of the National Health Research Institute (INS), the *in vitro* culture Laboratory of the National Institute for Agrarian Research (IIAM) and the Manhiça Foundation Laboratory are arguably the main representatives of the above-mentioned research groups. All these institutions are located in Maputo, which indicates an uneven geographical distribution.

The scarcity of competent and well-trained human resources is probably the major constraint to the development of biotechnology in Mozambique. Fewer than 100 researchers are involved in biotechnology activities. Of these, approximately 30 hold a master's degree and about 15 have PhDs. This scenario emphatically illustrates the urgent need for a specific human resources development plan for biotechnology, which should focus on the development of postgraduate training in Mozambique, the training of Mozambicans abroad and the mobilization of international expertise. In order to ensure sustainability to this process, it is essential that the human resources development plan is conceived in absolute compatibility with a general biotechnology institutional development frame, which should contain an effective staff retention strategy.

In general, the biotechnology research and development activities carried out in Mozambique take place in small labs which are usually fairly well equipped. The essential aspects to be taken into account in the upgrading of the labs are the improvement of quality of Internet connection, electric power supply, cold storage facilities and water supply. In order to increase competitiveness, substantial investment must be made in equipment in common use.

The existent research units in the area of biotechnology correspond to different organizational models. Both the Laboratory of Immunology (INS) and the *in vitro* culture laboratory (IIAM) are research units integrated in the main research institutes of the Ministry of Health and the Ministry of Agriculture, respectively. They are very specific in their range of activities. The Laboratory of Immunology has specialized in molecular epidemiology of viral and bacterial diseases, with a strong emphasis on HIV-AIDS. It is worth mentioning that this unit is responsible for quality assurance of the routine testing of HIV performed by the public lab network, whereas the *in vitro* culture lab is a unit responsible for micropropagation of important plant species. Presently, the lab deals mainly with important food crops such as banana and cassava, but is also establishing *in vitro* culture systems for medicinal plants like African potato (*Hypoxis hemerocallidea*).

The Manhiça Foundation Laboratory is an independent entity with a privileged relationship with the Ministry of Health. It is an extremely well-organized platform for micro-epidemiology, which has specialized in clinical trials and epidemiology studies. The laboratory applies extensive molecular methods for the study of a wide range of micro-organisms, their vectors and the complex relationships between them. This laboratory was recently involved in the trials of a malaria vaccine, trials that have comprised extensive scientific collaboration at national and international level. Over the years it has established a worldwide reputation in its area of research and is undisputably the Mozambican institution in this area with the widest range of international links.

The Biotechnology Center (CB-UEM) is a consortium formed by the Faculties of Agriculture and Forestry, Medicine, Science and Veterinary of the oldest and largest public university of Mozambique. The centre was created to boost the competence in molecular biology and genetic engineering through the development of a shared facility that could act as an incubator of a generation of researchers competent in biotechnology and aligned sciences. The researchers in the centre are biologists, biotechnologists, agronomists, veterinarians and medical doctors who work in mixed research teams that address transversal research topics such as molecular epidemiology and diagnostics, genetic characterization and biodiversity, molecular toxicology and environment. In the process of their research the different teams explore the striking communality existent among organisms at molecular level, hence developing a modern concept *in senso latum* of 'one health one medicine'. In the context of the National Program for Biotechnology the CB-UEM has the mandate for coordinating the training activities in the context of a specific human resources development plan for biotechnology.

## THE PARTNERSHIPS STATUS

The universities are playing a major role in promoting biotechnology in Mozambique. In this particular context, the participation of both Mozambican and foreign universities should be acknowledged. It is worth noting that their role goes far beyond the mere transfer of technological processes, assuring specialized training or engaging in defined scientific projects. The partnerships between universities are contributing decisively to the establishment of a scientific culture, which is a more sophisticated and sustainable stage of internalizing science.

Regarding biotechnology and in the context of Mozambique, Eduardo Mondlane University (UEM) has taken the strategic decision of assuming a pioneering and leading position in the field of biotechnology. The formal establishment of the Biotechnology Center in 2005 should be considered an objective indicator of institutional positioning in this area.

A strong partnership between Eduardo Mondlane University and a consortium of Italian universities composed of the Universities of Sassari, Pisa and Rome is securing the creation of an environment conducive to the establishment of the Biotechnology Center as an active unit in biotechnology research and development, and a leading player, nationally, in the promotion of postgraduate specialized training in this area.

There are also examples of efficient partnerships between universities and the research laboratories. The University of Barcelona, in close collaboration with the Mozambican Ministry of Health, has contributed decisively to the success of the model and the international reputation that Manhiça Foundation earns presently. In the same context, the collaboration between the Immunology Laboratory (INS) and the Federal University of Rio de Janeiro deserves to be mentioned. This partnership has contributed substantially to the quality of the present research output.

Moreover, a list of more than 50 small partnerships between foreign universities or international research organizations and the biotechnology research institutions in Mozambique could be compiled. Generally, these partnerships have as their sole object a specific scientific project, training of a student at postgraduate level, or both. This indicator could be used to ascertain the degree of internationalization of those institutions. The partnerships between biotechnology research institutions and international research centres, present a positive trend, which might be explained as an increase in the capacity for technology absorption by the national institutions. This trend will lead eventually to the consolidation of super-specialized niches in the Mozambican biotechnology institutions.

Partnerships are also developing between the university and other public sector institutions. At the Biotechnology Center, the combination of modern laboratory facilities with scientific expertise and a continuous flux of research have triggered a new type of personal and institutional demand. Mozambican students who are registered as postgraduate students abroad now have the opportunity to use the facilities to complete experimental parts of their studies. This new opportunity has led to the emergence of formal agreements between different public research institutions and UEM. Moreover, these agreements also include different forms of training and mentorship of the researchers from those institutions by the university.

The quantity and quality of the partnerships between the biotechnology research institutions and the private sector are still merely incipient. However, there are clear signs of development. A good example comes from the agricultural sector, where private enterprises and NGOs are in an advanced stage of liaison with the *in vitro* culture laboratory (IIAM) for the large-scale micro-propagation of banana and cassava plants. Another good example is the present demand from private companies for the university to develop and apply molecular diagnostic tools to prevent diseases in prawn and coconut industries respectively.

Recently the main cotton enterprise operating in the country has manifested its interest in embarking, together with the Faculty of Agriculture of the Eduardo Mondlane University, on the experimental introduction of a GMO (Bt-cotton). For reasons that are unclear this intention was not supported by the multinational supplier of agricultural seeds, Monsanto. A possible explanation for the refusal of this company to engage in such cooperation would be the limited market opportunities in combination with relatively tough labour laws.

Generally, the management of partnerships follows a formulaic model. A memorandum of understanding or an agreement establishing the object and the general principles, is signed by the parties involved and a steering committee or a board nominated by the parties takes care of the operational management. In some cases, more frequently in partnerships involving the private sector, the relationship is based on a contract with respect to a given service or a product.

The financial support to partnerships is diverse, the most common type being bilateral or unilateral responsibility for securing the funds. Less frequent is the case where a third party will secure financial support for the development of partnerships between two or more partners. This is especially the case for public funds or specialized donor agencies.

## INSTITUTIONAL POSITIONS AND PERCEPTIONS

In what concerns the promotion of biotechnology in Mozambique, a striking feature is the overwhelming political will and commitment of the Mozambican Ministry for Science and Technology leadership. The strong belief in the role of biotechnology as a development catalyst is not only clear in the political speeches of the ministry's senior officials, but it is also expressed in strategy documents and national development strategies.

The development of the 'Mozambique Science, Technology and Innovation Strategy' (MCT, 2006) is a landmark in terms of official positioning towards biotechnology. This document, which is considered the strategic backbone for science and technology in Mozambique, defines biotechnology together with Information and Communication technologies as 'strategic transversal areas', in recognition of their important impact in the promotion of research and development processes across a wide range of scientific disciplines. In this context the strategy gives particular emphasis to the development of common approaches to tackle different problems. Perceiving the dispersion of biotechnology activities by different small units as a potential problem for the rational use of resources, it is suggested that a strong leadership combined with a high level of coordination and communication would be the best preventing mechanism for duplication and misuse of resources. On the other hand, it is expected that the future organizational system would be the natural evolution and sophistication of the already existing institutional links among presently functional biotechnological units and centres. Although the functional relationships and hierarchies between different biotechnological institutions are only vaguely defined and analysed, the strategy advocates the development of a network of units of excellence as a reference centre. Moreover, it is envisaged that the reference centre would be under direct political control of the MCT, acting as a responding body to directives emanating from the government.

In this scenario of relationships between biotechnological research institutions and the MCT it is possible to predict several problems and eventually the distortion of some roles and functions. First, to secure competent leadership, MCT should consider the building of a strong specialized team to lead biotechnology research institutions effectively. In the present situation, where the scarcity of human resources is arguably the main constraint to biotechnology development, this approach might trigger a competition process between MCT and the research institutions for the more qualified researchers, which will lead eventually to a situation where scientific leadership is removed from areas where it is desperately needed. Furthermore, the crucial role of biotechnology research units and centres as specialized advisory entities and the need to establish and improve feedback mechanisms between research bodies and MCT is omitted from the text of the Mozambique Science, Technology and Innovation Strategy. This aspect deserves special attention, and adequate measures to correct this omission should be put in place.

The National Program of Biotechnology is a document commissioned by MCT and is in complete alignment with the Mozambique Science, Technology and Innovation Strategy. The document defines eight strategic objectives: the development of human resources and retention of human capital; the development of technical infrastructure; support to research, development and innovation in strategic areas such as agriculture, health, sea resources and fisheries, industry and environment protection; technology transference; establishment of a legal and regulatory framework for biotechnology; development of national and international networks; promotion of community participation and awareness; and the development of biotechnology enterprises. This document also contains the general principles and guidelines that should orient the main stakeholders in the biotechnology development process for the next ten years. An executive committee, composed of ministry officers and members from research institutions, has been appointed by the Minister of Science and Technology to lead the implementation of the programme.

The creation of the Inter-Institutional Group for Biosafety (GIIBS), presided by the Ministry of Science and Technology and composed of representatives from the Ministry of Agriculture, the Ministry of Environment, the Ministry of Health, the Ministry of Trade and Industry, The Ministry of Fisheries, the Ministry of Plan and Development and also representatives from academic and research institutions, constitutes a clear indication of the importance that the government of Mozambique attributes to biosafety matters. In this respect, it is worth mentioning that Mozambique has ratified the Cartagena Protocol in 2001. In line with the guidelines of the above-mentioned protocol GIIBS has developed a legal document, 'Regulations on management of genetically modified organisms and biosafety', which was approved by the ministers council and published in 2007 (Boletim da República, I série Número 17). This law on genetically modified organisms has as its main objective the regulation of all aspects pertaining to the import, export, transit, production, handling and use of these organisms.

The main legal responsibility of GIIBS is to provide advice on all biosafety aspects associated with GMOs. However, their legal mandate extends beyond biosafety in the issues concerning the preparation of updated technical reports on biotechnology and the promotion of public awareness activities at national level.

Similarly to what has occurred in areas such as water resources, energy, agriculture and health, the MCT has formally expressed the intention of establishing a 'National Council for Biotechnology'. This would be a committee of experts which would also have as its main function to advise on all matters concerning biotechnology, particularly on strategic options for its development in alignment with the country's general development policies and to explore national comparative advantages. It is not clear how this council will operate in parallel with GIIBS, considering the substantial overlap of their responsibilities. Another aspect that needs further clarification concerns the composition and selection procedure for the council. In the GIIBS case the representatives are nominated by the different member institutions. However, taking as a reference the process of formation of the other MCT councils, it is very likely that the members of the future National Council for Biotechnology will be selected directly by senior ministry officials, with the approval of the minister.

In order to evaluate and produce a prognosis on the relationship between the MCT and the primary research institutions it is advisable to briefly examine their genesis.

The inception of biotechnology research and application in the country was driven by Mozambican researchers who have had access to these technologies in the course of their postgraduate studies overseas. With the support of their universities and research institutes this embryonic stage has evolved to the formation of small research groups that matured to more structured laboratories with some degree of specialization. In other cases the establishment of research laboratories was mainly due to external motivation, particularly to the desire of research groups from developed countries to apply modern tools and technologies to relatively exotic problems.

Given the main characteristics of these research groups, the scarcity of senior researchers and the lack of a coherent development strategy, the definition of the research agenda relied upon quite arbitrary criteria. The experience of the researcher during his postgraduate period and the random opportunity of linkage with a foreign research group were arguably the essential elements to selecting a research topic.

More recently the government has increased the degree of assertiveness in the statement of its major socio-political goals. This change has triggered a more aligned and proactive attitude by government and donor agencies. In these terms, poverty reduction, gender balance and HIV-AIDS became essential background elements in the majority of the calls for research proposals. This position calls for a more applied and demand-driven type of research.

Anyway, the internal sustainability and the external competitiveness of the present biotechnology research institutions will essentially ride on the efficiency of human resources development and retention, and on the maximization of the common use of infrastructures and equipment. It is also important to achieve the right balance between political will and expectations, and scientific freedom and imagination.

## THE ROLE OF NON-GOVERNMENTAL ORGANIZATIONS (NGOs)

In Mozambique, the only records of direct involvement of NGOs in activities associated with biotechnology are connected with the support of specific agriculture programmes where there is often the need for distribution of improved plant varieties. In such cases there is a specific demand for *in vitro* micro-propagation of genetically improved plant varieties, which implies direct liaison with the *in vitro* culture Lab (IIAM), presently the only Mozambican institution with capacity for medium-scale production of *in vitro* micro-propagated plants. The national demand for genetically improved disease-free plants is increasing substantially, which indicates that it might constitute an important biotechnological interface area between NGOs and the public sector. Furthermore, this very same demand has the potential to trigger the development of micro, small and medium-scale biotech enterprises which will probably be more efficient in satisfying a demand that is consistently growing in agronomic and geographical complexity.

Some NGOs in Mozambique, particularly those operating in the area of agriculture and consumer rights and awareness, have been actively involved in the most polemic and politicized of the applications of genetic engineering – Genetically Modified Organisms (GMOs). Those NGOs were extremely proactive in bringing this controversy to the main political agenda, and in general their position was anti-GMO.

The inspiration for this behaviour could have been the strong anti-GMO campaign that swept Europe in the 1990s. Those movements, with a strong participation and commitment of certain segments of civil society, have mobilized the connivance and the alignment of most European governments and resulted in a long period of banning of genetically modified products from the major European markets. The similarity between the European position towards the GMOs and that assumed by some NGOs operating in Mozambique could be attributable to the dramatic financial dependence of the NGOs on European funds.

One of the major criticisms of the NGOs was related to multinationals' absolute control on GMO research and production and its effects. They pointed out that multinational owners of the GM business are leading the research process in this area with absolute control of research funds allocation. Seventy-five per cent of the total research funds for biotechnology research are generated from industries connected with agriculture activity. The industry of agriculture biotechnology is highly concentrated, with just four enterprises holding more then 50 per cent of the patents, and just Monsanto conducting two-thirds of all the trials with GMOs in the United States, spending more than USD2 billion in research and development.

The overwhelming dominance of multinationals was perceived as the main cause of the problems putatively attributable to this technology; the most common issue being the fact that the purpose of genetic modification was not primarily smallholder demand-driven, but mainly a strategy to maximize profit and to create further dependency on the multinational companies. The extreme vulnerability of poor peasants in this relationship is also frequently pointed out. Other frequent concerns in the GMO debate are the potentially detrimental effect on public health and the environment, where the possible reduction of biodiversity and the eventual emergence of super-weeds are the most commonly presented examples.

All this debate is developing against a background where the hard facts are that the first GMO crop appeared in the market 21 years ago, more than 80 million ha are used for GMO crops in the five continents of the planet, and hundreds of licences have been issued for new GMO trials involving a wide range of crops and a substantial number of intended modifications.

The position of some NGOs championing the anti-GMO cause was not enough to secure an alignment of the government, but was arguably instrumental in triggering the creation of an inter-institutional body to deal specifically with biosafety issues in Mozambique and has certainly played a substantial role in catalysing the process of development of the law on GMOs.

## THE ROLE OF DONORS

Research in general and particularly on biotechnology-related areas was never a topic that was included in the international community mainstream development aid agenda in Mozambique. With the exception of Sweden, and to a lesser extent, Italy, which have been historically associated with the consistent support of research at Eduardo Mondlane University, the main public university in Mozambique, the external support for research has been developed in the context of rather small bilateral or multilateral projects, oriented to the implementation of scientific projects focused on finding the solutions for narrow scientific questions and often associated with the attainment of postgraduate degrees.

In order to contextualize the type of donor involvement in Mozambique, it is important to revisit some essential elements of its recent history and associated social characteristics. The country became independent in 1975 and has adopted a socialist, centrally planned economy model. In 1980 a civil war started and spread all over the country causing approximately 1 million deaths and 3 million refugees. After the end of the war in 1992, Mozambique was considered one of the poorest countries in Africa with nearly 70 per cent of the population living below the poverty line with more than a 75 per cent illiteracy rate. Given this scenario it is easy to understand why aid, in this period, after a major focus on humanitarian aid, was mainly directed to poverty alleviation through increasing access to basic social services.

In the late 1990s some donors initiated a shift to a more programme-oriented aid modality, in the form of general and sectoral budget support and several forms of common funds. This change in the approach to aid provision was, for the case of general budget support, pioneered by four countries in 1997, now having the support of 19 donors represented in Mozambique.

One important consequence of this new approach to aid delivery is that it opens up the opportunity for government-owned development strategies and simultaneously increases the challenges and responsibilities for a more systemic process of developing such strategies, including the role of research. As mentioned before, a good example is the case of the recently formulated strategy by the Ministry of Science and Technology to grow its science sector. Completely in alignment with this strategy is the establishment in 2005 of the National Research Fund and the more recent engagement in fundraising for the implementation of the National Programme for Biotechnology.

A common observation in developing countries like Mozambique is that the majority of research funding is directed to commissioned research in areas prioritized for their potential to promote development, which often implies paramount urgency, from governments and donor agencies, for the attainment of immediate useful results. In fragile research environments, as in the case of biotechnology research institutions in Mozambique, this mindset could have potential detrimental effects on a long-term and sustainable research capacity which can only be achieved through a well-structured capacity-building process, where funding for human resources development, laboratory facilities and contextualized institutional development plans should be perceived as essential priorities.

## THE ROLE OF THE BUSINESS COMMUNITY

In spite of the negligible role of the business community in Mozambique towards biotechnology it is worth highlighting some issues that might shape future trends of this community in this specific area.

The large enterprises operating on cash crops in the agriculture sector, particularly the cotton enterprises, have openly stated their intention to embark on experimental-scale adoption of genetically modified cotton. This ambition has triggered a partnership with the Faculty of Agriculture-UEM, which has sought the intermediation of the United States Agency for International Development (USAID) to persuade the BT-cotton supplier to support the trial. For some reason that is not completely clear, they never obtained support for their intention from either USAID or Monsanto. The size of the potential market for cotton seed in the country, the relative lack of definition on legal issues and regional policy agendas could explain in part the low motivation of both parties with regard to the introduction of BT-cotton in Mozambique. However, it is realistic to predict that these companies will continue to put some pressure on the implementation of GMO introduction trials and hence they could play a major role in the introduction of this technology in Mozambique.

Conversely, there is a group of companies in which the main investors are from countries where the use of GMO products is still taboo. In this case the GMO-based agricultural production is not considered as a technological option and the fact that, with the exception of South Africa, the regional market (SADC) in general is not very positive towards GMOs is used to legitimate their position.

Apart from GMOs, the nascent demand for genetically improved and disease-free plant varieties has the potential to stimulate the establishment of private companies in this specific business niche. Furthermore, the dependence on South Africa for diagnostic services in human, animal and plant diseases and in the production of other biological products creates the opportunity for entrepreneurship focused on competitively replacing South African services and products.

## CONCLUSIONS

Biotechnology in Mozambique is in its early stages of development, as is indicated by the extremely low level of industrial application of classical biotechnology and the characteristics and the number of research groups involved in molecular biology activities in the universities and research centres. Moreover, there is virtually no link between the Mozambican research community and the industries where biotechnological processes are used.

However, there is enormous potential for biotechnology application in the promotion and preservation of local knowledge and culture. In this context, the transformation of the indigenous knowledge into defined biotechnological processes could lead to its application in new products, hence contributing to the economic promotion of the communities. Furthermore, the tremendous richness presently hidden in the billions of genes that constitute the genomes of living organisms that inhabit the rich tropical ecosystems is an asset that could be transformed through biotechnology in a major vector for socio-economical development in the country.

The existing research units in the area of biotechnology are located in Maputo and correspond to different organizational models. Some laboratories are integrated in ministries and respond to the specific mandate of those institutions. There is also an independent laboratory with strong international connections yet maintaining some links with the Ministry of Health, and there is the Biotechnology Center, which represents the university. This centre is the model with the highest level of knowledge integration, and besides research it has the mandate for training.

The scarcity of competent and well-trained human resources is probably the major constraint to the development of biotechnology in Mozambique. This situation demands a specific human resources development plan for biotechnology. In order to ensure sustainability to this process, it is essential that the human resources development plan is conceived in absolute compatibility with a general biotechnology institutional development frame, which should contain an effective staff retention strategy.

Partnerships involving a diversity of international and national actors are considered an essential element for the transformation of the present status of biotechnology in Mozambique. The universities, with emphasis on the main public university UEM, are playing a central role in partnership development. The partnerships between universities are contributing decisively to the establishment of sustainable processes of internalizing science and simultaneously enabling institutions as proactive actors in the universal scientific movement. The partnerships established between universities, universities and research laboratories, universities and research organizations, including International Research Centres are playing a major role in boosting the institutional dynamics in the area of biotechnology. Conversely, the quantity and quality of the partnerships between the biotechnology research institutions and the private sector are still in their early stages. However, there are clear signs of a positive trend.

In terms of macro-political environment, a striking political will and commitment of the Mozambican Ministry for Science and Technology (MCT) has been observed. This position is based on the strong belief in the role of biotechnology as a development catalyst.

An objective indicator of the politically proactive attitude of the MCT is the formulation of the Science, Technology and Innovation Strategy (MCT, 2006). In this document biotechnology is considered a strategic area with important linkages to research and development processes across a wide range of scientific disciplines.

Given the potential problem that the dispersion of biotechnology activities by different small units would pose to the rational use of resources, the strategy advocates a strong leadership combined with a high level of coordination and communication as a preventing mechanism for duplication and misuse of resources. Furthermore, the strategic view for the future organizational system would be the natural evolution and sophistication of the already existing institutional network of presently functional biotechnological units and centres. A weak point of the strategic institutional model is that the functional relationships and hierarchies between different biotechnological institutions are vaguely defined and analysed.

One operational issue that MCT will face in order to secure competent leadership is to build a strong specialized team to liaise effectively with biotechnology research institutions. This approach might trigger a competition process between MCT and the research institutions for the more qualified researchers, which will eventually lead to a situation where scientific leadership is removed from areas where it is desperately needed. Furthermore, the crucial role of biotechnology research units and centres as specialized advisory entities and the need to establish and improve feedback mechanisms between research bodies and MCT is omitted from the text of Mozambique Science, Technology and Innovation Strategy.

The implementation of the National Program of Biotechnology will constitute the first programmatic attempt to develop biotechnology in a coordinated and harmonized way. Nevertheless, the internal sustainability and the external competitiveness of the present biotechnology research institutions will reside essentially on the efficiency of human resources development and retention and in the maximization of the common use of infrastructures and equipment. It is also important to achieve the right balance between political will and expectations and scientific freedom and imagination.

In terms of regulatory and advisory bodies, the creation of GIIBS is an indication of the importance that the government of Mozambique attaches to biosafety. However, taking into account the mandate of GIIBS, the lack of an advisory body for strategic planning is quite obvious. This void will probably be circumvented by the creation of a National Council for Biotechnology. Attention should be paid to the design of the terms of reference for this council in order to avoid a possible overlap with GIIBS mandate.

Other important stakeholders in the area of biotechnology in the country are NGOs, donors and private companies. In Mozambique, the involvement of NGOs in activities associated with biotechnology is mainly in the scope of their support to specific agriculture programmes. In this context, the potential demand for genetically improved disease-free plants might constitute an important biotechnological interface area between NGOs and the public sector. Furthermore, this demand has the potential to trigger the development of micro, small and medium-scale biotech enterprises which will probably be more efficient in satisfying a demand that is consistently growing in agronomic and geographical complexity.

NGOs, particularly those operating in the area of agriculture and consumer rights and awareness, have been actively involved in the most polemic and politicized of the applications of genetic engineering – The Genetically Modified Organisms (GMOs). Those NGOs were extremely proactive in bringing this controversy to the main political agenda. The position of some NGOs championing the anti-GMO cause was not enough to secure an alignment of the government, but was arguably instrumental in triggering the creation of an inter-institutional body to deal specifically with biosafety issues in Mozambique (GIIBS) and has certainly played a substantial role in catalysing the process of development of the law on GMOs.

It is important in fragile research environments, as in the case of biotechnology research institutions in Mozambique, to find the right balance between the need for useful immediate results and the appropriate resources to enable the creation of a sustainable research capacity which can only be achieved through a well-structured capacity-building process, where funding for human resources development, laboratory facilities and contextualized institutional development plans should be perceived as essential priorities.

The contribution of the business community, with emphasis on large agricultural companies, towards biotechnology in Mozambique is presently negligible. However, it is worth mentioning that their crescent demands for genetically improved and disease-free plant varieties and for GMO technologies has the potential to stimulate public and private entrepreneurships.

In spite of the low level of development of biotechnology in Mozambique it seems that an environment where there is a combination of political will, appropriate strategic framework, institutional motivation, international support and social demand has the potential to take this technology from its present relatively marginalized position to an important factor of socio-economic development.

## REFERENCE

Ministério de Ciência e Tecnologia (MCT) (2006), 'Estratégia de ciência, tecnologia e inovação de Moçambique', MCT.

# PART III

# Asia

# 7. Appropriation of technology in universities: the case of biotechnology transfer in Vietnam

**Tran Ngoc Ca, Nguyen Phuong Mai, Tran Thi Phuong and Le Van Chuong**

## BIOTECHNOLOGY DEVELOPMENT IN VIETNAM AND THE ROLE OF THE UNIVERSITY SYSTEM

The development of biotechnology in Vietnam officially started when the government issued Resolution 18/CP (dated 11 March 1994), directed at Vietnamese biotechnology development up until 2010. Following this, biotechnology is considered an essential component in achieving national goals with regard to food, feed and fibre production, health care and environmental protection.

In practice, biotechnology development in Vietnam has focused more on the agricultural and forestry sectors. Research activities are concentrated mainly at government institutes and are heavily dependent on funding from the government. Despite the early start, due to many cumbersome barriers and shortcomings in management, it is not until recently that biotechnology development has taken off.

Internationally, Vietnam is lagging behind in biotechnology by many decades. Given the modest investments in biotechnology in the past 20 years, only about USD7.5 million in total (MOST, 2003), the early start did not produce a dynamic innovation capacity for the biotechnology sector. However, Vietnam managed to produce some modern biotechnology products in selected areas such as vaccines (Ministry of Health, 2003).

In general, the sector has several key weaknesses such as: (i) weak capability of R&D organizations and enterprises; (ii) lack of human resources and investments; (iii) shortage of new, modern equipment and facilities and knowledge about international standards and regulations; (iv) imperfect environment for the operation of competitive and innovative enterprises (MOST, 2003). To deal with the problems, the government approved a national programme for biotechnology development in the field of agriculture for a period of ten years from 2006 to 2015. This programme provides financial support for R&D organizations to conduct R&D and improve physical capacities, facilities and human resources. The budget for this programme is about USD63 million (from the state budget) and is managed by the Ministry of Agriculture and Rural Development. Similar programmes related to biotechnology in different areas, such as R&D (run by the Ministry of Science and Technology) in seafood production and so forth, have also been approved, but with a smaller budget, for the period of 2006 to 2010.

These programmes, together with the policy to reform public R&D organizations in general, are expected to encourage the development of biotechnology innovation in Vietnam. According to the reform aimed at increasing the autonomy of R&D organizations, public R&D organizations will receive less financial support from the state budget for their operations. These organizations will have to find new ways to earn income to cover staff salaries and other costs. The reform is expected to encourage public R&D organizations towards greater operational autonomy. Nevertheless, these policies and programmes are new, and it is too early to see their effects on the development of biotechnology in Vietnam. It is not only the public R&D organizations that are changing in order to cope with requirements of market demand; in the new context, the industry sector is also changing to work more for the domestic market and to exploit opportunities offered by new markets overseas. Enterprises should pay more attention to investments in technology innovation and learning. International standards and regulations, such as IPR regimes, Good Agricultural Practice (GAP) and Good Manufacturing Practice (GMP), are also being implemented.

Within this system, universities play an important role in providing human resources for biotechnology development. There are about 80 universities that have training activities relating to agriculture and/or pharmaceuticals and health care, which are related to biotechnology to differing extents. Also there are R&D centres/institutes, organized under universities, that could be listed as R&D organizations.

Universities in Vietnam in general are focused more on the training function than on research. The activities of biotechnology and bioscience mainly emphasize teaching. The Ministry of Education and Training, other ministries and S&T organizations have various training programmes for BSc, MSc and PhD degrees in biotechnology in several formats. Around 12 universities have created specific training programmes and centres for biotechnology.

Biotechnology research tends to concentrate in large cities such as Hanoi and Ho Chi Minh City. In Hanoi, the University of Technology is one example of a university providing training courses in biotechnology. The *training courses* focus on chemical biotechnology such as: physical chemistry technology; organic chemistry and petrochemical technology; technology of pharmaceutical chemistry; pesticides for agriculture; polymer technology; cellulose and paper technology; technology of inorganic chemistry and fertilizers; chemical engineering; chemical and food engineering and so forth. The university also conducts *research* (by the faculties and institutes under the universities) on several issues, such as: upgrading the quality of products from traditional materials; the effective use of natural resources in Vietnam; synthesis of fertilizers, pesticides for agriculture; pharmaceutical products; food technology and biotechnology.

There are other universities that provide training in biotechnology as well as carrying out research in this field. Biotechnology research is usually conducted by the Faculties of Biology and various major research centres at the relevant universities. The research areas are usually related to such issues as the development of animal germ cell technology; development of recombinant DNA technology for production of proteins/enzymes for medicinal and agricultural use; molecular characterization of oxidative stress; the investigation and characterization of natural bio-active compounds; application of germ cell technology of domestic animals for development of biotechnology in husbandry and veterinary areas; and the development of plant and cell tissue technology for the improvement of some fruit and crop varieties.

In Ho Chi Minh City there are a number of universities conducting both training and research on biotechnology, such as the University of Natural Science, Agriculture and Forestry University, and the University of Technology. In different regions and provinces there is a range of agro-forestry universities with faculties conducting research and training activities related to biotechnology.

In general, so far, training in biotechnology subjects has been conducted for more than 4000 staff, of which 1500 obtained bachelor degrees, 400 MSc degrees and 90 PhD degrees (Nguyen Phuong Mai and Tran Ngoc Ca, 2007).[1] But this training has not yet met the demand, especially not in the major fields of biotechnology such as genetic technology, cell technology, enzyme-protein technology and micro-organism technology.

As for overseas training and internship programmes, MSc and PhD training has been carried out with state budget funds of around USD7 million per year. Within the framework of national research programmes on science and technology, hundreds of biotechnology staff have received training via collaboration with colleagues from advanced countries through visits or research projects. In addition, scientists from overseas have been invited to Vietnam to train more than 200 groups of Vietnamese biotechnology scientists in research methodology. In recent years, Vietnam has also relied on overseas scholarship schemes to send students to the US, Europe, Australia and Japan for training. However, when these graduates return, they face the problem of not being able to find adequate professional jobs in Vietnam, because modern biotechnology is not sufficiently developed in Vietnam, and working facilities are scarce. Also, the budgets allocated for research activities are usually small (MOET, 2003).

In the past, universities were mainly public universities. Recently, the number of non-public universities has increased. However, these universities still usually pay more attention to, and invest in, training services rather than performing R&D. Discussing the role of the private sector, it should be noted that most of the biotechnology activities, both in training and research, occurred within state-owned organizations. There are a large number of state-owned manufacturing enterprises in Vietnam, including agricultural and pharmaceutical ones. Recently, these firms were equitized and turned into shareholding companies, but the state still owns a large number of equities of the firms (usually 51 per cent). In the pharmaceutical sector for instance, there are about 180 enterprises that produce modern medicine, and about 300 enterprises that produce traditional medicine. Small and medium enterprises dominate in relation to the total number of enterprises, by more than 90 per cent. Most of these enterprises are short of investment resources and lack R&D capacity (MOST, 2003).

As such, the role of enterprises, especially of private-sector enterprises in biotechnology research and application, is weak and unclear. The driving force for development and competitive capability of the state-owned enterprises in biotechnology has also been weak. The system linkages among actors in the biotechnological innovation system (e.g. enterprises, institutes and universities) have not been clearly formulated (Tran Ngoc Ca, 2003).

## A POLICY LANDSCAPE

Overall, the policy environment seems quite favourable for biotechnology development. In Resolution 18/CP, some policies for enhancing Vietnamese biotechnology development are set up as follows:

- Enterprises are encouraged to conduct innovation by implementation of R&D or import of advanced technology for processing new biological products, which are significant for socio-economic

development. The government provide incentives such as tax reduction, preferential interest rates, favourable procedures and other incentives to these enterprises.
- Products and technology which are manufactured domestically will not be imported; backward technologies using excessive energy or that are in other respects harmful for human health and environment are forbidden.
- Enterprises are encouraged to establish in-house R&D units to enhance continuous technology qualifications, and to make products with high competitiveness.

A National Committee on Biotechnology was established in 1997 in order to assist the Prime Minister in respect of implementing Resolution 18/CP on biotechnology development up to 2010. In order to implement the substance of Resolution 18/CP, the Ministry of Science and Technology (MOST) had built up the state-level research programme, titled 'Biotechnology for developing sustainable agriculture, forest and aquaculture, environment and human health protection' (1996–2000) and have been implementing the research programme, titled 'Science and technology for biotechnology development' (2001–2005). A range of biotechnology labs have been set up in various organizations and locations.

In terms of human resource training, the Ministry of Education and Training (MOET) has promoted human resource training in biotechnology fields. With the cooperation of MOST, by implementing research programmes, many government staff in biotech fields have received training. Training human resources with high professional qualification and building infrastructure are two important factors for promoting Vietnamese biological development. In addition, MOET is running a programme that provides funding for selected students to study abroad (Programme 322). About USD6 million is spent on this programme every year. Thanks to Programme 322, about 30 students and 200 government staff have received funding to study overseas on the specialization of biotechnology (Nguyen Manh Quan, 2004).

In 2000, MOST and other ministries formed a strategy for developing biotechnology up to 2010. This strategy is an important part of the national S&T strategy. In December 2003, the government approved a national S&T strategy, including the biotechnology component.

In 2006, the Prime Minister issued Decision 11/2006/QD-TTg, approving a programme on biotechnological development and its application in the domain of agriculture and rural development up to 2020. This programme aims to create new plant varieties; new animal breeds; micro-organic strains and agricultural biotechnological preparations of high yields; high quality and economic efficiency in order to serve economic restructuring in agriculture and rural development. Another objective is to raise the quality and competitiveness of commodity farm produce and the proportion of processed agricultural, forestry and aquatic products in service of domestic consumption and export. USD62.9 million will be spent to develop and apply biotechnology in agriculture and rural development over the next 15 years. Funding will be provided for scientific research, trial productions and specialized training for the programme.

In addition, in 2005, important laws and regulations related to technology development in general such as the Law on Intellectual Property Rights and the Law on Technology Transfer were enacted; these laws facilitated further development of biotechnology in Vietnam. Vietnam became a WTO member at the end of 2007, which made the country more exposed to international trade rules. There are also WTO incentive regimes available, such as the 'green box' policies, which should be utilized more.

To promote international cooperation in biotechnology, the government has encouraged organizations to use international experts or Vietnamese individuals living abroad with high professional qualifications in order to act as consultants or to take part directly in building development plans, training processes and R&D activities. In some specific cases, the government has encouraged and financed the bilateral and multilateral cooperation between Vietnam and international organizations in the field of biotechnology.

Apart from legislation as mentioned above, government policies have been promulgated on crop and animal varieties; state-supported agricultural contracts for high-technology agriculture; forest and aquaculture development programmes. Policies for encouraging the investment in all economic components relating to agricultural high technology will continue to be formulated, such as infrastructure investment, tax reduction, import tax exemption for imported equipment and so forth.

In general, it can be said that in order to implement the policy under Resolution 18/CP, much work has been done for the development of biotechnology in Vietnam.

Despite these efforts, the current policies are unclear and do not appear to be specific enough to make enough of an impact on biotechnology development. The results of biotechnology development efforts are still limited, due to insufficient investments in this field, by the state as well as by other actors. Also, the biotechnology sector has not developed as expected because it does not have a workable innovation system: the linkages between actors in the system are very weak, while actors themselves have limited capabilities. Although new programmes for biotechnology development, especially in the fields of agriculture, health care and pharmaceuticals, are expected to have a positive impact on the behaviour of actors and organizations, at this moment it is too early to assess the impact. Despite a range of policies, laws and regulations, their implementation and enforcement has proved very difficult. An example of this is the Law on Intellectual Property Rights. In addition, Vietnam still lacks laws and regulations relating to biosafety and traditional knowledge, which are important for the development and operation of the biotechnology sector.

Broadly, it is the overall policy scene (including the innovation policy mindset), related to S&T and innovation development in general, and to biotechnology R&D and transfer in particular, that are the more crucial issues. To move from a traditional S&T policy mindset to one more informed by innovation policies requires more decisive action and certainly more resources, including time.

## THE TWO CASES: HUE UNIVERSITY AND CAN THO UNIVERSITY

The study on universities has been undertaken using case studies. Two units were chosen for in-depth study: the Institute of Resources, Environment and Biotechnology (IREB) under Hue University; the Faculty of Agriculture and Applied Biology (FAAB) and the Institute for Research and Development of Biotechnology (IRDB), both under Can Tho University.

### The Institute of Resources, Environment and Biotechnology (IREB), Hue University

IREB is a subsidiary of Hue University in the Central part of Vietnam, created as the Center for Resources, Environment and Biotechnology by decision No. 3283/GD-DT on 13/09/1995 of Minister of MOET and with permission to operate under Decision No. 445 on 17/01/1996 of Ministry of Science, Technology and Environment (now the Ministry of S&T). By the decision of Hue University, the centre was turned into IREB in 2007, with functions to carry out R&D and technology transfer in respective fields. IREB has two centres: (i) Center for Environmental Monitoring and Sustainable Development and (ii) Center for Researching and Transferring Biotechnology.

During the period 1995–2003, the Center for Researching and Transferring Biotechnology (the 'Biotechnology Center') worked mainly on environmental issues and carried out virtually no biotechnology activities. In 2004, under the framework of the Education Reform project of the World Bank for a number of universities under the MOET, an investment of USD760000 was made to set up a biotechnology lab in the Biotechnology Center, and since then, the Biotechnology Center has begun more work on actual biotechnology research. The overall design of the Biotechnology Center's activities adhered to the conceptual design of the World Bank project.

At that time, IREB mobilized experts and lecturers in Hue University to conduct research in biotechnology. A PhD graduate in biotechnology who graduated from Korea has been appointed new director of the Biotechnology Center.

Now IREB is cooperating with several countries in the world, such as Japan, Belgium and Singapore, in the biotechnology field. Thanks to this cooperation, the infrastructure of IREB in general and that of the Biotechnology Center in particular has improved considerably. The equipment is more modern; the research capabilities of the experts are also improved. Besides, IREB also cooperates with other universities and research institutes throughout the country, such as the Hanoi University of Agriculture; the Institute of Biotechnology belonging to the Vietnam Academy of Science and Technology; the Genetics Institute belonging to the Ministry of Agriculture and Rural Development; the University of Sciences, Vietnam National University, Hanoi; the University of Technology and University of Education of Danang University; the University of Sciences, Vietnam National University, Ho Chi Minh; Can Tho University; the Institute of Biotechnology and Environment; and Ho Chi Minh Nong Lam University.

As of the end of 2007, the total number of staff working at IREB was 35. This included two associate professors. Of the staff members, one held a doctorate, four held master's degrees, six are research students and post-graduate students, eight are engineers and 19 hold bachelor's degrees. The Biotechnology Center was staffed by eight people. Additionally, invited lecturers and experts in Hue University also participate in activities of the Biotechnology Center. Most Heads of Faculty and the managers of labs in the Biotechnology Center hold doctorates and have conducted post-doctoral studies in countries which have a more developed biotechnology sector such as Japan, Germany, Denmark, Korea and Singapore.

The IREB has activities within seven subjects, including:

- applied DNA technology
- biologically active elements technology
- molecular medicine and biology
- food and agricultural biology technology
- environmental evaluation

- environmental informatics and geographical information systems
- environmental technology.

IREB has a total of 12 labs with 300 comprehensive and modern pieces of equipment which can serve well for scientific research and training; there are three computer rooms with 50 high configuration computers. Each computer room is capable of meeting the demands of research and training. The Biotechnology Center has eight labs for (1) immunity; (2) animal cell technology; (3) gene technology; (4) protein technology; (5) genetics technology; (6) microorganism technology; (7) flora cell technology; and (8) secondary compounds.

Recently, the Biotechnology Center has been used as a laboratory supporting seven doctoral theses, 21 master's theses, 22 graduation papers of engineers and bachelors from member universities under Hue University, such as the University of Nong Lam University, and the University of Sciences and the University of Medicine. Additionally, it is used as a laboratory supporting biotechnology experiments carried out by postgraduate students of the University of Technology and the Danang University of Education.

Now IREB has to rely on self-financing mechanisms for its operations. Pursuant to the new regulations,[2] university R&D organizations are to become more independent under the university system. They have more autonomy on finance, human resource management and other operations, while still retaining linkages with the mother university. As a result, the main income of IREB comes from the services it provides in carrying out environment impact assessment and other contracted research topics and projects. The income of the Biotechnology Center is only enough to pay for staff salary and materials for experiments (about 400 million VND a year). The other annual allowances are contributed by the income of the Center of Environmental Monitoring and Sustainable Development also under the IREB.

**Biotechnology development: strategies and activities**

Based on the policies of Hue University on giving priority to the development of biotechnology, the Biotechnology Center is using gene and cell technology to preserve, multiply and improve plant and domestic animal genes. It is using molecular, medical and biological techniques to diagnose diseases in humans, cattle, poultry and aquatic products; and using molecular medical techniques to analyse biological diversity. Additionally, the Biotechnology Center is using these techniques to research and produce biological elements from growing flora cells; to research and produce recombined protein used in food and medicine; to rehabilitate the environment; to research and apply informatics and biology in order to manage and exploit the gene and protein database.

The Biotechnology Center's strategy is that of strengthening relationships with partners inside and outside Hue University, expanding relationships with new partners from foreign countries such as Belgium, Japan and Korea and so on. This strategy will help to improve the research capability of staff in the Biotechnology Center and to attract investment capital.

The Biotechnology Center has been researching many issues in accordance with the orientation and development strategy of IREB. The applied research projects account for 40 per cent of the projects, compared to the basic research projects. Most of the published results in the biotechnology field emanate from basic research.

In IREB's research portfolio a range of notable applied research projects can be found:

- Research on three varieties of rice which can withstand drought; have a shorter growth cycle; avoid the flood season; have better quality compared to its previous generations and which are also suitable to the conditions of the region where the Biotechnology Center is located. IREB has now undertaken a project to produce the rice, and this project has been approved by Hue University. Currently, the Biotechnology Centre is waiting for funding to transfer the results to the Agriculture Department, to the Agriculture Extension Center, to agricultural cooperatives and to farmers, for production on a large scale.
- Research to produce asparagus by tissue culture. This research project has been completed and has produced results, but the Biotechnology Centre has not found partners to which it can transfer the research result.
- Research to produce orchids by growing tissue. This project has already produced results. Although the market demand for this product is very high, the lack of investment capital for pilot production makes it difficult to apply the result to practical production.
- Research using enzymes to hydrolyze shrimp heads to create chitin for the production of sauce. The experiments produced valuable results. This project uses the waste of an aquatic product as a material for sauce production, leading to reduction of production cost and environmental pollution.
- Research on food vaccine. The nature of this project is to modify the genes of plants to create new plants which have higher quality and productivity. However, the regulations regarding genetically-modified products in Vietnam are not clear and detailed.

- Diagnose mental deficiency in embryos using PCR (Polymerase Chain Reaction) techniques. At present, Hue Central Hospital does not have PCR technology, so there is no technology available to carry out diagnosis of mental deficiency in embryos. Even though this project is not finished, the available results are satisfactory. If this research is successful, it will play an important role specifically for families and for society in general.

In general, the research results in biotechnology of IREB are only the beginning, and yet have contributed to business and production of some partners in the area. However, there is still a gap between these results and the research carried out at large research centres. Most research cycles require a pilot phase, something that is lacking in IREB's methods, and it is difficult to make a final product for end users out of this unfinished research cycle.

There are several issues that hinder the development of biotechnology R&D at IREB. First to be mentioned is funding. In general, total funding available for R&D at the university is quite limited, while MOET provides funding for only five ministerial projects of Hue University (the funding is at this level of around USD20 000 each on average). At the university level of R&D project, the funding level is, on average, only USD200 per project.

The autonomy of R&D institutions is another issue to be mentioned. So far, IREB has relied mostly on its income from environment-related projects. In the context of moving R&D institutions toward more autonomy, young and new organizations like IREB may have more problems in facing an uncertain future, without the support of public funds for operation, the reason being that biotechnology research cannot generate as much income compared to activities like environmental services projects.

Within Hue University, there is a lack of systems to deal with the transfer of technology and commercialization of research results, such as a Technology Licensing Office, or a Technology Transfer Office. This makes it difficult for research activities to reach the market. University staff lack market and business management knowledge, and thus, even for some useful technologies, it is hard to turn them into marketable products.

**Policy issues**

Biotechnology development requires substantial resources. Therefore, in order to promote biotechnology, research must be *funded* properly and R&D must be coordinated with transfer activities. In order to link R&D and the transfer of biotechnology, *intermediate stages* such as pilot production are required, together with appropriate marketing strategies for products and suitable promotion policies. Finally, the research orientation must be based on market demand. However, up to now, most of the research conducted by IREB in biotechnology has merely been done in the lab.

Although the Biotechnology Center's research staff were trained in advanced biotechnology countries such as Korea, Germany and so on, and the infrastructure and the equipment is sufficient to meet the demand of applied research, IREB has not yet produced many products that can be transferred to companies, enterprises or customers. The main reason is that the *research agenda setting* is done by researchers, and based on individual preferences or research competences, instead of on any real societal demand, and without appropriate information channels to companies and people. Moreover, with the need to become autonomous in terms of financing, which requires quick return in terms of technology transfer and contract-based services, a long-term basic research type of project may not be favoured by the staff. The interests of basic research as opposed to applied-oriented research for commercialization come strongly into conflict in this context.

Even though there are many national policies to promote the development of biotechnology such as Decision number 50/CT-TW on 4 March 2005 of the Communist Party of Vietnam with regard to promoting the development and application of biotechnology in industrialization as well as the Prime Minister's programme of developing and applying biotechnology in agriculture and rural development up to 2020, these policies have barely contributed to the transfer of IREB research result in biotechnology. Staff at the Biotechnology Center seemingly lack knowledge of related *policies* of the government on biotechnology. They have *difficulties in accessing government R&D programmes and projects* on biotechnology. To take part in these programmes seems beyond their reach. Many university staff consider the process of research agenda-setting, selection of projects, and funding allocation not sufficiently transparent and too complicated for them to join.

Therefore, Vietnam needs more suitable policy measures from the government so that biotechnological research results, new processes and products may be transferred and exchanged easily between scientists, enterprises and farmers.

*Linkages* between IREB and related organizations in biotechnology R&D activities also seem inadequate. They have little cooperation even with the S&T departments of nearby provinces such as Hue or Quang Tri. It is interesting to note that each of these S&T departments has a lab for tissue culture funded centrally by the Ministry of S&T, but the activities of these labs are limited and linkage with the IREB is poor.

The local government and Hue University claim that there is a lack of knowledge about biotechnology and suitable policies. Therefore, there is a need for more practical and detailed policies to support the development and transfer of biotechnology in general and the development of IREB in particular. We need *financial policies* providing incentives through reduced tax for projects which invest in production using biotechnology or other policies which encourage and create a good environment for individuals, domestic and foreign organizations to invest in this field. In addition, we need policies for encouraging the enterprise sector to use biotechnology in production. Then, following success in production, companies will have to invest in further developing biotechnology to make competitive products in the market. When biotechnology becomes a real force to take into account, enterprises will not only carry out research by themselves but also invest in additional biotechnology labs through research contracts for new technologies or products.

On the part of IREB, it needs to change its R&D strategy with regard to market orientation. For example, agriculture research needs information from many sources, especially from the agricultural extension system, to find the real demand of farmers so that scientists can define the agenda for their research. Besides, the IREB needs to expand cooperation and relationships with biotechnology research centres in the north and the south of Vietnam, and with international organizations. They also need to approach governmental programmes and projects to obtain more funding for pilot production.

In general, IREB can be seen as the first effort of Hue University in the development of biotechnology scientific research. The Biotechnology Center under IREB is one of the efforts of Hue University to develop a new research area under the support of the government and international donors, in order to strengthen the profile of the university towards becoming a research university proper. Some of these efforts have paid off in terms of research results and linkages with farmers, firms and society at large. However, the overall policy environment that was supposed to be supportive for R&D in general and for biotechnology in particular has still not materialized. The policy intention to make R&D organizations more responsible for their own funding may transform them into mere service providers rather than research organizations. There is a strong need to classify these organizations, including university research types, into various kinds of financially autonomous regimes. Apart from this, many specific policy measures are far from the knowledge and implementation capacities of universities. It appears that biotechnology research efforts at this specific university are still in their infancy, with many underdeveloped activities. The policy environment must be improved significantly at both central and provincial levels. Government organizations such as MOST and MOET should have more active roles to play.

### The Faculty of Agriculture and Applied Biotechnology (FAAB) and The Institute for Biotechnology Research and Development (IBRD), Can Tho University

Can Tho University (CTU) was founded in 1966. Currently, CTU has nine faculties, two research institutes, seven training and service centres and ten administrative departments. The first main mission of CTU is to train the 16000 students located at CTU and the 15000 students located in associate facilities in other provinces of the Mekong River delta. In addition, there are 1015 postgraduate students in 6 doctoral programmes and 25 master's programmes. The second mission of CTU is research, of which the most notable programmes are researching applications of new techniques in planting, processing agro-products, and applied biotechnology research for agriculture. To support its activities, CTU has developed strong cooperative ties with many international organizations. CTU has organized various research centres or institutes to deal with the need for training, research and serving communities.

In 1997, via the Joint Financing Programme for Cooperation in Higher Education (MHO) with the Netherlands government, CTU began training in biotechnology. For the first phase (1995–99), funding received was USD13 million and USD9 million for the second phase (2000–2004). CTU also received support from the Japanese government (USD23 million) under the programme to upgrade the Faculty of Agriculture. In addition under a research cooperation programme with Belgium, CTU received funding of EUR 3 million for 1998–2002 and EUR 3 million for 2003–2007. During 2005–2006, MOET invested around USD500000 to improve the capabilities of the Institute of Biotechnology Research and Development (IBRD). Biotechnology activities under CTU mainly take place in two organizations described in more detail below.

#### The faculty of agriculture and applied biology (FAAB)

FAAB has ten units, including nine divisions and one administrative unit. The Faculty has 290 staff (of which 132 are female). For the academic year 2006–2007, the Faculty taught ten specialized subjects to 2432 students and via ten centres for 2032 more students. The Faculty also has 350 masters' students and 37 doctoral students. In addition to training, the Faculty has been involved in various technology transfer activities, and in linkages with local communities, related to matters such as integrated pest management (IPM), agricultural extension network, safe vegetable production and garden farming.

**The Institute for Biotech R&D (IBRD)**

IBRD has two main functions: teaching and research. There are five divisions working in the areas of molecular biology, biochemistry, and micro biotechnology. IBRD has 33 staff members and has trained more than 100 master's students, under joint training programmes with certain Dutch universities. Since 2004, it has also trained a couple of doctoral students every year. Starting in 2006, the Institute began training students in undergraduate programmes in biotechnology (around 100 students) and advanced biotechnology (around 30 students). IBRD focused its activities on areas such as gene technology, molecular marker, isozyme and gene mapping, and so on.

**Biotechnology development: activities**

FAAB has a number of projects on biotechnology research. In 2007 it had 14 ministerial projects and two key projects in the area of agricultural genetics, receiving funding of around 100000 USD. In addition, FAAB conducted three incubation projects for the Ministry of Education and Training. The Faculty has many joint projects with different provinces in the Mekong River delta. Projects include issues of biodiversity; application of high-technology biotechnology to produce seedless oranges and kumquat of high quality; production of mangoes, lotus seeds and flowers; production of green and clean fertilizers (trichoderma); creation of bio-products from the nuclear polyhedrosis virus (NPV) to produce clean and safe vegetables.

FAAB has conducted several international cooperation projects. On average, during a year, FAAB receives around 40 international delegations and has been conducting more than ten joint research projects and programmes together with partners from the US, Germany, Japan and Sweden. These projects focused on issues of agricultural biotechnology, such as sustainable rice production, plant protection, and plant and food processing technology.

As for IBRD, it works on issues of protection of genes; the production of medicines, enzymes in agriculture, using applied micro-organisms in agriculture; the production of environmentally-friendly micro-organism fertilizers; the biodiversity of crops and plants in the Mekong River delta. IBRD also conducts research on molecular biotechnology to identify diseases in food products to serve as a basis for food safety. To conduct these research activities, IBRD has a national-level research programme on biotechnology with funding of around USD2 million. During the period of 2006–2008, IBRD had seven projects receiving USD85000 in funding.

These two organizations mainly focused their biotechnology activities on the applied side of R&D, in having contracts with farmers and local communities in order to transfer to them specific technologies. There has not been much basic research in the biotechnology area. Many activities are dealing with the practical needs of farmers in the provinces and have contributed significantly to the socio-economic development of the communities.

**Policy issues**

The organizations concerned complain mostly about financial policy and its specific funding mechanisms for state programmes. To supplement these sources and to avoid the complexity of disbursement procedures, these organizations have to promote their other activities, such as cooperative contracts with users in production sectors; activities to serve the needs of farmers; increasing work with those organizations that receive funding for the development of key economic zones in the Mekong Delta.

Apart from funding from provinces and local governments, international cooperation is another important source of funding that these organizations should further develop. *Procedures* and preparation for *international cooperation* are not always appropriate. The state via MOET also needs to have appropriate *allocation policies* in funding R&D activities within the university system. The current situation is considered untenable as the state budget for most university-level research projects is almost negligible, making it practically impossible to do any kind of serious large-scale research that the region needs.

Biotechnology R&D and knowledge transfer in the two organizations has begun to contribute to the needs of society in their respective provinces. However, due to constraints in funding and in available human resources, these contributions have mainly been minor, and have been mostly restricted to applied research activities.

To make the outreach of the two organizations more feasible, the research staff of FAAB and IBRD always attempt to close the gap between their research activities and the actual needs of end users during different periods. Information on trends of production, markets and socio-economic development in general of the Mekong Delta region has to be priority input for the research agenda-setting of these organizations. Vice versa, trends in science and technology development must guide the overall development strategy and orientation of the region in utilizing available resources. FAAB and IBRD recognized that the development of the Mekong Delta and their own operations have close mutual impact.

There are several state organizations within the region and they have a need to link different research organizations and university faculties in joint research programmes, and in implementing research results in practice. Technology transfer for provinces in the Mekong Delta region is a strength of CTU in general, and FAAB in particular. Joint technology transfer activities for the purpose of economics and social services is a type of linking channel, and FAAB and IBRD both consider that they need to tap more into this possibility, exploiting their existing strengths. Still, the *mechanisms of linkage and joint technology transactions* within a technology market are not fully developed.

## EMERGING ISSUES

In their own context, all the organizations in this study mentioned plans and strategies to develop biotechnological R&D by the respective universities and higher-level organizations such as the MOET and the MOST. Still, study participants complained about the weak capability of their institutions in terms of financial resources, human resources and insufficient physical facilities such as equipment and laboratories.

To fully utilize the strengths of biotechnology, it is essential to invest sufficiently in this area. Furthermore, there is a required set of intermediary activities such as pilot production, testing facilities and marketing. Most importantly, research agenda-setting should be based on the actual needs of society. The organizations involved in the study to various extents began to carry out R&D in biotechnoloy-related matters and tried to turn them into useful products for end users, communities and farmers. Despite these efforts, so far, many research results remain in the lab. One of the main reasons has been that the research priorities that were set tended to rely on the existing expertise of scientists, and not always because of real needs. There is no mechanism for funding the pilot phase to turn end-of-lab results into more fully developed final products for transfer. Financial regulations and systemic barriers have been named as key ingredients for this insufficiency. This reflects the fact that the financial policy is not conducive enough for research and innovation activities.

Institutional and policy issues are among those that emerged most strongly in this study. All organizations concerned mentioned the national government's grand vision and well-documented policies. But when it comes to implementation of these policies, there are always problems. Despite the fact that there are many macro policy documents promulgated, such as Resolution No. 50/CT-TW of 4 March 2005 of the Party Central Committee, or the Prime Minister's Programme on Biotechnology in Agriculture, the impact of these policy statements remains to be seen.

One of the notions coming out of this study is that most of the national biotechnology policies are far from the university domain. Resources and capacities for performing biotechnology R&D are beyond the reach of many universities. This is a dominant feature especially for those universities that are located in remote and regional areas, outside the metropolitan areas of Hanoi and Ho Chi Minh City. One example is that these institutions (like IREB) have no or only negligible information about or contact with the national biotechnology R&D programmes that are being promoted by ministries. Usually, information about these national programmes reaches the provinces via provincial departments for S&T or for agriculture and rural development, which are branches of the central ministries. These departments, from case to case, may or may not have not close linkages with universities in the same provinces or cities. These loose links make the interactions in the innovation system of the respective localities quite weak. Even if the links exist, they tend to be unsustainable.

A factor that hinders the appropriation of technology from the university context to the market context is that not many universities have offices or units dedicated to commercializing their R&D results. The necessary intermediary technology organizations are still missing links in this context. Contact between scientists and potential users of their research results is limited. The technology market is still in its infancy.

Lastly, the policy change to turn universities and R&D institutions into commercial-type entities or enterprise-type operations, without being accompanied by further specific plans of actions, could be too much, and too soon. This might turn them away from carrying out necessary research, without contributing to the manufacture of products of interest to the market.

## CONCLUSION

In the past few years, biotechnology has made a number of contributions to Vietnam's development. Initial infrastructure, organizational and institutional framework and personnel training for biotechnology have been established, leading to an improvement in research capabilities and technological development. Its application to production areas has contributed to the improvement of the quality of agricultural and aquacultural products, as well as the production of vaccines and medical products. Still, in spite of starting relatively early, Vietnam is only now taking the first steps towards the development of modern bioscience and biotechnology. Biotechnology in Vietnam lags behind several other countries in the region and cannot yet meet the increasing demand for socio-economic development and an improved living standard. This is due to problems in the operation of the overall innovation system, including the legal, institutional and policy framework.

A number of issues related to the general situation of biotechnology development have been highlighted. First, although the Strategy for Vietnamese S&T development up to 2010 was approved, a policy statement in which the *priorities for biotechnology* and bio-industry were defined, these strategy objectives need to be specified. It is necessary to further narrow priority fields in R&D and biotechnology application to make action plans more feasible. Biotechnology is an expensive activity for economies, and one cannot do everything. In the case of Vietnam, it is wise to focus on a few activities and to find a niche where the country's bioscience and biotechnology capabilities can best address it needs.

Second, *international cooperation* is crucial for starting research. This must be accompanied by government policies that provide a conducive environment to enable manufacturing to be transferred from the research sector. No less important is the market incentive regime that stimulates entrepreneurship among scientists. In this regard Vietnam is introducing policies to attract multinationals, first for industrial production capacity, then hopefully for related R&D capacity. The Technology Transfer Law and the High Technology Law recently passed by the National Assembly will contribute to a better incentive environment. The effort to restructure legal regulations, as well as policies for joining the WTO, is another factor that would affect the sector, and the recently enacted Law on Intellectual Property Rights and other regulations on biosafety is expected to ensure greater reliability for stakeholders.

Third, *learning* is a key ingredient for any success. Conditions such as sufficient and focused investment for learning from international partners, and internal learning from training, should be in place. One of the measures to address the fragmentation of investment, build-up of an innovation system, as well as to break down institutional barriers is the creation of Centers of Excellence in bioscience and biotech. Already, several advanced centres for biotechnology have been set up in some cities.

The specific experiences of the university cases show the attempt to create new centres for R&D excellence in biotechnology. Still, their experience reveals several weak aspects of the research and innovation in the university system. There is still a long way for universities in the country to go to become credible research universities. In addition to the existing hurdles that the overall S&T and innovation system in Vietnam faces, the research efforts within a university domain confront additional problems. The second mission of the universities, which is to carry out research, is not well developed, while the third mission regarding reaching out to the wider society is in a shaping phase. Government, as well as local authorities and the respective universities, should discuss and agree on specific policies conducive to biotechnology development and transfer.

There should be concrete steps in the policies of tax exemption, land allocation, or other incentives provided to organizations and persons engaged in biotechnology. Policies with a focus on promoting firms involved in biotechnological activities should be strongly emphasized.

University organizations should change their strategy on R&D priority setting, to rely more on the actual needs of users. For instance, in agriculture, the needs of farmers and the community should be top priorities in considering the research and technology development agenda. These organizations should also expand linkages with biotechnology centres located in big cities and international organizations, to tap into the network of excellence for biotechnology funding and expertise in order to serve the needs of local communities.

## NOTES

1. According to other sources, the number of people working in biotech with university degrees and above is only just over 2000 (Nguyen Van Ngu, 2006).
2. Decree 115 of the Government states that R&D organizations should become more autonomous in terms of securing funding, managing human resources and operating activities.

## REFERENCES

Le Van, Chuong and Tran Ngoc Ca (2007), 'Biotechnology transfer at Faculty of Agriculture and Applied Biotechnology (FAAB) and Institute for Biotech Research and Development (IBRD), Can Tho University (CTU)', case study.

Ministry of Education and Training (MOET) (2003), 'Training and human resource for biotechnology industry', conference report, National Conference on Biotechnology, Hanoi.

Ministry of Health (2003), 'Application of Biotechnology in diagnostic, therapeutics and pharmaceuticals', conference report, National Conference on Biotechnology, Hanoi.

Ministry of Science and Technology (MOST) (2003), 'Review of the biotechnology development in Vietnam', conference report, National Conference on Biotechnology, Hanoi.

Nguyen Manh, Quan (2004), 'Policies and mechanism to develop bio-technology and bio-industry in Vietnam up to 2010', NISTPASS.

Nguyen Phuong, Mai and Tran Ngoc Ca (2007), 'Biotechnology innovation system in Vietnam', report for the project of UNU MERIT-IDRC on Biotech innovation systems in Asia and Africa in comparative perspective.

Nguyen Van, Ngu (2006), 'Planning the training of high-level science and technology staff, and training in countries with advanced level of S&T', research project

report, Institute of Education Strategy and Programs, Ministry of Education and Training.
Tran Ngoc, Ca (2003), 'International cooperation in science and technology: some critical issues for Vietnam', paper presented at MOST-IDRC roundtable on international cooperation in science and technology, Hanoi, October.
Tran Thi, Phuong (2008), 'Technology transfer of Biotech at Institute of Resources, Environment and Biotechnology (IREB), Hue University', case study.

# 8. Biotechnology transfer and application in China: background and case study

**Wang Haiyan and Zhou Yuan**

## INTRODUCTION

Biotechnology is of great importance to China's economic and social progress. The rapidly growing biotechnology industry accelerates industrialization, internationalization and knowledge accumulation. Much attention is attached by policy-makers to the development of modern life sciences, as they believe the application of biotechnology will help China solve some of its most urgent problems, such as population growth, food supply, health care and protection of the environment. Science and education hold the key to a prosperous future of biotechnology. In this chapter, we describe in detail the Chinese research institutions and the human research capacity in biotechnology.

### Domestic Biotechnology Research Institutions

A couple of programmes that prioritize the development of biotechnology have been carried out by the Chinese government since the late 1970s. The programmes cover a wide spectrum such as basic research programmes, hi-tech research and development programmes, key science infrastructure programmes, as well as the establishment of key laboratories and engineering centres.

An ambitious scheme to promote biotechnology research was started in the beginning of the 'Seventh Five-Year Plan' (1986–1990) when the first comprehensive National Biotechnology Development Policy Outline was issued. Under this outline, a number of high profile technology programmes have been launched from the mid-1980s. Some of the most significant programmes include the '863 High-Tech Plan', the '973 Plan', the Initiative of National Key Laboratories on Biotechnology, Special Foundation for Transgenic Plants Research and Commercialisation

(SFTPRC), Key Science Engineering Program, Special Foundation for High-Tech Industrialization (or Commercialisation), Bridge Plan, and so on (Huang and Wang, 2003).

The '863' Plan, also called the National High-Tech Research and Development Plan, was approved in March 1986 as a result of the recommendation from four leading scientists in China. The '863' Plan supports a large number of applied as well as basic research programmes. Biotechnology is one of seven supported areas under the plan. The National Basic Sciences Initiative, also named the '973' plan, another high-tech research plan, was initiated in March 1997. That plan is complementary to '863' and many other national initiatives on high-tech development, as it exclusively supports basic research. Life science, with biotechnology as a priority, constitutes one of the key programmes under the plan. In order to strengthen the national research and industrialization of China's agricultural biotechnology, the Ministry of Sciences and Technology (MOST) initiated a new programme, SFTPRC, in 1999. This is a unique foundation to promote both research and commercialization of transgenic plants.

MOST and SDPC (the State Development and Planning Commission) jointly led the Key Science Engineering Program (KSEP), a national programme to promote the fundamental construction of research, and initiated a biotechnology sub-programme in the late 1990s. Moreover, the State Council passed a new Agricultural S&T Development Compendium in 2001. The compendium re-emphasizes the importance of agricultural biotechnology in improving the nation's agricultural productivity, food security, and farmers' income, which has led to a new decision to further raise research budgets for biotechnology development.

By 2007 there were about 1000 laboratories at national and local level, located in more than 300 research institutes and universities across the country working on agricultural (plant and animal) biotechnology. Over the last two decades, China established more than 30 National Key Laboratories (NKL). Among these NKLs, twelve are working exclusively on agricultural biotechnology while three others have major activities in that area (Huang et al., 2001). Besides NKLs, there are numerous key biotechnology laboratories and programmes under various ministries and local provinces.

## Human Research Capacity-Building and Investment

The funding from central government for R&D has been increasing continuously over the past 30 years at a pace considerably higher than the already impressive economic growth rate of the country as a whole. China has adopted preferential policies in venture capital and taxation for

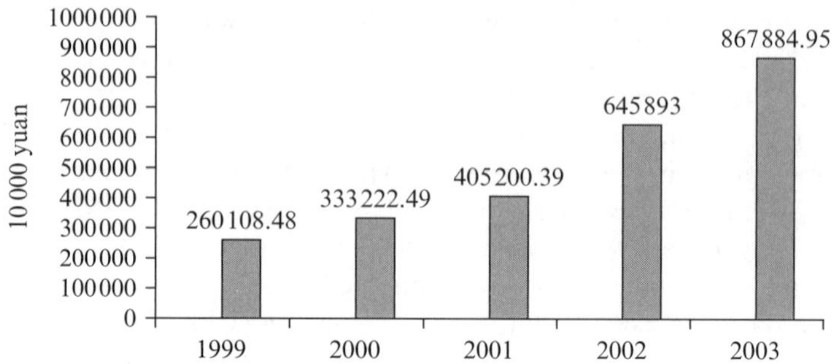


*Source:* China Bio-industrial Report (2004)

*Figure 8.1 The R&D investment of main biotech corporations in China between 1999 and 2003*

research institutes and high-tech companies. Many R&D centres and facilities for biotechnology have been set up and more and more scientists with PhD degrees are returning home after being trained in western countries. There are more than 456 universities, research institutes and companies and a total of over 20000 scientists and researchers involved in biotechnology, including about 6800 professionals. More than 2000 biotechnology PhD students receive their degree each year (Wang, 2007). The number of research papers published internationally is increasing rapidly. According to statistics, more than 100000 overseas Chinese students undertook projects in life science and related research. The number of plant biotechnology researchers more than tripled in the past two decades. If we exclude the animal sector, the number of researchers in agricultural biotechnology may be more than 4000, which probably is one of the largest in the world. Moreover, a remarkable improvement has occurred in human resources to conduct biotechnology research. Among professional staff, the share of researchers with PhD degrees increased from 2 per cent in 1986 to more than 20 per cent in 2000 (Huang et al., 2001; Department of *High-Tech* Industry of the State Development and Reform Commission, 2007).

This share is expected to keep rising in the future as the capacity to run PhD programmes in biotechnology has been strengthened. Biotechnology research in China is predominantly financed and undertaken by the public sector. The growth in biotechnology research investment in the public sector has been substantial. The estimated investment in biotechnology research was only 2.601 billion yuan in 1999 (1 USD = about 6.5 yuan, 2010) (see Figure 8.1). The investment grew to 8.679 billion yuan in 2003: 3.332 billion yuan in 2000, 4.052 billion yuan in

2001, and 6.459 billion yuan in 2002; the increase in 1999–2003 represents an annual growth rate of 58 per cent. The investment in biotechnology research continued to grow in the first few years of the twenty-first century. Nearly all investments in biotechnology in China come from government sources. Public investment accounted for 94 per cent of the total biotechnology budget in 1999 (Huang et al., 2001). That share has increased to about 98 per cent in 2003 (Li, 2004). Biotech research in animal and plant agriculture is estimated to account for 15 per cent of China's agricultural research, much higher than the 2–5 per cent share typically devoted to biotech research in developing countries. China accounts for an estimated 10 per cent of the world's public expenditures on agricultural research and development.

With the support of the government, the scientific community and industry, China has made significant progress in biotechnology development.

- The total sales of biotechnological products in China have increased 50 times during the past ten years. In 1997, it was about 13 billion RMB (USD1.6 billion). In the year 2000, it reached more than 20 billion (USD2.5 billion). The total output and sales were about 507.38 billion yuan and 479.089 billion yuan (Zhang, 2008).
- A total of 18 bio-pharmaceutical products have been commercialized, including recombinant medicines and vaccines; 30 more are at the stage of clinical trials.
- Crop breeding techniques are generating huge economic wealth in China: the yield of super hybrid rice is as high as 100kg/ha per day during the rice growing period.
- China was one of the earliest countries to carry out gene therapy for patients and Chinese scientists have found several disease genes including the neural deaf gene (Huang et al., 2004).

## THE ROLE OF UNIVERSITIES IN BIOTECHNOLOGY R&D IN CHINA

Modern biotechnology study and biotechnology industry started late in China, but have developed comparatively fast over the last decade. Universities, corporate R&D institutions and public research institutions are the three main executive institutions of biotechnology R&D in China. According to a rough estimate, at least 95 per cent of the biotechnology R&D funding and personnel in China are working in these three institutions. At present, the official statistics in China do not offer detailed data on the distribution of biotechnology R&D resources between these three institutions, so we can only collect sporadic data from related reports and papers to approximate the situation.

First, the biotechnology R&D funding obtained by universities has increased substantially in China over the last decade. The funding for biotechnology R&D in China mainly comes from corporations and the government; corporate R&D institutions received a majority of the corporate funds, while most of the government funds traditionally flowed to public research institutions but not to universities. Therefore, for a long time, universities were the most underfunded of the three executive institutions for biotechnology R&D in China. However, over the past ten years, following the popularization of the idea of research universities, the Chinese government began to pay more attention to university R&D. Both the central government and local governments at all levels have issued a large number of policies aiming at promoting university R&D, and have also set up multiple R&D funds for universities. As a major support object of the Chinese government, biotechnology has benefited particularly from this development. Moreover, universities have also deepened their cooperation with corporations, and have become the second most important recipient of corporate biotechnology funds after corporate R&D institutions.

Secondly, the number of biotechnology R&D personnel in universities is not as high as that in corporate R&D institutions, but its predominance in terms of quality of personnel is obvious. The university itself is a place to cultivate research talent, so the average educational qualifications of the R&D personnel in universities outclass that of corporations and public research institutions. In recent years, judging from the application and licensing of biotechnology patents in China, universities are also more active in development of biotechnology than public research institutions, and second only to corporations.

Lastly, compared to the other two kinds of institutions, especially to corporate R&D institutions, biotechnology R&D in universities also has unique characteristics:

1. The project working period is different. The biotechnology R&D period in universities is comparatively longer, while corporate R&D institutions pay more attention to short-term projects for the recovery of funds.
2. Universities pay more attention to public welfare research, and tend to perform basic research and public technology, while corporate R&D institutions only focus on proprietary technology that has market potential and is not easy to imitate.
3. The present performance evaluation system in universities mainly takes published books and papers as performance indicators, so

the research in universities emphasizes advancing new theories and exploring new research methods.

## THE DEVELOPMENT OF THE CHINESE BIOTECHNOLOGY INDUSTRY IN 2007

The higher costs of upstream raw material and lower growth in investment in biological medicine affect the sustainable development of the biotechnology industry. As it is expected that the biotechnology industry will speed up its development, there is a need to quickly establish investment and financing channels and enhance the financing and taxation policy guidance and support, in order to create a good market environment. Below we will go through the main characteristics of the biotech industry in 2007 in five areas; biological medicine; biological agriculture; biological energy; biological manufacturing and biological service, areas which exhibited rapid development in 2007.

### Rapid Biological Pharmaceutical Production and Revenue Growth

The industrial output of biological pharmaceutical industries was 507.38 billion yuan in 2007 with a year-on-year growth of 25.87 per cent. Its growth rate was 5.18 per cent higher than the overall growth of the high-technology industry from January to October in 2007. Among them, biological and chemical products achieved an output of 44.135 billion yuan with a year-on-year growth of 23.16 per cent. Sales accounted for 479.089 billion yuan with a year-on-year growth of 25.02 per cent, 4.85 per cent higher than high-technology industrial growth. The profit was 33.118 billion yuan from January to August with a year-on-year growth of 48.76 per cent. Among them, the profit of chemicals and Chinese patent drugs increased 50.77 per cent and 52.73 per cent respectively. Biomedical engineering continued to expand. The output of medical equipment and instrument manufacturing was 45.667 billion yuan with a year-on-year growth of 21.64 per cent between January and October of 2007. The year-on-year profit growth achieved in the first eight months was 52.32 per cent and investment of 6.443 billion yuan in the first nine months yielded a year-on-year growth of 65.29 per cent (Department of High-Tech Industry of the State Development and Reform Commission, 2007).

The revenue of the biological pharmaceutical industry in 2007 in China kept rising for a variety of reasons: first, the industry experienced a downturn in revenue in 2006 but recovered strongly in 2007. Secondly, several measures boosted the pharmaceutical market growth in 2007, such as the active promotion of a new rural cooperative medical system, the expansion of the scope of planned immunization, and introducing pilot basic medical insurance for urban residents. These measures are estimated to have created a pharmaceuticals market of 40–50 billion yuan. This greatly boosted the development of the medical industry. Thirdly, the drug price policy was relatively stable, which created a good market environment for the development of the medicine industry.

### Steady Growth of Biological Agriculture

The planting area of agroforestry further expanded and the share of high-yielding agricultural products increased. In 2007 more than 136 varieties were examined, including 52 varieties of rice, 36 varieties of corn and 25 varieties of soybean. The planting area of hybrid rice was 16 million hectares (2.4 million acres) in 2007. The cotton planting area was 5.68 million hectares (85.23 million acres), including more than 4 million hectares (60 million acres) transgenic pest-resistant cotton. The market scale of biological products for animal uses expanded gradually, animal vaccine renewal speeded up, industrial scale grew swiftly and the year-round output value increased to exceed 60 billion yuan. Feed additives kept a steady growth, with annual sales estimated at 1 billion yuan. The application scope of biological pesticide and biological fertilizer expanded further.

### Rapid Growth in Investment in the Production and Marketing of Biological Energy

Production of fuel ethanol progressed smoothly. The output of four designated ethanol production enterprises was about 1.3 million tons, with an output value of 6.5 billion yuan. The sales of fuel ethanol decreased in 2007 because of lower gas prices and higher grain prices. Biodiesel kept growing steadily and the investment increased rapidly. According to the statistics of the Chinese petroleum chemical industry association, the capacity for biodiesel production was about 1 million tons, with actual production about 0.2–0.3 million tons, and with an additional future production capacity of about 1.8 million tons coming from projects presently under construction. In addition, a set of straw-fired cogeneration projects was under construction.

### Speeding up the Industrialization of Biological Manufacturing

China has constructed polylactic acid factories with a production scale of 5000 tons as well as acid ester (mainly hydroxyl silane to gather hydroxyl butyrate acid ester) factories with a capacity of 2000 tons of production. Some big companies, such as COFCO, have invested in the construction of the base material of biological projects to be put into production. According to incomplete statistics, the total consumption of current domestic biological base plastic, including polylactide and hydroxyl silane acid ester, was about 100 000 tons. Among them, the consumption of polylactic acid was about 5000 tons, the consumption of acid ester (mainly silane hydroxyl PHBV) was about 6 million tons, and the consumption of starch-based polymer was about 800 000 tons. The consumption of other cellulose and chitin was about 15 000 tons and wood and plastic about 100 000 tons. China will construct a market for biological base plastic with a production of over 200 000 tons per year, and a wood and plastic product market with a production capacity of 300 000 tons in three years' time.

**Rapid Growth in Internationalization and Expansion of the Service Outsourcing of Biomedical Research**

A great number of international biological pharmaceutical giants, such as AstraZeneca, GlaxoSmithKline, Roche, Vince and American Pharmacopoeia, set up R&D centres or regional headquarters in China in 2007. Transnational companies continued to outsource biological pharmaceutical R&D to China. The biological pharmaceutical outsourcing industry in Shanghai and Beijing continued to grow at an impressive speed. According to statistics, the output value of biological pharmaceutical service outsourcing for Shanghai Pudong was 0.22 billion yuan in 2004, 0.7 billion in 2006 and more than 1 billion yuan in 2007 (Department of High-Tech Industry of the State Development and Reform Commission, 2007).

**Rapid Development of Domestic Biological Industry Parks**

From January to October 2007, 12 domestic biological industry parks earned 26.27 billion yuan with a year-on-year growth of about 21 per cent, with a profit of 28.8 billion yuan and with a year-on-year growth of about 33.6 per cent, as shown in Table 8.1. The total number of enterprises in these parks is over 4000. Domestic biological industry parks in Beijing form the hub of an industrial organization system, with a core of large companies such as Tong Ren Tang. Relying on the advantage of focusing scientific research institutes, the system includes a complete chain of research and development, incubation and production.

*Table 8.1 The development of China's domestic biological industry base in 2007*

| Base | Total industrial output value | | Profits | | The number of bio-technology enterprises in base |
|---|---|---|---|---|---|
| | Output value/ 100 million yuan | Year-on-year growth | Total amount/ 100 million yuan | Year-on-year growth | |
| Beijing (Medicine) | 192 | 22% | 28 | 43% | 200 |
| Shijiazhuang | 280 | 18% | 16 | 70% | 270 |
| Changchun | 216 | 30% | 16 | 33% | 175 |
| Shanghai | 293 | 9% | 27 | 51% | 370 |
| Guangzhou | 225 | 23% | 47 | 20% | 350 |
| Wuhan | 220 | 18% | 33 | 30% | 328 |
| Changsha | 242 | 30% | 26 | 30% | 310 |
| Chengdu | 351 | 9% | 37 | 28% | 500 |
| Chongqing | 226 | 23% | 19 | 23% | 513 |
| Kunming | 43 | 25% | 4 | 19% | 155 |
| Qingdao | 73 | 26% | 12 | 28% | 182 |
| Shenzhen | 266 | 23% | 23 | 28% | 752 |
| Total | 2627 | 21% | 288 | 33.6% | 4105 |

*Source:* Annual report on bioindustry in China (2007)

## The Constant Improvement of Development Environment of Biological Industry

In April 2007, the General Office of the State Council promoted the 'Biological Industry Development under Eleventh Five-year Plan' edited by the State Development and Reform Commission. For the first time, the biological industry was taken as an important strategic industry for the domestic economic deployment. In order to bring the financing advantage of the state development bank into play and increase the financing credit support of biological industry development, the State Development and Reform Commission and the China Development Bank drew up 'The biological industry in China' to jointly promote the funding of the sector. Moreover, a medium-to-long-range programme for renewable sources of energy has been issued. China's first conference on the biotech industry was held in Shijiazhuang. The new rural cooperative medical care and social security system for urban residents is rapidly advancing. Some local governments attach great importance to the development of biological industry and have formulated programmes and supplementary policies and measures to promote industrial development (Department of High-Tech Industry of the State Development and Reform Commission, 2007).

## DECISION-MAKING PROCESS OF BIOTECHNOLOGY R&D PROJECTS IN CHINA

### The Proposal and Confirmation of General Projects

Compared to other subjects, the particularity of biotechnology in China is mainly embodied in safety issues and ethical issues. Therefore, the proposal and confirmation of biotechnology R&D projects in China mainly deals with three aspects: public safety, ethics and funding.

For qualified research institutions, the principal of the executive unit can, in principle, independently approve all research projects whose safety level is at 'moderate risk'. However, due to the R&D funding system, the approval right for this kind of project in practice actually belongs to the fund supplier.

At present, there are many public R&D funds in China; some are directly managed by the State Department of PRC, and some are provided by the ministries and commissions of the State Department or the local government where the research institutions are situated. Generally speaking, the administrative power is vested in the government at all levels. The application and approval procedure is generally carried out annually, and the fund management department issues the appropriate funding plan and the approval standards to the related research institution and researchers. Then, the researchers form project groups and fill in the application according to the plan and standards. Each application should first be approved by the academic committee of the institution that the applicant belongs to, and then be handed over to the fund management department. The academic committee of the fund management department has the final approval right.

Compared to university and public research institutions, the corporate R&D institution, which mainly relies on corporation funds, is more convenient for the proposal and confirmation of general projects, and the project approval right basically belongs to the corporations themselves. However, if they hope to get financial aid from the government, or enjoy the preferential policy that the government has set up for corporation R&D activities, such as tax exemption, the related government department must approve the R&D projects involved. In China, except for a few national high-tech projects or large-scale technology projects, the decision-making rights of most corporation R&D projects belong to local government, especially at the municipal level, and the specific management method is also developed by the local government itself.

**Public Safety and Ethics Issues**

Compared to general projects, the approval procedure of projects related to safety and ethical issues is stricter.

At present, the management of biological safety and ethics by the Chinese government mainly focuses on transgenic research and medical biology research. Over the past 20 years, the Chinese government has particularly strengthened the management of transgenic safety. According to the 'Security Management Method for Genetic Engineering' issued in December 1993 by the original National Commission on Science and Technology (now Ministry of Science and Technology), the security of genetic engineering in China is uniformly managed by the National Committee on Genetic Engineering Security organized by the National Commission on Science and Technology. In addition, transgenic research related to special fields such as agriculture and medication, needs to accept the management of the related authority and provisions. Projects deemed to carry moderate risk should be carried out after being approved by the administrative department in charge, while high-risk projects can only be approved by the National Committee on Genetic Engineering Security.

The management of medical biology research is similar. Besides the medical research related to genes, medical biological security mainly focuses on the issue of whether or not the study involves highly pathogenic micro-organisms. Research that carries moderate risk needs to be approved by the provincial health administration, while high-risk research or research that is not on the categories list needs to be approved by the Ministry of Health.

The management of ethics in biological research mainly focuses on the field of medicine. In 2007, the Ministry of Health issued 'The Approval Method for Biomedical Research Ethics involving Human Beings', which prescribed that each institution that carries out biomedical research and related technology application activities involving human beings, must set up its own ethics committee, and that the Ministry of Health and provincial health administration respectively should set up a National Medical Ethics Committee of Experts and local ethics review, guidance and advisory organizations. The general biomedical research involving human beings should be carried out after being approved by the ethics committee of the institution, while important R&D projects need to accept the ethical review by the local ethics review, guidance and advisory organization, or even by the National Medical Ethics Committee of Experts.

## THE GENERAL SITUATION OF APPLICATION AND AUTHORIZATION FOR CHINESE BIOTECHNOLOGY INVENTION PATENT BETWEEN 1997 AND 2006

The Patent Office of the PRC was founded in 1980 as a component of Deng Xiaoping's globalization programme. China revised its Patent, Trade Mark, and Copyright laws in 1992 and 2000, and they are now similar to those in the European Patent Convention. An invention is clearly defined as patentable if it possesses novelty, inventiveness, and industrial applicability. The revisions also made it easier to determine the court jurisdiction of an infringement lawsuit and raised the maximum intellectual property remedies to 500000 yuan. Genetic technologies rely on main patent protection, while new species of plants and animals fall under special protections.

These new intellectual property protections were established to promote investment in the Chinese market, improve international collaboration, and reward technical innovation. More than 250000 applications were filed between 1985 and 1999, mostly for Chinese herbal medicines, foodstuffs and pharmaceuticals (Lui, 2001). Since 2003, China's patent agency has reviewed nearly 3 million applications, many of which are for new genetic engineering techniques and methods for modernizing traditional Chinese medicine (TCM). As a result of these changes, foreign companies are finally starting to trust China's life science intellectual property systems.

The application of biotechnology invention patents in China kept a steady growth from 1997 to 2006, from 1161 invention patents in 1997 to a high of 3839 in 2003. During these years, the domestic application of 15589 patents amounted to 63 per cent, whereas overseas applications amounted to 9129 patents or 37 per cent, as shown in Table 8.2. The total number of applications for patents was 24718, of which pending patents accounted for 41 per cent (10108), granted patents for 34 per cent (8489), and rejected, withdrawn or invalid patent applications for 25 per cent (6121) (Department of High-Tech Industry of the State Development and Reform Commission, 2007).

If we look at the geographical distribution of the domestic applications of biotechnology patents we find that Beijing and Shanghai accounted for 4061 and 2970 respectively, or 32 per cent and 24 per cent of the national total. They were followed by Jiangsu with 1140 patent applications (9 per

*Table 8.2 The general situation of invention patent application and authorization for Chinese biotechnology between 1997 and 2006*

| Home and abroad | Year of patent application and authorization | | | | | | | | | | Total |
|---|---|---|---|---|---|---|---|---|---|---|---|
| | 1997 | 1998 | 1999 | 2000 | 2001 | 2002 | 2003 | 2004 | 2005 | 2006 | |
| Domestic | 290 | 442 | 853 | 2246 | 1611 | 1957 | 2582 | 2535 | 2181 | 892 | 15589 |
| Overseas | 871 | 962 | 1162 | 1249 | 1134 | 1153 | 1257 | 1189 | 121 | 31 | 9129 |
| Total | 1161 | 1404 | 2015 | 3495 | 2745 | 3110 | 3839 | 3724 | 2302 | 923 | 24718 |

*Source:* Annual report on bioindustry in China (2007)

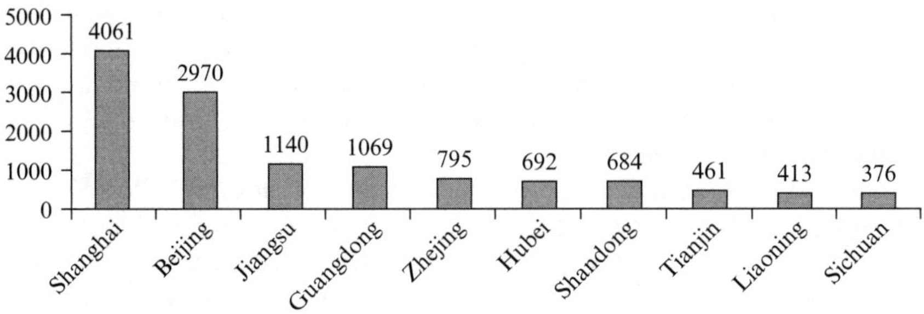


*Source:* Annual report on bioindustry in China (2007)

*Figure 8.2 The area distribution of domestic biotechnology patent applications*

cent). In addition, there were also a high number of patent applications in Guangdong, Zhejiang, Hubei, Shandong, Tianjin, Liaoning and Sichuan province, as shown in Figure 8.2. Also in terms of granted patents, Beijing, with 1318 granted patents, and Shanghai, with 815, accounted for the highest levels. In addition, other provinces, such as Guangdong, Suzhou, Hubei, Shandong, Zhejiang, Tianjin, Sichuan and Yunnan accounted for more than 150 patents each. The domestic patent applicants in biotechnology are mainly composed of colleges and research institutions; they accounted for 31 per cent and 26 per cent respectively of total applications; applications from enterprises ranked third, amounting to 25 per cent, while applications from individuals accounted for 17 per cent of total patent applications (Department of High-Tech Industry of the State Development and Reform Commission, 2007).

## THE ISSUE OF INTELLECTUAL PROPERTY RIGHTS PROTECTION

Innovation fuels the knowledge-based economy. Protection of intellectual property rights will stimulate and accelerate technology transfer. We have to keep a good balance between the free information exchange and protection of inventors' interests. To take the human genome project as an example, China is a part of the consortium, and insists that all of the basic human DNA sequence information should be available to the public and used freely. In the meantime, China agrees to the patent protection of advanced techniques and new products, especially in functional genomic research.

China has been adopting a policy to protect patents and intellectual property rights. We have gradually established various laws and the regulations regarding patents, trademarks and copyrights. The Chinese government formally accepted the international consensus on patenting and intellectual property rights protection. China has been working hard to create an environment to recognize patents and protect the interests of patent owners. As a result, the number of patent applications in China has increased significantly during the past few years.

## CASE STUDY: BIOTECHNOLOGY R&D AND TRANSFER CONDITIONS OF TIANJIN UNIVERSITY OF SCIENCE AND TECHNOLOGY

Tianjin University of Science and Technology (TUST) is a key comprehensive university supported by Tianjin government, with 130 professional biotechnology researchers, and having a strong R&D capability and technology transfer experience in the field of biotechnology.

### R&D Situation

Some of the biotechnology R&D projects in TUST are commissioned projects from government organizations, which are generally called vertical projects; others are commissioned projects from corporations or public research institutions, which are generally called cross-cutting projects. Both the number and funds of vertical projects have experienced a sharp increase, obviously a policy-pulling effect (see Figures 8.3 and 8.4). Although the number of cross-cutting projects is not increasing as fast as the vertical projects, it is still increasing steadily (see Figures 8.5 and 8.7).

The funding for cross-cutting projects has basically stabilized at between

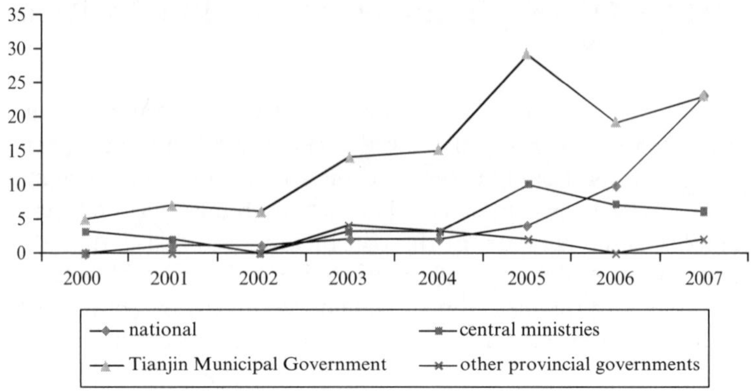


*Source:* Science and Technology Department of TUST

*Figure 8.3 The number of biotechnology vertical projects in TUST, divided by commissioned units*

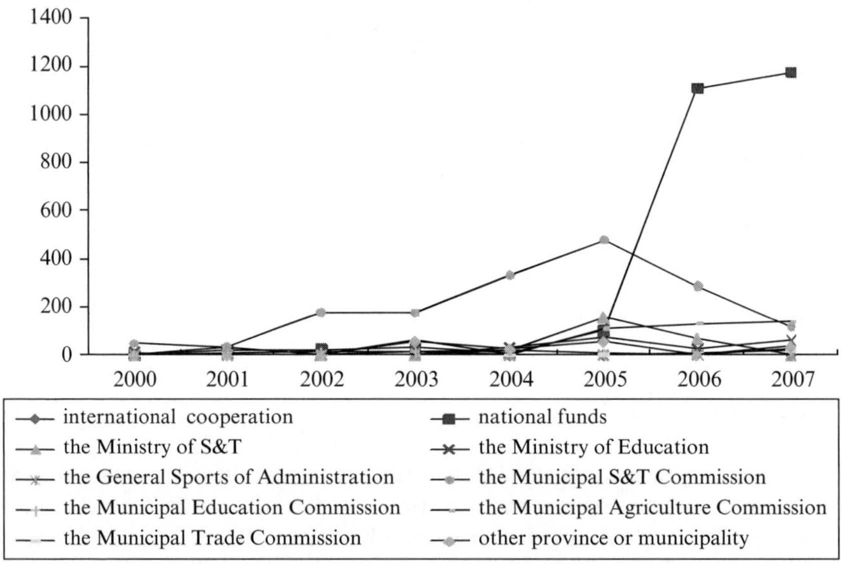


*Source:* Science and Technology Department of TUST

*Figure 8.4 Funding of biotechnology vertical projects in TUST, divided by commissioned units*

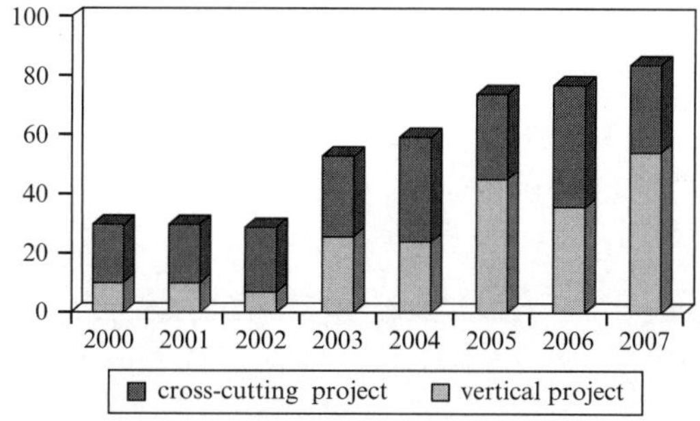


*Source:* Science and Technology Department of TUST

*Figure 8.5 The number of biotechnology projects in TUST, divided by source*

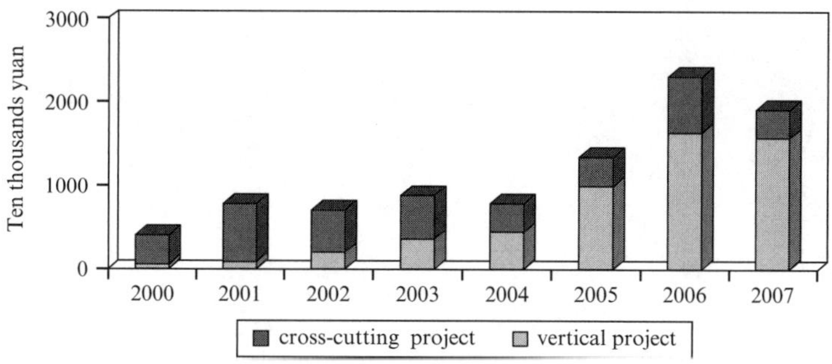


*Source:* Science and Technology Department of TUST

*Figure 8.6 Funding of biotechnology vertical projects in TUST, divided by source*

3 million and 7 million per project (Figures 8.6 and 8.8). Corporations are the main sources of biotechnology cross-cutting projects in TUST, which indicates that TUST has already built a long-term cooperation relationship with a group of corporations in biotechnology, and has had comparatively steady funding sources for cross-cutting projects.

Lastly, from the point of view of the ownership of the corporation that has cooperated with TUST, the percentage of national corporations is the biggest and the number is also stable (see Figure 8.9). This indicates that the cooperation of the two parties is effective and long-term. Since 2004, along with the continuous growth of private S&T corporations and their increasing demand for technology transfer, the cooperation of TUST and private corporations has increased rapidly in biotechnology R&D.

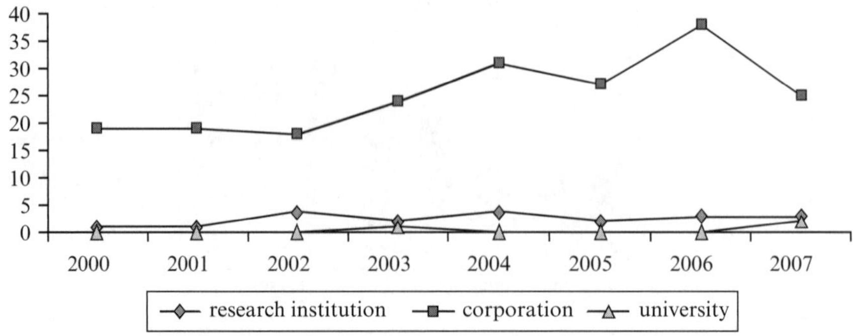


*Source:* Science and Technology Department of TUST

*Figure 8.7 The number of biotechnology cross-cutting projects in TUST, divided by commissioned units*

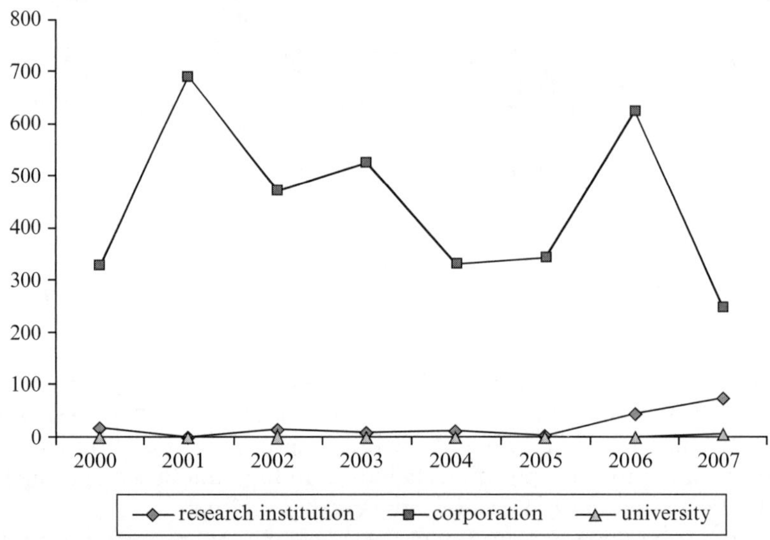


*Source:* Science and Technology Department of TUST

*Figure 8.8 The funding of biotechnology cross-cutting projects in TUST, divided by commissioned units*

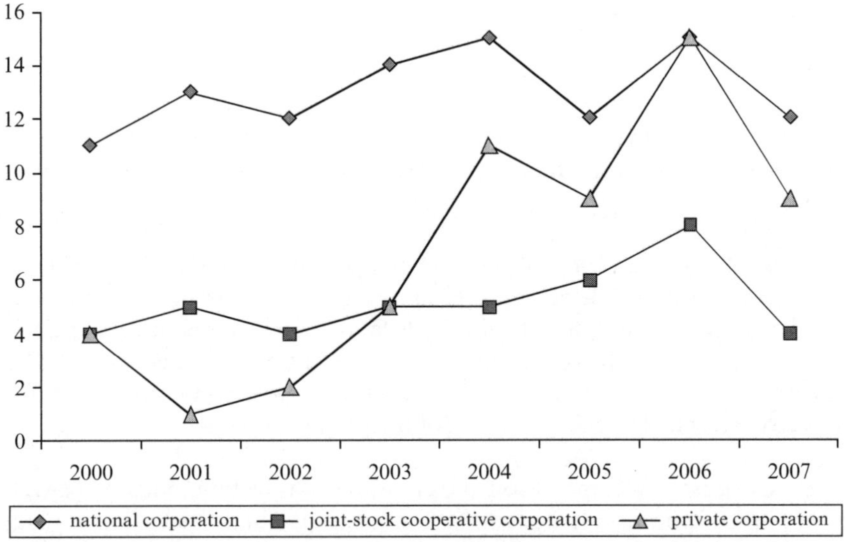


*Source:* Science and Technology Department of TUST

*Figure 8.9 The classification of cooperative corporations*

## Transfer Conditions

### Main modes of biotechnology transfer in TUST

In recent years, universities have paid more and more attention to biotechnology R&D and utilization of research results. By using the technology transfer experience at home and abroad as a reference, TUST has gradually formed and improved some common technology transfer modes.

*Transfer implemented by the state* This technology transfer mode is organized directly by the state, and mainly exists in vertical projects. It is applicable for those important original innovations or major technology innovations that relate to the national economy and the people's livelihood or the strategical development of national industry, and is also applicable for the technology transfer relating to national security and national defence. The characteristics of these projects are: mainly funded by the state, with comparatively abundant funds, covering multiple industries and subjects.

*Transfer inside the university* After a successful R&D project, the university builds a corporation to develop the product or carry out industrial production. The advantage of this kind of transfer mode is that the inventor of the technology can participate directly in the establishment of the corporation, which enables improvements and upgrades of the technology after transfer; and its disadvantage is that the university-run corporation generally lacks the systematic management experience for the whole process of production, and the progress towards large-scale production is slower.

*Export-oriented market transfer* The successful cooperation of TUST with corporations effectively accelerates the combination of industry, academy and research and raises the level of scientific research in TUST, and also promotes the development of university S&T industry as well as the regional economy. At the same time, the corporation itself has gained positive effects, particularly in the following three forms:

- Corporation commissioned technology. According to its own development needs, a corporation commissions a university to carry out the R&D or innovation of a certain technology. The corporation devotes R&D funds, and the university devotes R&D resources and completes the R&D task according to the demand of the corporation, and finally delivers it to the corporation.
- Cooperative research. The university and the corporation join together to carry out the R&D and innovation of a certain project or technology. Specifically speaking, the corporation devotes funds, the universities and the corporation form a research group together, and the two sides share risk and benefit. This kind of user-oriented close cooperation research mode makes the transformation of the research achievement very effective, since the investor or user participates fully in the R&D work all the time.
- Collaborative construction. This is a kind of long-term strategic S&T cooperation, such as collaboration to build a research centre, research institute or laboratory, which has strategic significance for the development of the cooperative parties. The two parties can complement, promote and develop mutually.

**The main channels of biotechnology transfer in TUST**

In the practice of biotechnology transfer, the three parties of government, university and corporation all make great efforts to strengthen the construction of technology transfer platforms, to exchange information, promote cooperation, and ensure effective technology transfer channels. Corresponding to the three modes of biotechnology transfer, there are also three channels of biotechnology transfer in TUST.

*Technology transfer with the assistance of government* The university–industry research cooperation under the organization and coordination of government is an important channel of biotechnology transfer. It is mainly applicable for technology transfer projects implemented by the state and parts of projects established by universities and corporations together. In recent years, taking the scientific research plans issued by governments at all levels as a guideline, TUST has developed university–industry research cooperation with many domestic corporations such as Tianjin Emperor Wine Co. Ltd and Shandong Fu Feng Group, and has carried out many successful biotechnology transfer activities.

*Technology transfer by the S&T department set up by universities* In order to offer a platform for technology transfer, TUST specifically set up an S&T department as a window for consultations with corporations. Not only does it take over the tasks of technology service, technology consultation and the R&D of new products delivered by corporations at home or aboard, it also presents the R&D achievements of universities to corporations. The majority of cooperative projects with corporations or commissioned projects from corporations in recent years are indebted to the work of the S&T department. At present, through this platform, TUST has taken over biotechnology projects such as the progress design of the full set of a product line for condensed pineapple juice from Indonesia.

*Direct cooperation of universities and corporations by public information platform* Because TUST is well known in the national biotechnology field, some corporations will take direct contact with the related research group to solve their problem. At the same time, the biotechnology researchers in TUST often issue their latest technology achievements on the Internet to seek potential cooperation. Most of the successful projects through this channel are commissioned projects from corporations.

## CONCLUSION

### Problems Reflected in Biotechnology R&D and Transfer

- At present, the government-driven mode of biotechnology transfer is comparatively monotonic, mainly relying on direct investment and in need of improvement. From the above analysis of biotechnology transfer projects and fund sources in TUST we can see that, since 2005, the funding from all levels of government for vertical projects account for three-quarters or even more of the total funding obtained by TUST. Although these projects can increase the research resources of the university and improve the performance level and technology transfer of research results in the short run, they often need support from government after completion before they can be used in production by corporations. This increases the transfer period and reduces the efficiency of the technology transfer.

- In the course of technology R&D and transfer, the funds for 'development' are quite insufficient. In the field of biotechnology, it is absolutely necessary to carry out a medium-scaled 'test' production between lab research and industrialization. In China, however, there are strict limitations of funds and equipment in this phase at present. In the drastic academic competition for funding, universities prefer to devote their funds to basic research and to the purchase of experimental equipment, rather than testing equipment. Due to economic realities, corporations only devote small funds to testing equipment. Therefore, there is a lack of settings and resources for carrying out testing of research achievements.
- The building of a technology transfer platform needs to be strengthened. At present, there is a serious problem in the course of research, development, transfer and production of biotechnology, which is that the university researchers who have R&D capability often lack practical experience on commercialization and industrialization. Their research achievements often do not reach the market, or are in need of further development after their transfer to corporations has been completed. Therefore, this kind of cooperation needs not only an agency that simply offers the information of research achievements to the corporations, but also a platform with better functions that can support a long-term cooperation between corporation and university.
- Currently, there are a lot of problems regarding intellectual property protection in China. The intellectual property protection is an issue concerning both the university and corporations dealing with biotechnology R&D. For example, the invention patent or research achievement obtained through the cooperation between university and corporation will involve the benefits of university, project teams and corporation, and there are still no specific provisions in law on how to coordinate the benefits of these three parties in technology transfer, which would positively influence the project teams and corporations in technology transfer.

## Policy Suggestion

The Chinese government plays three main roles in the promotion of science and technology development. First of all, the government should strengthen the support for basic research. It is the basis of any new knowledge and methodology. Secondly, favourable policies and necessary conditions should be created for technology innovation, transfer and industrialization. This requires a stable macro-economy, a healthy financing system, an educated labour force, a competitive market environment, and so on. Thirdly, the innovation system should be improved. Innovation no longer relies on companies or institutes alone, but on integrated efforts from both sides. Obstacles for cooperation and networking should be removed and close connections should be set up through networking among universities, public organizations, industrial companies and financial institutions. In order to promote integration of science, technology and the economy, the following policies have been implemented in China: (1) setting up high-tech incubators for biotechnology; (2) adopting favourable taxation policies for biotechnology development; (3) supporting SMEs in biotechnology; (4) encouraging venture capital investment and technology transfer; (5) setting up specific zones for high technology development; (6) recognizing intellectual properties as capital investment.

The industrialization of university biotechnology achievements is a complex and risky process. In order to solve the problems above, it needs the cooperative endeavour of the community, university, corporation and government departments. According to the problems and experience of TUST in biotechnology transfer, we put forward the following suggestions:

- The continuous increase of government support should rest on the adoption of various and more flexible support measures. Since 2000, the investment of government to biotechnology R&D and transfer has obviously grown considerably, much in line with the global trend. However, if we look at the direction of the investments, apart from direct investment to university, the funds should be dispersed in a more flexible strategy. For example, more specialized funds should be devoted to corporations, in order to encourage the cooperation of corporations and university. Or some indirect support measures could be chosen, such as tax credits to replace direct investment. And finally, a multi-channel, multi-level financial investment system should be built, with the government as the leading part and corporations as the main body.

- Middle and small-sized biotechnology corporations should be developed, which give priority to the transfer of technology achievements and technology services. The establishment of this kind of corporation alleviates problems in the testing phase caused by the lack of testing centres. Moreover, it would help to connect technology application corporations and technology research institutions through technology outsourcing services, to accelerate the effective transfer of biotechnology results.
- In today's global economy, scientific collaboration has become a prerequisite for development. China has made great progress in the national economy during the past 20 years and believes that international cooperation will help to achieve its goals. Up till now, the Chinese government has established active scientific cooperation and exchange programmes with 152 countries throughout the world. Agreements have been signed with 96 countries for research collaboration. The cooperation has different formats, including the collaboration on joint investigations, basic research and industrialization. The Chinese government will continue to implement an open-door policy and economic reforms. It will continue to initiate research programmes for international collaboration. The government has approved the setting up of a specific fund to support international cooperation during the tenth 5-year plan.
- There is a common saying, 'Everything has its own advantages'. For scientific cooperation, we should establish equal partnership, and the cooperation should benefit all parties. We will follow international rules and fully protect intellectual property rights. We are, and will always be, a responsible member of the international family.
- Risk investment funds should be established and financing channels enlarged. Due to the great risk, few of the loan operations in the finance department involve biotechnology transfer projects. Therefore, the risk prevention mechanism of S&T loans should be improved, and the support of the finance department to biotechnology transfer projects should be promoted.
- Testing and pilot production should be encouraged. Government departments at all levels should encourage pilot production projects, strengthen their support for testing projects in biotechnology, and give special treatment in related policies such as tax offers to technology service corporation that engages in product testing and pilot production.
- Establishing some Biotechnology achievement introduction institutions. From the experience of TUST, the specialized technology

transfer institution organized by the university is helpful in accelerating technology transfer. This kind of institution can not only offer a common technology transfer intermediary service, but can also take charge of the organization and communication work for strategic association of 'industry–academy–research institution', facilitate the connection of university to biotechnology corporation and public research institution, be the bridge for the industrialization of university biotechnology achievement, coordinate and promote the cooperation between university and industry, and accelerate the industrialization of university biotechnology achievement.

In the age of global economy, we strongly believe that the development of modern life science and biotechnology will bring us a brighter and more prosperous future. A worldwide coorporation will certainly help us to realize our bright prospects. The resolving of technology transfer problems is also the resolving of the inner relation problem of technology and economy in the field of biotechnology. The present experience indicates that in the commercialization of biotechnology achievement, the regulation and support of the government is still vital. Perfecting technology transfer systems, improving risk investment systems, establishing effective market agencies, and strengthening the function of corporations in technology transfer, are the main problems to be resolved in the biotechnology sector of the future.

## REFERENCES

Department of *High-Tech* Industry of the State Development and Reform Commission (2007), *Annual Report on Bioindustry in China: 2007*, Chemical Industry Press, pp. 18–32.

Huang, J. and Q. Wang (2003), 'Agricultural biotechnology development and policy in China', *AgBioForum*, **5**, 1–15.

Huang, J., R. Hu, C. Pray, and S. Rozelle (2004), 'Plant biotechnology in China: public investments and impacts on farmers, "New directions for a diverse planet"', *Proceedings of the 4th International Crop Science Congress*, 26 September – 1 October, Brisbane, Australia, available at www.cropscience.org.au, pp. 1–10.

Huang, J., Q. Wang, Y. Zhang and J. Zepeda (2001), 'Agricultural biotechnology development and research capacity in China', working paper, Centre for Chinese Agricultural Policy, Chinese Academy of Sciences.

Li, X. (2004), *China Bio-industrial Report*, Central Documentary Press.

Lui, J. (2001), 'Patenting biotechnology inventions in China', *Nature Biotechnology*, **19**(1), 83–4.

Wang, C. (2007), 'The development of Chinese biological industry under the background of globalization', *Chuangxinkeji*, no. 8, pp. 14–19.
Zhang, Y. (2008), 'The situation and suggestion on biotechnology in 2007', *China Venture Capital*, no. 3, pp. 32–6.

# PART IV

# Europe

# 9. Biotechnology in Europe: background information on biotechnology industry characteristics and policy environment in Denmark, Germany, Latvia, Russia and Sweden

**Thomas Reiss, Ralf Lindner and Ulrich Schmoch**

## INTRODUCTION

This overview report aims at providing some general information on the policy environment related to biotechnology and on the biotechnology industry characteristics in Denmark, Germany, Latvia, Russia and Sweden. Thereby a common starting point for the five country case studies presented in separate reports will be set.

An important requirement for the comparative country analysis is the availability of a common understanding of the field of investigation. In the case of biotechnology such an approach faces a number of specific difficulties which are related to the fact that biotechnology per se is not a business sector covered by standard industry statistics provided, for example, on an international basis by the OECD. Rather, biotechnology is a cross-cutting technology which can be applied in a wide range of industrial sectors ranging from pharmaceuticals through to mining. Therefore, the statistical challenge is not only to identify business and public activities which can be considered as core biotechnology, but also to identify those business activities in the various application sectors that to some extent use biotechnology.

More than five years ago the OECD developed a definition of biotechnology based on the work of a large group of experts in biotechnology and in survey design. The resulting definition is a compromise between the need for a workable definition of biotechnology that is suited to survey

### BOX 9.1 LIST-BASED DEFINITION OF BIOTECHNOLOGY TECHNIQUES

**DNA/RNA:** Genomics, pharmacogenomics, gene probes, genetic engineering, DNA/RNA sequencing/synthesis/amplification, gene expression profiling and use of antisense technology.

**Proteins and other molecules:** Sequencing/synthesis/engineering of proteins and peptides (including large molecule hormones); improved delivery methods for large molecule drugs; proteomics, protein isolation and purification, signalling, identification of cell receptors.

**Cell and tissue culture and engineering:** Cell/tissue culture, tissue engineering (including tissue scaffolds and biomedical engineering), cellular fusion, vaccine/immune stimulants, embryo manipulation.

**Process biotechnology techniques:** Fermentation using bioreactors, bioprocessing, bioleaching, biopulping, biobleaching, biodesulphurization, bioremediation, biofiltration and phytoremediation.

**Gene and RNA vectors:** Gene therapy, viral vectors.

**Bioinformatics:** Construction of databases on genomes, protein sequences; modelling complex biological processes, including systems biology.

**Nanobiotechnology:** Applies the tools and processes of nano/microfabrication to build devices for studying biosystems and applications in drug delivery, diagnostics etc.

research, the need to differentiate between modern and traditional biotechnologies and the need to retain the flexibility to cover new developments in a rapidly changing field (Patel et al., 2008).

The OECD recommends the use of a combination of a single definition of biotechnology and a list-based definition (Van Beuzekom and Arundel, 2006).

Accordingly, the single definition defines biotechnology as 'the application of science and technology to living organisms as well as parts, products and models thereof, to alter living or non-living materials for the production of knowledge, goods and services'. The accompanying list-based definition of biotechnology techniques is show in Box 9.1.

The single definition also includes more traditional biotechnology and is intentionally broad in order to incorporate new discoveries that are not yet foreseen in the list. The list-based definition, on the other hand, limits the definition more to modern biotechnology.

The OECD list-based definition, has so far been used in biotechnology surveys in 15 OECD countries (Van Beuzekom and Arundel, 2006). Some countries also included the category 'other' in the list-based definition in order to capture additional activities not yet foreseen. The general advantage of the list-based survey approach is that companies can quite easily decide by themselves whether their business activities fall under one of the listed items and accordingly can be considered as biotechnology. Also some of the main sources used for the five country case studies use the list-based definition. These include in particular:

- The Biopolis project (Enzing et al., 2007) which was funded by the European Commission under FP6 and aimed at providing an inventory and analysis of public policies that stimulate biotechnology research and exploitation in Europe. The Biopolis project not only provides an overview of biotechnology policies in 32 European countries, but also comprises a performance analysis of these countries using various science, technology and innovation output indicators.
- The BIO4EU project (Reiss et al., 2007) which elaborated an evaluation of the consequences, opportunities and challenges of modern biotechnology for Europe in terms of economic, social and environmental aspects. The evaluations within the BIO4EU project are based as much as possible on quantitative indicators. The project goes back to an initiative of the European Parliament which had requested the European Commission to carry out an assessment of modern biotechnology.
- The sector report 'Monitoring, analysing trends and identifying challenges in biotechnology' within the Europe INNOVA initiative of DG Enterprise which has been completed in 2008 (Patel et al., 2008).

In addition to these public sources a recent private survey on biotechnology was also used as a source. This concerns the Europa Bio-Critical I surveys from 2005 and 2006 (Critical I, 2005, 2006) which provide internationally comparative data of economic activities in biotechnology. These surveys cover a wide range of European countries and explicitly describe their methodology for identifying biotechnology firms and gathering data. In this respect the Europa Bio-Critical I surveys are different from many other private surveys which usually do not give sufficient information for judging the methodological quality of the respective research.

In the next section of this chapter, some general background information on the five countries involved is given. Following that, the general policy environment is described with relation to biotechnology. Next, country specialization and the biotechnology industries are characterized, using, among others, quantitative data provided by the above-mentioned surveys. Finally, we draw some preliminary conclusions which will be further elaborated in five country case studies described in Chapter 10–13.

## BACKGROUND INFORMATION ON THE FIVE COUNTRIES

In order to put the respective biotechnology activities in the participating countries into a broader context, the following gives some basic information on the economic situation in the five countries.

### Denmark

With a population of 5.4 million, Denmark has a modern market economy with extensive government welfare initiatives, a stable currency (Danish Krone, DKK) pegged to the euro, and excellent standards of living. According to Eurostat (2006), at the beginning of 2002, the gross domestic product (GDP) per capita measured in standard purchasing power (SPP) was 21.4 per cent greater than the European average (EU25). In 2002 economic growth slowed down to a growth rate of 0.5 per cent (measured in terms of real GDP). However, Denmark experienced an economic recovery in the period 2002–05, reaching 3 per cent growth rate and (GDP) per capita 24 per cent greater than the EU25 average.

Investments in research and development (R&D) amounted to 2.56 per cent of GDP in 2005. This level was above the European average (EU25: 1.86 per cent). Moreover, 59.9 per cent of the investment in R&D was carried out by industry. According to the Danish National Research Foundation (2005), in the last decade most of the increase in the financial resources directed to research and development activities was caused by increased private investment, while public investment remained stable at a lower level.

### Germany

In 2005 Germany was the largest economy in the European Union (EU) in terms of GDP, and the most populous country in the EU. According to Eurostat,[1] with a GDP of EUR 2 244 000 million and a population of

82.5 million inhabitants, in 2005 Germany's GDP per head was 7 per cent greater than the EU25 average. Between 2000 and 2003, economic growth in Germany slowed down, reaching −0.2 per cent growth rate of its real GDP in 2003. The economy recovered slowly to reach 1.2 per cent growth rate in 2004, but fell again in 2005 to 0.9 per cent. With this development Germany has become one of the slowest-growing economies in the European Union (EU25 average in 2005 2.3 per cent). With an unemployment rate of 9.5 per cent in 2005, policy activity is very much dominated by labour market and welfare system issues. Moreover, the weak economic development in eastern Germany is still a main political concern. The relatively large public deficit share as percentage of the GDP (over the 3 per cent threshold in 2003) is an important obstacle to undertaking public investments to try to overcome the economic stagnation. The outstanding export performance is the mainstay of Germany's economic growth.

Despite the weak macroeconomic performance in the last decade and major structural problems of the German economy, the efforts in research and development activities (R&D) grew continually between 1996 and 2003. In 2004 R&D investment declined slightly. Nonetheless, with a GERD/GDP[2] ratio of 2.49 per cent, in 2004 Germany was the fourth largest EU country in terms of R&D investment as a percentage of its GDP, behind only Sweden, Finland and Denmark.[3] The private sector plays a major role in the German innovation system. In 2004 industry financed 67.1 per cent of R&D activities. This was the second largest national rate behind Finland.

## Latvia

Latvia became independent from the Soviet Union in 1991 and developed into a parliamentary democracy. In 2004 Latvia joined the European Union and NATO. With a population of 2.32 million people (2004) Latvia is the second largest of the three Baltic States behind Lithuania (Eurostat, 2006).[4]

The Latvian economy enjoyed one of the highest growth rates in recent years within the EU. Between 2001 and 2005 GDP grew at an average of 8.1 per cent annually and reached a growth rate of 10.2 per cent in 2005 (Ministry of Economics of the Republic of Latvia, 2006). Production volume increased by 7.9 per cent in 2004 (European Commission, 2005a) and the employment situation improved continuously in Latvia as indicated by a decreasing unemployment rate from 8.5 per cent in 2004 to 7.4 per cent in 2005 (Ministry of Economics of the Republic of Latvia, 2006).

While such relative indicators paint a very favourable picture of the development of the Latvian economy, in absolute figures the Latvian economy is among the lowest performing economies in the EU. In 2005 the GDP reached EUR 11 800 million (Eurostat, 2006). On a per capita basis this is only 43.7 per cent of the EU25 average (European Commission, 2005a). Also labour productivity per person employed is only 43.6 per cent of the EU25 average (European Commission, 2005a).

The R&D performance of Latvia presently also ranks far below the EU25 average. In 2003 the GERD amounted to EUR 39 million, corresponding to an R&D intensity (GERD by GDP) of only 0.39 per cent, which ranks Latvia close to the bottom of the European member states. Only Cyprus (0.33 per cent) presented a lower R&D intensity rate (Eurostat, 2006). One reason for the low R&D intensity of Latvia is the low contribution of the business sector to total R&D investment. In 2002 the share of the business sector was only 21.7 per cent of total research investment (Eurostat, 2006). Accordingly, one of the main innovation challenges in Latvia is the mobilization of a stronger business involvement in innovative development.

## Russia

During the last 15 years Russia has been trying to overcome a major macroeconomic and political crisis. Radical changes took place in the country, including the emergence of new fundamental economic and political institutions such as private property, democracy and federalism, and the development of vital economic institutions and mechanisms such as taxation, budgeting, real estate and land management, intellectual property and technology transfer, a deregulation system aimed at lowering administrative barriers hindering business activities, integration processes in research, education and industry, and so on.

The growth of the Russian economy – which only began in 1997 after the post-socialist recession – was interrupted in 1998 due to the radical deterioration of the global market situation and a flight of capital from many emerging markets, including Russia. In 1999 the economy started to grow again. Generally, in 2001–06 the country's GDP grew by 48.7 per cent (in 2000 prices). Increased economic activity was based on the growth of the domestic market, with investments growing at a higher rate than consumer demand. Another important characteristic of the country's economic development was imports growing faster than domestic production. The high demand for imported goods was mostly based on an exceptionally favourable situation regarding global energy and raw material markets. Diversification of the Russian economy, reducing its dependency on the situation on raw materials markets, remains a major strategic problem that Russia must deal with.

As for research and development activities, it should be noted that Russia's domestic R&D expenditures as share of GDP grew from the second half of the 1990s until 2003 (from 0.85 to 1.28 per cent, except during the 1998 recession when this figure dropped to 0.95 per cent), but in 2004 the R&D expenditures started to fall again (to just 1.07 per cent of GDP in 2005). In 2007 this indicator grew to 1.12 per cent, still below the level of the last pre-reform year of 1991, when the share of GDP spent on research and development activities was in excess of 1.4 per cent (Indicators of Science, 2008).

An international comparison shows that Russia's R&D expenditures as share of GDP are below the average for EU27 (1.07 versus 1.77 per cent in 2006). In 2007 Russia was behind almost all EU15 countries except Greece, as well as behind several former socialist countries including the Czech Republic, Slovenia and Estonia (OECD, 2008).

Russia is one of the countries where public funds amount to a larger share of R&D expenditures (about 60 per cent in 1995–2006, 61.1 per cent in 2006) than the corporate sector's investments (less than one third of the total expenditures during the period in question: 30 per cent in 2006). This makes the government a serious player in the country's S&T sphere, and at the same time highlights the rather weak interest of the business community in implementing national research projects.

A breakdown of domestic R&D expenditures by R&D sectors reveals that the entrepreneurial sector's organizations are the most active ones: the sector's share never went below 66 per cent during the last 12 years, the average figure being 68.8 per cent. This sector's share of the total Russian R&D expenditures (66.7 per cent in 2006) is comparable with the relevant values for such countries as Austria, Belgium, Denmark and Germany, and exceeds the EU27 average. The public sector's organizations' share of total R&D expenditures during the same period varied between 24.3 and 28.2 per cent (27 per cent in 2006) – much higher than in most of the EU member states. Russian educational institutions make only a very small contribution to research and development: their share is just 6.1 per cent of the total R&D expenditures, about three times lower than in many European countries.

The breakdown of domestic R&D expenditures by type of activity (basic, applied research and development) (69.3, 15.3 and 15.4 per cent in 2006, respectively) remained practically unchanged throughout the period in question. This allows us to conclude that in Russia, R&D organizations are mostly oriented towards short-term projects, due to lack of funds, specific structure of funding (relative shares of various funding sources) and an archaic institutional structure of the S&T sphere.

**Sweden**

Sweden has a population of 9.4 million people, with a GDP per capita of EUR 37 500 (current) in 2009–10. Swedish manufacturing production is relatively large and knowledge-intensive by international standards, and its share of the high and medium-high-technology manufacturing industry (relative to the size of the economy) has ranked among the top five countries in the OECD for many years (Marklund et al., 2004).[5] Moreover, the Swedish high-technology manufacturing sector has a very broad scope, embracing most segments of telecommunications and pharmaceuticals. The relative weight of knowledge-intensive services is slightly lower relative to some leading OECD economies. Two other distinctive features of the Swedish economy are the dominant role played in manufacturing employment and R&D by a small number of large national industrial groups, and the fact that the Swedish economy is largely driven by exports (which in 2002 accounted for 43 per cent of GDP), with non-electrical machinery, motor vehicles, pulp and paper and telecommunications as the leading export sectors in terms of volume (Marklund et al., 2004).

Over the period 2001–03, growth rates of GDP have been low, causing a fall in GDP per capita (in PPS) relative to EU25, from 119 in 2000 to 114 in 2003, though recovery in 2004, with a GDP growth rate of 3.5, has pushed the ratio up to 116. Despite the fact that Sweden remains well above OECD average levels of GDP per capita, some concerns have been raised due to the fact that over the period 1970–2003 Sweden has fallen from fourth to fifteenth place in the OECD's rankings (Marklund et al., 2004). This weak long-term economic performance has led to questions about whether the returns on Swedish R&D investments have been lower than expected and whether R&D institutions in Sweden are able to transform effectively the knowledge generated into activities that generate growth.

Regarding its R&D efforts, Sweden is outstanding for a variety of reasons. First, Sweden has steadily increased its gross domestic R&D expenditures (GERD) as a percentage of GDP since the early 1990s (2.7 per cent in 1991, rising to 4.27 per cent by 2001), and has been the top EU country for GERD as a percentage of GDP since the early 1990s. Second, Sweden has also led OECD rankings in terms of business expenditures on R&D (BERD) relative to GDP, and the percentages grew from 1.9 per cent in 1991 to 3.3 per cent in 2001. The business sector in Sweden plays a key role in the country's R&D efforts, responsible for 72 per cent of GERD by source of funding, while the public sector contributes only 21 per cent (OECD, 2005). This contrasts with EU25 figures, where the percentages are 55 per cent and 35 per cent, respectively. Thirdly, Sweden has a comparatively large percentage of business R&D expenditures from foreign-owned firms, indicating that Sweden is an attractive location for large multinational manufacturing groups (Marklund et al., 2004). Finally, the current structure of the Swedish innovation system is largely dependent on a few large corporations that are responsible for nearly 70 per cent of the business R&D investment (Ministry of Industry, Employment and Communications, 2004). This feature makes the Swedish innovation system relatively vulnerable to the international strategies of local groups regarding their R&D investments.[6]

## POLICY ENVIRONMENT

In this section a brief overview of the European biotechnology context will be given, followed by a summary of the biotechnology policy landscape in the countries considered, including information on public perception of biotechnology and regulatory frameworks. Finally information on biotechnology funding will be compared in Denmark, Germany, Latvia, and Sweden, as elaborated within the Biopolis project (Enzing et al., 2007).

### EU Policy Context

Biotechnology is considered to have the potential to contribute to key EU policy goals. In particular, biotechnology is expected to promote growth, employment and competitiveness in the EU. The European Council and the European Parliament have recognized the importance of biotechnology, and accordingly the Commission has proposed an action plan to address the challenges and opportunities of biotechnology. In 2002 the Commission adopted a strategy for Europe in the field of life sciences and biotechnology (European Commission, 2002). The European strategy proposed a 30-point action plan involving the Commission, other European institutions and other stakeholders. The strategy runs until 2010. Originally, the scope of the strategy was very broad, in order to cover all possible relevant policy issues and facilitate the uptake of the technology in a wide range of sectors. Action was proposed under four headings: harvesting the potential; promoting governance; answering global challenges; and assuring coherence between the full range of policies concerned.

In 2007 a mid-term review of the strategy was carried out. This review was largely based on the BIO4EU study, which presents an exhaustive picture of possible applications with concrete examples and assesses their impact from the economical, social and environmental point of view, including comparative data on the situation in third countries. The BIO4EU study was finalized in April 2007 (Reiss et al., 2007; Zika et al., 2007).

The mid-term review of the European strategy for life sciences in biotechnology (European Commission, 2007) concluded that the potential of biotechnology to support EU policies is real and has been proven by many practical examples. In consequence it was considered vital to continue promoting the development of life sciences and biotechnology in the EU, in particular by increasing research activities and promoting competitiveness. While the original design of the strategy was wide in scope, to give an initial mapping of the situation and to identify the full range of linked policy areas, the mid-term review tried to re-focus the strategy in order to improve its output by 2010. Specifically, five main interdependent themes were identified:

- Promote research and market development for life sciences in biotechnology applications and the knowledge-based bioeconomy (KBBE).
- Foster competitiveness, knowledge transfer and innovation from the sciences base to industry.
- Encourage informed societal debates on the benefits and risks of life sciences and biotechnology.
- Ensure a sustainable contribution of modern biotechnology to agriculture.
- Improve the implementation of the legislation and its impact on competitiveness.

Accordingly, the original action plan of the European strategy for life sciences and biotechnology was re-focused under these five headings. Following the mid-term review, the Commission concluded that it will continue the implementation of the action plan of the strategy up to 2010, that biotechnology will be included in the implementation of the innovation strategies and that in cooperation with member states and stakeholders the implementation of the strategy will be improved.

**National Situation**

The following summaries of the national situation in biotechnology in the selected countries are based on the national reports of the Biopolis project (Enzing et al., 2007).

**Denmark**

Denmark has a long experience of designing and implementing public programmes to promote biotechnology. The Danish research councils have been funding biotechnology research with specific initiatives since 1987. In the 1990s public support for biotechnology research activities combined institutional funding and competitive funding. Both these types of funding were implemented via bottom-up initiatives (e.g. FØTEK programmes targeting industrial research) and top-down approaches (e.g. BIOTEK programmes). Moreover, these biotechnology initiatives traditionally targeted research collaboration between university and industry actors (Assouline, 1999).

With regard to the public acceptance of biotechnology research and applications, according to a survey of public attitudes to new technologies in Europe (European Commission, 2005b), 72 per cent of Danish respondents believe that developments in biotechnology and genetic engineering can positively affect our way of life over the next 20 years.

The Eurobarometer survey explores public attitudes to several applications of genetics. Interestingly, in most cases the percentage of Danish citizens who would never approve these applications is higher than the EU average:

- 67 per cent would never approve of the use of genetic tests on children to identify talents and weaknesses (EU average 54 per cent);
- 57 per cent would never approve of the use of genetic treatments to prolong our expected life span by 25 years (EU average 42 per cent);
- 45 per cent would never approve of developing genetic tests for everyone to identify diseases they might contract (EU average 34 per cent);
- 40 per cent would never approve of genetic treatments to get rid of bad habits like smoking or alcoholism (EU average 33 per cent);
- 26 per cent are totally opposed to using genetic testing to produce a child that could be a bone-marrow donor for a sibling with a life-threatening disease (EU average 31 per cent);
- 12 per cent are totally opposed to storing everyone's genetic data so that criminals could be caught more easily (EU average 21 per cent);
- 16 per cent would never approve of storing the population's genetic data to study the causes of human disease (EU average 17 per cent).

The Eurobarometer survey also explores public attitudes towards genetic modification. In this application field Danish citizens are not as critical as the EU average when it comes to the development of genetically modified crops to increase the variety of regionally grown food: 35 per cent of the citizens would never approve of this application of biotechnology (EU average 37 per cent).

In the case of environmental applications, Danish citizens are more open to biotechnology applications than the EU average. Only 11 per cent of the respondents would never approve of developing genetically modified bacteria to clean up after environmental catastrophes (EU average 19 per cent).

The legal framework conditions pertaining to traceability, labelling and marketing of biotech products are regulated in Denmark in accordance with EU regulations. Moreover, Denmark is a strong proponent of common EU regulations on coexistence. In June 2004, the Danish parliament passed legislation on the coexistence of biotechnology and non-biotechnology crops (including organic agriculture) (GAIN, 2006).

Stem cell research and cloning are regulated by the Act on Artificial Fertilisation (10.6.1997/460). Stem cell research is only allowed on surplus embryos, while reproductive cloning is not allowed (Norden, 2006).

**Germany**

As early as the 1970s, the federal government was showing its concern about developing the biotechnology knowledge base. Since the mid-1980s, three thematic framework programmes have been launched by the federal government, explicitly targeting the development and industrial application of biotechnology in Germany:

- Applied biology and biotechnology (launched in 1985);
- Biotechnology 2000 (launched in 1991);
- Biotechnology Framework Programme (launched in 2000).

Accordingly, national support for biotechnology research has been very much determined by these biotechnology-specific programmes. The last two programmes involved a wide range of instruments targeting biotechnology-related activities, such as collaborative research between industry and academia and network formation. Moreover, the initiative BioRegio, launched by the federal government in 1995 within the framework of the Biotechnology 2000 Programme, prepared the ground for the establishment of regional biotechnology clusters in Germany. This policy instrument aimed to develop the regional potentials in biotechnology, to establish biotechnology companies and networks exploiting regional knowledge and regional industrial resources. The success of the initiative raised great awareness of the potential of biotechnology among regional policy-makers. According to Giessler and Reiss (1999), besides the measures at the federal level, the regional policies designed and implemented by the 16 *Länder* in the period 1994–98 included a number of measures promoting biotechnology. All in all, the support for biotechnology in Germany has been an important priority for policy-makers in the last two decades.

According to a survey of public attitudes towards new technologies in Europe (European Commission, 2005b), in Germany 65 per cent of the respondents believe that development in biotechnology and genetic engineering can positively affect our way of life over the next 20 years. This exactly matches the European average. However, in terms of specific applications, German citizens are quite distrustful of the use and development of biotechnology.

The survey explores public attitudes to several applications of genetics. Interestingly, in every case the percentage of German citizens who would never approve these applications is considerably higher than the EU average:

- 60 per cent would never approve the use of genetic tests for children to identify talents and weaknesses (EU average 54 per cent);
- 60 per cent would never approve the use of genetic treatments to prolong our expected life span by 25 years (EU average 42 per cent);
- 43 per cent would never approve the development of genetic tests for everyone to identify diseases they might contract (EU average 34 per cent);
- 38 per cent would never approve genetic treatments to dispense with bad habits like smoking or alcoholism (EU average 33 per cent);
- 45 per cent are totally opposed to using genetic testing to produce a child that could be a bone-marrow donor for a sibling with a life-threatening disease (EU average 31 per cent);
- 16 per cent are totally opposed to storing everyone's genetic data so that criminals could be caught more easily (EU average 21 per cent);
- 31 per cent would never approve of storing the population's genetic data to study the causes of human disease (EU average 17 per cent).

The survey also explores public attitudes to genetic modification. Germany is stricter than the EU average when it comes to the development of genetically modified crops to increase the variety of regionally grown food: 43 per cent of citizens would never approve this application of biotechnology (EU average 37 per cent). This critical attitude is persistent in the case of environmental applications: 24 per cent of respondents would never approve of developing genetically modified bacteria to clean up after environmental catastrophes (EU average 19 per cent).

In general terms, the results of the survey suggest a low acceptance for most potential applications of biotechnology and genetic engineering.

**Latvia**

Latvia has a long tradition in chemical research, which is also relevant for biotechnology. Further, biotechnology has been identified as a national research priority since 2002. In particular, a focus of biotechnology research should be on biotechnology, biomedicine and organic synthesis, with a particular emphasis on gene therapy and new technologies for the synthesis of biological active compounds (Bundule, 2006). So far, however, there is no specific biotechnology research programme in Latvia. Rather, there is mainly project funding for various biotechnology activities in the framework of generic measures.

An important initiative in Latvia is the Latvian Genome Programme, which was established in 2001 (Pīrāgs and Grēns, 2002). The goal of the programme is to establish a genome database of the Latvian population and to compare genomic data with clinical information. The long-term goal is to develop individual therapies based on genetic characteristics of patients. The project is coordinated by the Biomedical Research and Study Centre at the University of Latvia. All in all, the project is planned for a period of ten years. So far, funding comes from the Latvian Council of Science. In addition, it is planned to use EU structural funds in particular for building up the required advanced infrastructure and equipment for sequencing and data processing/storing activities (Antonovs, 2005).

Finally, it is worth mentioning that Latvia is also a member of the ScanBalt, which comprises a network of networks in the Scandinavian region focusing on the advancement of biotechnology in that region (see www.scanbalt.org).

Information about the public perception of biotechnology in Latvia can be derived from the latest Eurobarometer survey 64.3 from 2005 (Gaskell et al., 2006), which for the first time included the new member states. Among the new member states, Latvia belongs to the more sceptical countries. For example, the optimism about the potential of biotechnology to improve our way of life in the future is rather modest in Latvia. Among the ten new member states, Latvia ranks in sixth place, with an index value of 60. The most optimistic member states are Malta (index 81) and Estonia (index 79). In line with this rather sceptical attitude, the support for four specific biotechnologies (nanotechnology, pharmacogenetics, gene therapy, GM foods) is rather low in Latvia. The country ranks 22nd among the EU25, clearly below the European average. Support is particularly low for GM food where Latvia, together with Cyprus and Greece, is most sceptical; also the approval of embryonic stem cell research is comparatively low in Latvia. In contrast to this sceptical attitude towards biotechnology, confidence and trust in the governance of science is rather high in Latvia. For example, a clear majority, which is above the European average, favours the governance of science and technology according to the principle of scientific delegation. Such a system is based primarily on scientific evidence about the risks and benefits involved and relies mainly on the advice of experts. Further, public confidence in the key actors and institutions involved in biotechnology is also rather high in Latvia. This includes confidence in university scientists, scientists and industry, the European Union, the industry in general and the government.

**Russia**

Biotechnology is generally considered to be among the so-called 'technologies of the future'; it is one of the major priority areas regarding development of science and technology in Russia, designated 'Live Systems'. It should be noted that this priority area (the list of which is reviewed every four years by means of expert analysis) currently includes bioengineering technologies; cellular technologies; biomedical and veterinary technologies for life support and protection of people and animals; genome and post-genome pharmaceutical technologies; biocatalysis, biosynthetic and biosensor technologies; and bio-information technologies.

A number of programmes to support the development of priority biotechnologies are being implemented in Russia. For instance, the Federal Target S&T Programme 'Priority S&T Development Areas for 2002–2006' included funding of research in such areas as human genome, biological diversity, gene diagnostics, gene therapy and so on. The Federal Target Programme 'National Technological Basis in 2002–2006' also included a section to support development of biotechnology. Another Federal Target Programme, 'Priority Development Areas for the Russian S&T Complex in 2007–2012', also includes funding of biotechnology research and development.

A comparison of data obtained during a representative polling of the adult Russian population, conducted by ISSEK of SU-HSE in 2006 and 2007, reveals that in 2006 respondents have been much less enthusiastic about new technologies than in 2007; compared with similar surveys in European countries (European Commission, 2005b), Russians generally demonstrated higher loyalty and trust in scientific progress.

Thus in 2006 the share of positive expectations regarding improvement of living standards in the next 20 years by development and implementation of biotechnologies amounted to 40 per cent; by use of stem cells, to 39 per cent; and through introduction of genetically modified products, only to 9 per cent of the respondents. Note that the population feels deeply prejudiced about the latter: 40 per cent of the respondents had negative expectations about the consequences of introduction of such products. Note also that negative expectations regarding use of biotechnologies were more often expressed by older respondents, by people with higher-level education and by residents of large cities; negative expectations regarding stem cell research were more often expressed by Muscovites and respondents with higher education; regarding use of genetically modified products, by middle-aged respondents, people with higher levels of education and belonging to higher income groups, as well as residents of large cities, especially Moscow and St Petersburg.

It should be stressed that the attitude towards genetically modified products serves not so much as an indicator of the population's attitudes towards innovations as it shows the influence of mass media on the public opinion. Despite the fact that most of the respondents (77 per cent) didn't know what genes are (much less what genetically modified products were), 79 per cent of the respondents felt able to make a decision regarding whether or not they should use them. Two-thirds of the respondents wouldn't plant genetically modified vegetables in their land plots (gardens), despite the scientists' assurances that genetic modification only served to improve storage potential, protect vegetables from diseases and make them taste better. Just 1 per cent of the respondents were ready to trust the scientists completely and plant genetically modified varieties, while 13 per cent were willing to test the scientists' assertions personally by planting common and genetically modified plants alongside each other (these were mostly rural residents). At the same time researchers established that every second European (54 per cent of those surveyed) believed genetically modified products to be dangerous, and only one in every seven (14 per cent) didn't think so (European Commission, 2005b).

It should be noted though, that during the 2007 survey Russians gave more tolerant answers and showed more trust regarding acceptability of practical implementation of genetic research; this shift may have been due to the fact that by then mass media had largely switched to covering nanotechnology development. As for the actual results of the survey, the following were among the findings.

- genetic testing to determine susceptibility to various diseases including incurable ones (testing without any limitations whatsoever; only under public or government supervision; or only in exceptional cases) was deemed acceptable by 32, 40 and 15 per cent of respondents, respectively;

- use of genetic techniques to cure harmful addictions such as smoking or alcoholism was deemed acceptable by 31, 39 and 17 per cent of respondents, respectively;
- establishment of a genetic data bank to research genetic causes of diseases was deemed acceptable by 26, 44 and 15 per cent of respondents, respectively;
- creation of genetically modified bacteria capable of cleaning up the environment after ecological disasters was deemed acceptable by 26, 42 and 16 per cent of respondents, respectively;
- genetic treatment techniques which would allow life expectancy to be extended for more than 25 years were deemed acceptable by 25, 28 and 18 per cent of respondents, respectively;
- maintaining genetic data banks in order to facilitate catching criminals was deemed acceptable by 21, 43 and 17 per cent of respondents, respectively;
- genetic testing of children in order to identify their talents and shortcomings was deemed acceptable by 14, 28 and 23 per cent of respondents, respectively;
- development and dissemination of genetically modified grains with high crop capacity, highly resistant to diseases and with other superior qualities was deemed acceptable by 13, 31 and 16 per cent of respondents, respectively;
- genetic experiments to give birth to a donor child for an incurably sick brother or sister were deemed acceptable by 12, 27 and 23 per cent of respondents, respectively.

Also, changing the nature and wording of questions, switching from generalized and, to many of the respondents, abstract concepts such as 'biotechnology' and 'stem cells' to specific examples of practical application of such technologies and their useful results, resulted in a more favourable rating of the importance of developing biotechnology and nanotechnology in Russia. In particular, the vast majority of respondents accepted that the following technologies were to some degree important (very important or rather important);

- understanding mechanisms of normal cells turning into cancerous cells (very important: 52 per cent and rather important: 32 per cent of the respondents, respectively);
- treatment of common diseases (e.g. heart attacks, apoplexy) with

stem cells taken from the patients themselves or from their relatives (48 and 33 per cent of the respondents, respectively);
- transplanting nerve stem cells from the patient's spinal marrow to help recovery from paralysis (45 and 34 per cent of the respondents, respectively);
- miniature sensors to monitor critical medical parameters (40 and 34 per cent of the respondents, respectively);
- nano-preparations for targeted delivery of drugs to sick organs, e.g. to treat cancer with minimum side effects (48 and 33 per cent of the respondents, respectively);
- nanobiosensors (nanodevices) to identify viruses, bacteria and malignant cells (46 and 34 per cent of the respondents, respectively);
- implanting microchips into the brain of deaf and blind people to restore their hearing and sight (46 and 33 per cent of the respondents, respectively);
- nanomaterials to replace damaged tissues in human organs, e.g. cartilage, bones, skin (45 and 35 per cent of the respondents, respectively);
- understanding mechanisms of ageing and finding ways to prolong life (40 and 33 per cent of the respondents, respectively).

**Sweden**

Since 2001, the creation of The Swedish Governmental Agency for Innovation Systems (VINNOVA) has provided the main policy support for biotechnology. While the overall mission of VINNOVA is to promote sustainable growth by means of problem-oriented research and effective innovation systems, VINNOVA has established a strategic plan for the period 2003–07 in which 18 priority growth areas have been identified. A growth area is defined by VINNOVA as a sectoral innovation system in which initiatives can have a significant impact on growth (VINNOVA, 2002). Of these priority growth areas, at least five are intimately linked to the field of biotechnology, while the others correspond to ICT, services, manufacturing, materials and transport. All VINNOVA's initiatives, regardless of whether they involve the development of national or regional innovation systems, focus primarily on the prioritized growth areas.

The relevance of biotechnology within VINNOVA's strategic plan is also shown by its Biotechnology Department, in charge of initiatives in the area of life sciences and biotechnology. In addition, VINNOVA explicitly considers biotechnology to be a knowledge platform, since it is a generic technical field that is expected to play an important role in a number of growth areas and industrial sectors.

The five prioritized growth areas in relation to biotechnology are indicated below. VINNOVA's initiatives in all five areas are focused on problem-oriented research that brings together medicine, biology, IT and engineering, and in all cases target the development and growth of small and medium-sized R&D-intensive companies.

- *Pharmaceuticals and diagnostics* initiatives focus on the development of new analytical tools for exploiting post-genomics, the development of strategies and structures for bioinformatics and systems biology, initiatives in medicine technology, and initiatives to stimulate growth in small and medium-sized enterprises.
- *Biotech supply – biotechnologies for research and production* initiatives focus on the development of bio-analytical methods, bioinformatics, IT integration and new technologies for biotechnical production ranging from the research level to industrial scale.
- *Biomedical engineering* initiatives focus on stimulating R&D cross-cooperation between scientific disciplines within the fields of implant biomaterials and tissue engineering, and to promote and encourage new developments in biomedical technology linked to IT and needs in extended home health care.
- *Innovation in foods* initiatives focus on the promotion of cross-disciplinary scientific research aimed at improving the use of medical, biological and biotechnological knowledge in the food industry and food research institutions.
- *Green materials from renewable resources* initiatives focus on stimulating multi-disciplinary research in order to establish platforms within the area of composite and fibre technology, enzyme technology and biopolymers.

VINNOVA also considers that strong regional innovation environments and close cooperation between the research community, business and government are essential prerequisites for sustainable national growth. Consequently, VINNOVA has launched a programme – VINNVÄXT – to promote growth in a number of regions and is committed to give support to a new generation of Competence Centres, VINN Excellence Centres.

A survey of attitudes to new technologies in Europe Eurobarometer (2005) found that the Swedish public is slightly more receptive (70 per cent) than the EU25 average (65 per cent) in expectations about the positive effects of biotechnology and genetic engineering on our way of life in the next 20 years. However, the Swedish public is less receptive to most applications of cloning than the EU25 average, except for cloning human stem cells to make cells and organs to treat people with diseases. Sweden would approve it if highly regulated and controlled (55 per cent), compared to the EU25 average of 41 per cent. It also had a higher disapproval rate than the EU25 average for various applications of genetics (including genetic testing, genetic treatments or storing the population's genetic data to aid either catching criminals or to study the causes of disease).

In Sweden, surplus embryos from fertility projects may be used for human embryonic stem (HES) cell research, though it requires explicit consent from the gamete donors, as well as regulatory approval. Sweden forbids the creation of embryos for research purposes, including the procurement of HES cells. However, in Sweden, long-held policy positions with respect to embryo research are being debated and new regulatory regimes may contemplate permission for HES cell research to move forward. For instance, in early 2003, the Committee on Genetic Integrity published a consultation document that included the recommendation that the creation of embryos by fertilization or by somatic cell nuclear transfer (SCNT) for research purposes would not be specifically prohibited, though it recommended that reproductive cloning be unequivocally banned. In short, Sweden, where more liberal policies are being instituted, is likely to join the UK and Israel in terms of a permissive regulatory environment for HES cell research (Knowles, 2004).

## Comparison of Public Funding Activities

In this section funding activities of the countries under review will be compared using data obtained in the BIOPOLIS project (Enzing et al., 2007). Since so far such data has not been compiled for Russia, the comparison is restricted to four countries.

### Public biotechnology budgets 2002–05

In the BIOPOLIS project information on biotechnology funding was collected for the period 2002–05. Table 9.1 summarizes information on total biotechnology budgets and information on generic funding and regional funding.

By far the largest public biotechnology budget was provided in Germany, followed by Denmark and Sweden. This is not only the case for the absolute funding budget but also for the specific budget which has been corrected for the size of the countries considered. While in Denmark, Latvia and Sweden so-called generic funding instruments, which do not have a specific biotechnology focus, but which are allocated to biotechnology after

*Table 9.1 Public biotechnology budgets in the period 2002–05*

| | Total budget (mio. €) | Specific budget: (€/capita) | Share of generic instruments [1] ( per cent) | Share of regional funding ( per cent) |
|---|---|---|---|---|
| Denmark | 166.0 | 30.7 | 77 | 0 |
| Germany | 4575.0 | 55.5 | 21 | 35 |
| Latvia | 0.7 | 0.3 | 100 | 0 |
| Sweden | 145.5 | 16.3 | 72 | 0 |

*Note:* 1. relates to policy-directed funding; we differentiate between biotechnology-specific instruments and generic instruments which have no specific technology focus

*Source:* BIOPOLIS project (Enzing et al., 2007).

bottom-up procedures are more important; in Germany biotechnology-specific funding instruments contribute the largest share to the budget as a whole. In addition, regional funding is rather important in Germany. All 16 German states (*Länder*) have implemented some sort of biotechnology funding, which in total contributes almost one third to the total budget.

**Coverage of policy goals 2002–05 as indicated by share of policy-directed funding**

Biotechnology funding can be differentiated according to specific policy goals. In the BIOPOLIS project budgets were classified according to ten different policy goals (see Table 9.2).

Biotechnology funding in Latvia mainly concerns two policy goals: to support a high level of biotechnology research and to facilitate knowledge flow and collaboration among scientific disciplines. Thus, mainly creating a science base is in the focus of respective policy initiatives. The policy goals followed in Denmark, Germany and Sweden have many common items, such as support of a high level of biotechnology research, and support of industry-oriented research or knowledge transfer, just to name a few. However, there are also some interesting differences in the respective funding profiles. Firstly, in Denmark and to a lesser extent in Sweden, there is also a strong focus on support of the adoption of biotechnology for new industrial applications. Germany, on the other hand, is the only country where supporting firm creation receives a considerable share of the public funding budgets. Sweden and Denmark have instruments in place to support business investment in R&D. Taking care of the availability of human

*Table 9.2 Allocation of biotechnology funding between 2002 and 2005 to ten different policy goals (per cent)*

| Policy goals | 1 | 2 | 3 | 4 | 5 | 6 | 7 | 8 | 9 | 10 |
|---|---|---|---|---|---|---|---|---|---|---|
| Denmark | 23 | 27 | 5 | 0 | 18 | 11 | 0 | 4 | 11 | 0 |
| Germany | 28 | 22 | 6 | 3 | 17 | 2 | 15 | 0.2 | 4 | 3 |
| Latvia | 80 | 0 | 20 | 0 | 0 | 0 | 0 | 0 | 0 | 0 |
| Sweden | 28 | 12 | 10 | 18 | 10 | 9 | 1 | 0.1 | 11 | 2 |

*Notes:*
Policy goals:
1 High level of biotechnology research
2 High level of industry-oriented (and applied) research
3 Knowledge flow and collaboration among scientific disciplines
4 Availability of human resources
5 Transfer of knowledge from academia to industry and its application to industrial resources
6 The adoption of biotechnology for new industrial applications
7 Firm creation
8 Social acceptance of biotechnology
9 Business investment in R&D
10 Bio-safety, risk assessment

*Source:* BIOPOLIS project (Enzing et al., 2007)

resources is another policy goal receiving considerable funding shares in Sweden.

**Thematic priorities of biotech funding in terms of budget allocation to different biotechnology areas between 2002 and 2005**

In the BIOPOLIS project (Enzing et al., 2007) nine thematic priorities have been defined for biotechnology funding. Table 9.3 summarizes the allocation of biotechnology funding to these different areas.

All four countries have in common that the highest budget is allocated to health biotechnology. However, concerning the other thematic areas, the four countries exhibit different specialization. In the case of Latvia health biotechnology, plant biotechnology, food biotechnology, and basic biotechnology are equally important. Denmark is different from all other countries with respect to support of animal biotechnology, which comprises more than one fifth of the total budget allocation. In Germany, besides the strong focus on health biotechnology, the support for industrial biotechnology is relatively high compared to all other countries. In the case of Sweden a rather high significance of support for basic biotechnologies is striking.

*Table 9.3 Thematic priorities of biotechnology funding in terms of budget allocation (per cent)*

| Thematic priorities | 1 | 2 | 3 | 4 | 5 | 6 | 7 | 8 | 9 |
|---|---|---|---|---|---|---|---|---|---|
| Denmark | 15 | 22 | 0 | 35 | 10 | 11 | 7 | 0 | 0 |
| Germany | 9 | 1 | 0 | 52 | 2 | 26 | 0 | 3 | 7 |
| Latvia | 25 | 0 | 0 | 25 | 25 | 0 | 25 | 0 | 0 |
| Sweden | 10 | 1 | 8 | 39 | 5 | 15 | 20 | 2 | 0 |

*Notes:*
Biotechnology areas:
1 Plant biotechnology
2 Animal biotechnology
3 Environmental biotechnology
4 Health biotechnology
5 Food biotechnology
6 Industrial biotechnology
7 Basic biotechnology
8 Ethical, legal, social aspects of biotechnology
9 General

*Source:* BIOPOLIS project (Enzing et al., 2007)

## COUNTRY SPECIALIZATION AND BIOTECHNOLOGY INDUSTRY CHARACTERISTICS

In the following section we will first provide some information on the biotechnology profiles of the five European countries participating in the project. The analysis will be based on patent indicators. In the second part of this chapter the biotechnology industry in the five countries will be characterized using publicly available data. This description will draw largely on the information provided by the BIO4EU project (Reiss et al., 2007).

### Biotechnology Profiles in Different Countries

#### Approach

The basis for deriving biotechnology profiles is an analysis of the number of patent applications in different sectors which are defined by sub-classes of the International Patent Classification (IPC). The IPC primarily describes technologies and products, but not sectors. However, it is possible to define sectors by their major products. In the real world, enterprises often manufacture products which, in principle, do not belong to other sectors. So the assignment of products/technologies to sectors is not perfect. Nevertheless, the definition of sectors by IPC codes is possible with reference to broader samples such as all enterprises of a country.

A classification with 19 sectors, with a focus on technology-intensive sectors, was used. Furthermore, so-called transnational patent applications which comprise International Applications (PCT applications) and European Applications were examined. As many International Applications are transferred to the European Patent Office, it is necessary to exclude double counting.

Figure 9.1 shows the quantitative distribution of all transnational applications on the 19 sectors for the period of 2003 to 2005, in terms of years of first application. We analysed the three-year period 2003 to 2005 for achieving sufficiently large samples of applications per sector, disaggregated by countries. The computation was made in the online database WPINDEX of the host STN. The number of applications per sector is quite different, with telecommunications and pharmaceuticals the largest and machine tools the smallest. The consumer goods sector proves to be quite large, although consumer goods are generally less technology-intensive. In this sector a broad variety of different sub-sectors is comprised, covering games as well as furniture. The search was made by all IPC symbols of the applications considered. It is possible to classify an application by more than one IPC code, so that a certain overlap between the sectors exists.

In addition to the sectors we searched for the total number of biotechnology-linked applications (total biotechnology). Furthermore we computed the applications with no reference to a specific sector separately (basic biotechnology). According to these analyses, most biotechnology applications are associated to specific sectors. Only 19 per cent of all biotechnology applications can be qualified as basic ones.

As a next step, patent applications in the different sectors with co-classification in biotechnology were retrieved. On this basis the share of biotechnology applications in each sector was determined, as shown in Figure 9.2. In order to avoid statistical artefacts in the following analyses, the description of biotechnology profiles was focused on sectors with a minimum share of biotechnology applications of 1.8 per cent.

The share of biotechnology in 'Computers' with 2.9 per cent seems to be quite high. The patent applications primarily refer to special computer programs analysing and conceiving large biotechnology structures (biocomputing). In the sector 'measurement' (23.8 per cent), the majority of patent applications concerns the analysis of blood for medical purposes.

In the sector for medical equipment, a variety of different fields, for example introduction of medical substances into the body, or artificial organs including tissue engineering, are considered. In the case of basic

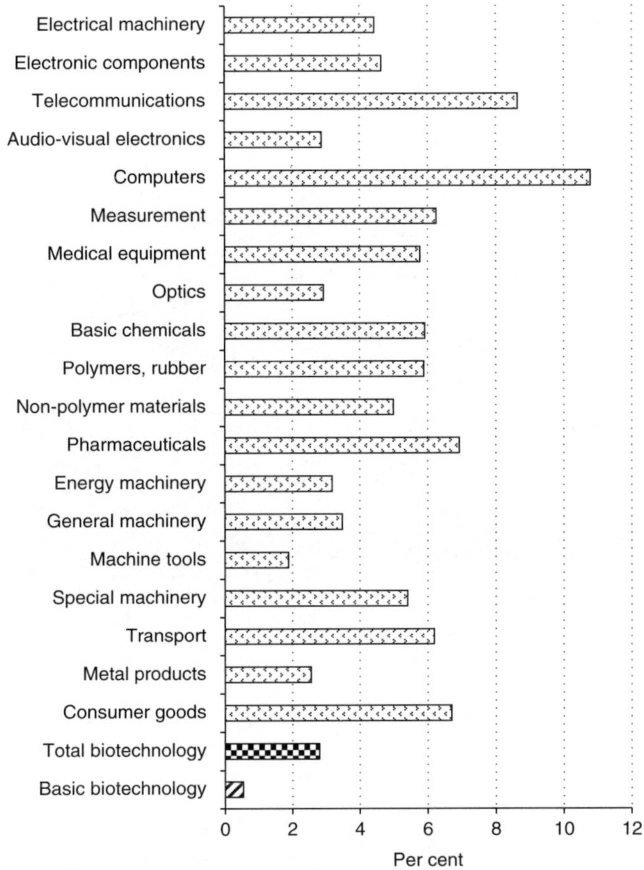


*Source:* WPINDEX (STN), computation by Fraunhofer ISI

*Figure 9.1 Distribution of all transnational patent applications on sectors, 2003–05*

chemistry and polymers, biotechnology primarily refers to the improvement of processes. As expected, the largest share of biotechnology applications can be found in pharmaceuticals (38.1 per cent). As to special machinery, biotechnology is again used for the support of specific processes, for instance in the production of food. The rather high share of biotechnology in consumer goods is primarily due to the classification of new plants in this sector.

A disadvantage of the sector approach for the present purpose is that typical fields of application, in particular water cleaning and pesticides and

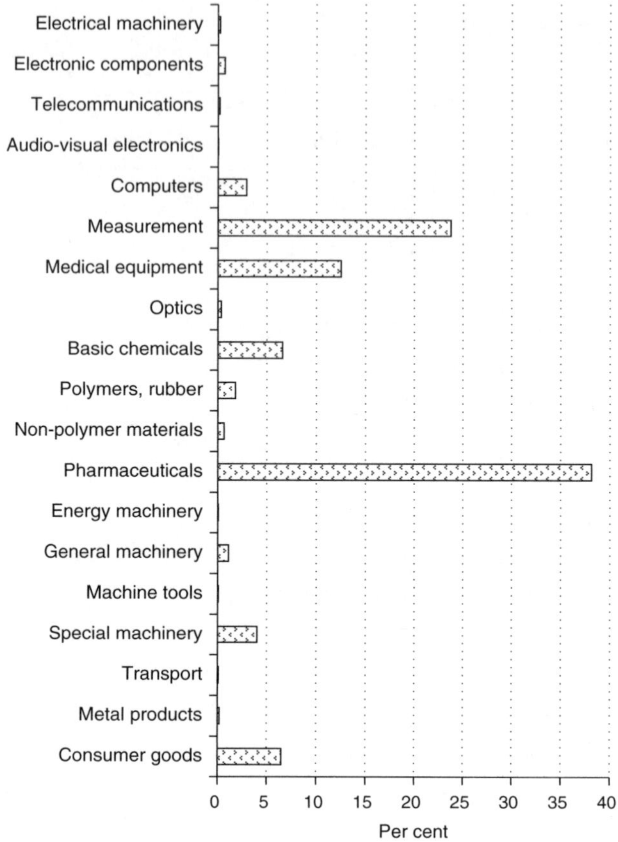


*Source:* WPINDEX (STN), computation by Fraunhofer ISI

*Figure 9.2 Share of biotechnology applications in different sectors for all transnational applications, 2003–05*

herbicides are not displayed separately, but as part of basic chemistry. The major advantage is that it is possible to compare the profile in biotechnology with the general country profile.

For the country profiling, specialization indices were used. Such an index is the RPA (Revealed Patent Advantage). It is defined as follows:

$$RPA_{ij} = 100 \tanh \ln [(Pat_{ij} / \Sigma_i Pat_{ij}) / (\Sigma_j Pat_{ij} / \Sigma_{ij} Pat_{ij})]$$

Therein, *i* refers to the country, and *j* to the sector. The index compares the share of a country in a specific sector to the general share of this sector

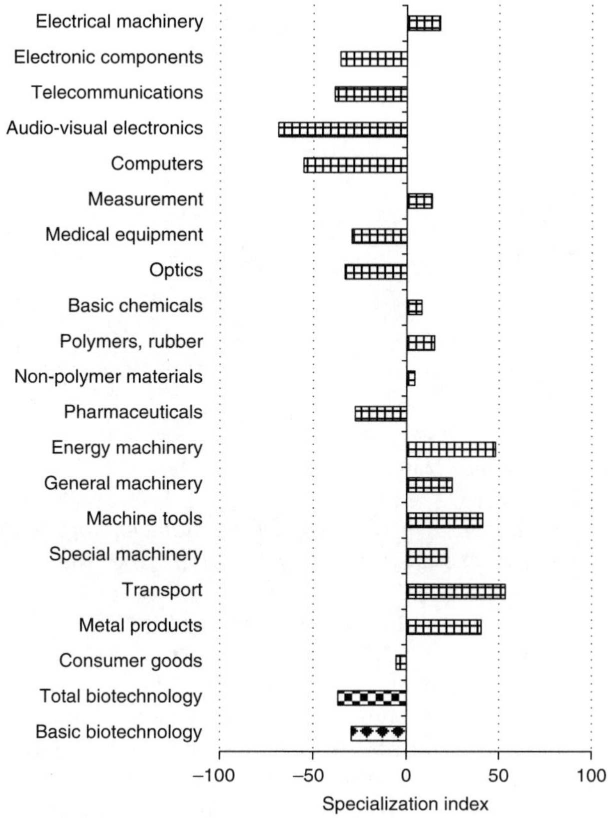


*Source:* WPINDEX (STN), computation by Fraunhofer ISI

*Figure 9.3 Specialization profile of Germany as to sectors, calculated on the basis of transnational applications, 2003–05*

in the total patent applications for all countries. Thus, if this share is higher than the worldwide average, we assume a specialization above average of this country in this sector. The index is conceived in a way that the indices for specialization above average are positive, and below average are negative. The maximum and minimum values are +100 or −100.

**Country profiles**

*Germany* The first part of a country analysis is the calculation of a specialization profile for all sectors, and in addition for total and basic biotechnology. The graph for Germany is depicted in Figure 9.3. The

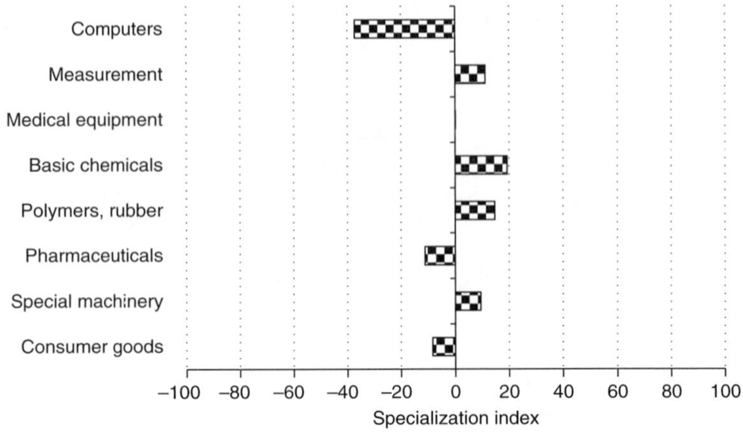


*Source:* WPINDEX (STN), computation by Fraunhofer ISI

*Figure 9.4 Specialization of Germany within biotechnology, calculated on the basis of transnational applications, 2003–05*

figure shows a distinct orientation on all fields of mechanical engineering, whereas in electrical engineering Germany is specialized below average, with the major exception of electrical machinery. The specialization is very low in pharmaceuticals, the major application field of biotechnology. In consequence the specialization in total biotechnology as well as in basic biotechnology is clearly below the world average.

It is important to see that specialization indices are relative measures. In Germany the absolute level of transnational applications is very high, with 334 applications per year and 1 million inhabitants. So even in fields with low specialization, the activity is still quite relevant.

The second step is the calculation of the specialization within biotechnology (Figure 9.4), so the references are the German and worldwide totals in biotechnology. According to this perspective, Germany is less specialized in computers and pharmaceuticals, and above average in measurement, basic chemicals, polymers and special machinery. The specialization in basic chemicals and polymers supports the notion that Germany has a good position in the field of white biotechnology, which refers to the use of biotechnology for improving/modifying chemical processes. So the specialization in biotechnology reflects the general specialization of Germany.

*Sweden* In Sweden the total level of transnational applications is at a level of 326 annual applications per 1 million inhabitants.

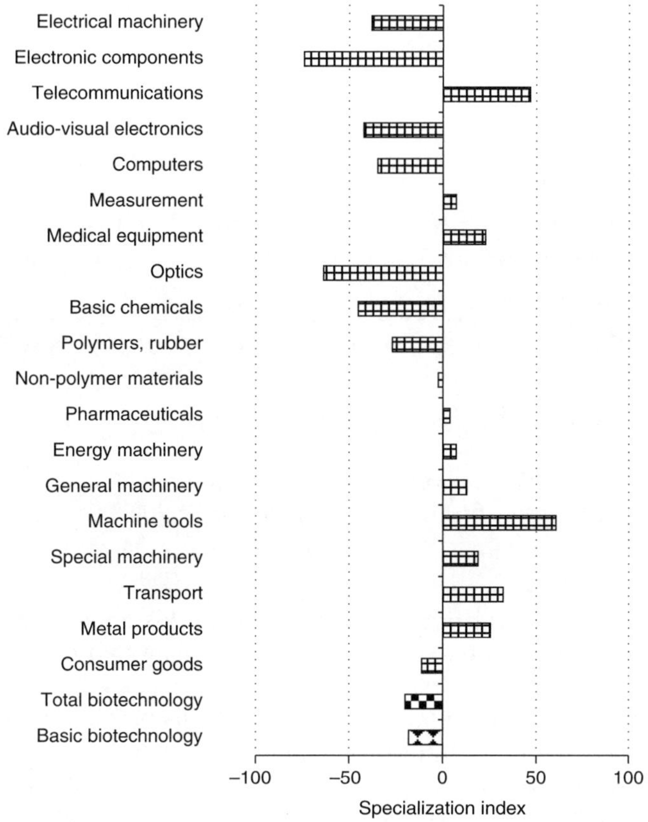


*Source:* WPINDEX (STN), computation by Fraunhofer ISI

*Figure 9.5 Specialization profile of Sweden as to sectors, calculated on the basis of transnational applications, 2003–05*

In the general profile, a strong specialization in telecommunications and machine tools can be observed and still clear specializations in special machines, transport and metal products (Figure 9.5).

The specialization in biotechnology does not reflect directly this overall specialization (Figure 9.6). But in measurement, pharmaceuticals and consumer goods, a specialization in biotechnology at an average level takes up average specializations in these sectors in the overall profile. The orientation of biotechnology activities on basic chemicals and consumer goods indicates a special orientation of biotechnology on agriculture in addition to the use of biotechnology in pharmacy.

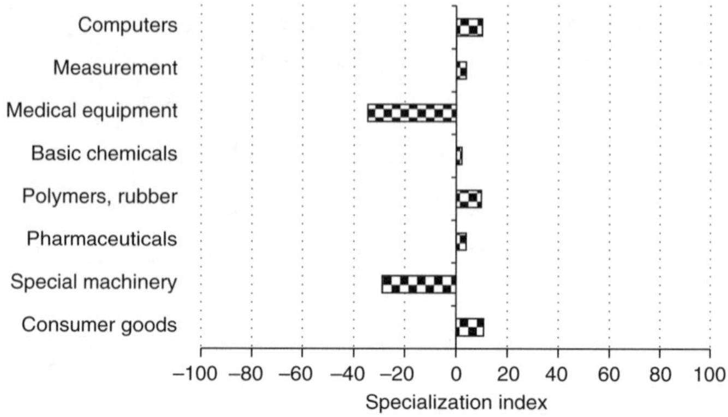


*Source:* WPINDEX (STN), computation by Fraunhofer ISI

*Figure 9.6 Specialization of Sweden within biotechnology, calculated on the basis of transnational applications, 2003–05*

*Denmark* The general level of Denmark is at 258 annual applications per 1 million inhabitants. This activity is lower than that of Germany or Sweden, but still quite high in an international comparison.

In the general profile (Figure 9.7), a clear focus on pharmaceuticals is visible, and in addition on medical equipment, energy machinery, metal products and consumer goods. Furthermore, the specialization on total and basic biotechnology is extremely high.

In the biotechnology profile (Figure 9.8), the specialization in pharmaceuticals is slightly above average, but the analysis primarily points to strong specializations in basic chemicals, which reflects Denmark's strong position in enzymes, and consumer goods, a sector with close linkage to agriculture.

*Russia* The general patent level of Russia is at 47 annual applications per 1 million inhabitants, a typical level for a catching-up country regarding international trade. In this context, it has to be noted that patents do not exclusively reflect technological strength, but are always linked to market expectations. So the technological strength of Russia in many fields is well known; however, the transfer of these competencies into marketable products can be improved.

In the overall profile (Figure 9.9), a high specialization in basic chemicals, non-polymer materials, general machinery and special machinery is obvious. In basic biotechnology, there is a strong specialization highly above average.

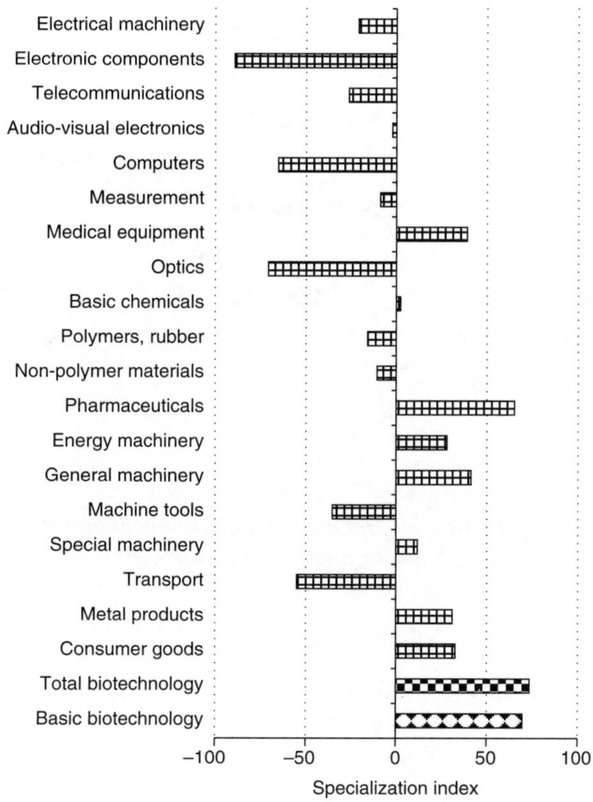


*Source:* WPINDEX (STN), computation by Fraunhofer ISI

*Figure 9.7 Specialization profile of Denmark as to sectors, calculated on the basis of transnational applications, 2003–05*

Within biotechnology (Figure 9.10), a distinct orientation on basic chemicals, polymers, pharmaceuticals and consumer goods emerges, so only in basic chemistry is an agreement with the general profile visible. However, the strong specialization in basic biotechnology shows that the research activities are still at an early, basic stage, and the linkage to specific sectors will come later.

*Latvia* The absolute level of Latvia is at 12 annual applications per 1 million inhabitants. This does not mean that there are no or very few technological activities. Rather, the low level indicates that the country is still in an early catching-up stage with a high relevance of imitation,

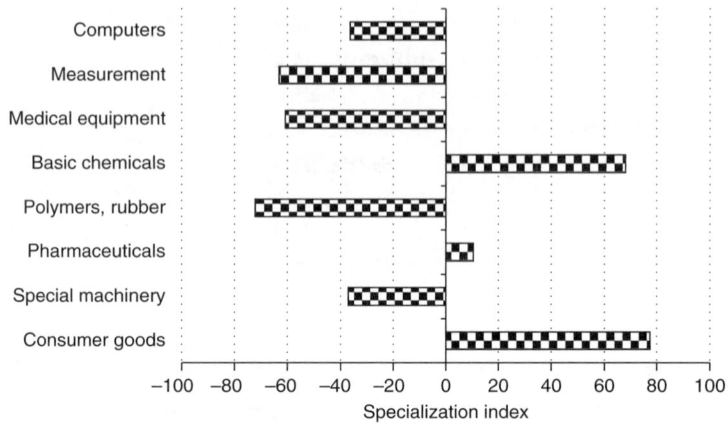


*Source:* WPINDEX (STN), computation by Fraunhofer ISI

*Figure 9.8 Specialization of Denmark within biotechnology, calculated on the basis of transnational applications, 2003–05*

and that the strict legal requirements for patents are generally not yet achieved.[7]

In the overall profile (Figure 9.11), a focus on non-polymer materials, pharmaceuticals and consumer goods becomes obvious, whereas the orientation on biotechnology is clearly below average. As the number of biotechnology patents is very low, the calculation of a meaningful specialization profile within biotechnology is not possible. However, the clear focus on pharmaceuticals in the overall profile indicates that Latvia has achieved an internationally competitive level. This suggests that the biotechnology activities should be linked to this sector.

## Comparison of Biotechnology Industry Characteristics

In the following section an overview of the biotechnology industry characteristics in the countries considered will be given. Unfortunately, comparable data from the same sources as used for Germany, Denmark and Sweden is not available for Latvia and Russia. Accordingly, the discussion will be restricted to the former three countries.

The following indicators will be used: The number of dedicated biotechnology firms (DBF) in absolute terms and in relative term adjusted by the size of the country. 'Dedicated biotechnology firms' are firms which have biotechnology as their main commercial activity. According to the OECD such firms are also called 'core biotech firms'. 'Bio-active

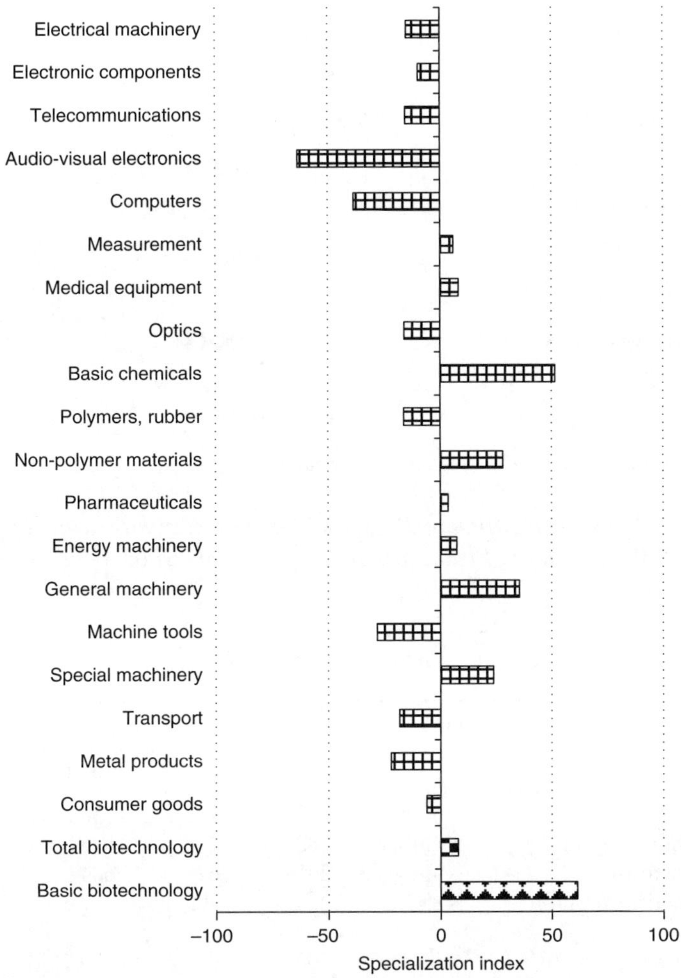


*Source:* WPINDEX (STN), computation by Fraunhofer ISI

*Figure 9.9 Specialization profile of Russia as to sectors, calculated on the basis of transnational applications, 2003–05*

firms', on the other hand, are all firms that report some activities in biotechnology. The scope of commercial and R&D activities of the firms considered is captured by the indicators 'capital raised', 'revenues', 'R&D expenditure' and 'employment', each calculated per firm (see Table 9.4).

Accordingly, in absolute numbers Germany has the strongest biotech

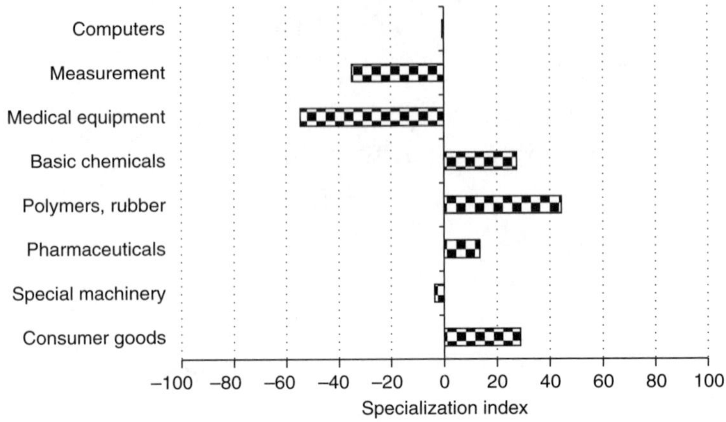


*Source:* WPINDEX (STN), computation by Fraunhofer ISI

*Figure 9.10 Specialization of Russia within biotechnology, calculated on the basis of transnational applications, 2003–05*

*Table 9.4 Biotechnology industry characteristics*

| Indicator | Unit | Year | Country | | |
|---|---|---|---|---|---|
| | | | Denmark | Germany | Sweden |
| DBF | Number | 2004 | 117 | 538 | 138 |
| DBF/million capita | Number | 2004 | 21.7 | 6.5 | 15.4 |
| Bioactive firms | Number | 2004 | 267 | 604 | 216 |
| Bioactive firms/million capita | Number | 2004 | 49.5 | 7.3 | 24.1 |
| Capital raised per DBF | Million € | 2004 | 1.10 | 0.75 | 0.28 |
| Revenue per DBF | Million € | 2004 | 46.1 | 5.4 | 6.19 |
| R&D expenditure per DBF | Million € | 2004 | 8.50 | 2.80 | 2.66 |
| Total employment in DBF | Number | 2004 | 18461 | 16094 | 3942 |

*Note:* DBF: Dedicated Biotechnology Firms

*Source:* BIO4EU project (Reiss et al., 2007; Ciritical I, 2006; OECD, 2006)

industry. However, considering the different sizes of the countries, Denmark is outstanding. In particular in terms of capital raised, revenues, R&D expenditure and total employment, Denmark is more advanced compared to Germany and Sweden. The large employment figure for

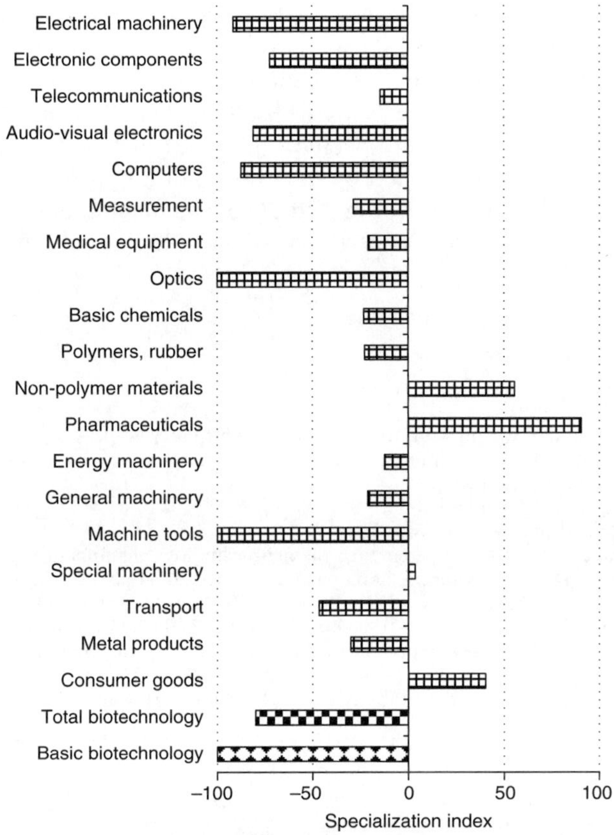


*Source:* WPINDEX (STN), computation by Fraunhofer ISI

*Figure 9.11 Specialization profile of Latvia as to sectors, calculated on the basis of transnational applications, 2003–05*

Denmark indicates that compared to the other countries some larger biotech firms are included in the sample, which also correlates with the figures for revenues and R&D expenditure.

Another interesting difference between Denmark on the one hand and Germany on the other, is the relation between bioactive firms and dedicated biotechnology firms. In Denmark and Sweden the relation between bioactive firms and dedicated biotechnology firms is much higher compared to Germany. This seems to indicate that a broader diffusion of biotechnology to firms, which besides biotechnology have also a number of other activities, has taken place in Denmark and Sweden.

Finally, the specific volume of capital raised per dedicated biotechnology firm is much lower in Sweden compared to Denmark and Germany. This seems to indicate that the private capital market is more difficult to access for biotechnology firms in Sweden. On the other hand, this difference might also reflect specific difficulties in Sweden in the year 2004. This notion is supported by the observation that the total volume of capital raised in the preceding year 2003 was much higher in Sweden, while in the other two countries a similar level could be achieved (Critical I, 2005).

## NOTES

1. This paragraph draws on data from Eurostat (see http://www.eurostat.org).
2. GERD/GDP: gross expenditure on research and development expressed as a percentage of gross domestic product.
3. Only data for 2004 available.
4. Eurostat (see http://www.eurostat.com).
5. For instance, by 2000 the percentage of high-technology manufacturing employment relative to total manufacturing employment was around 44 per cent in Sweden, above that of the US, Japan, France and the Netherlands (Marklund et al., 2004, p. 14).
6. For instance, over the period 2001–03 business sector R&D fell by nearly 10 per cent, but mainly occurred in the telecommunications sector. Other sectors maintained their R&D investments (Sandgren, 2005).
7. A basic requirement for patentability is novelty on a worldwide scale, not only at the national level.

## REFERENCES

Antonovs, Aleksandrs (2005), 'Development of Latvian genome database', paper presented at 'Shaping Best Practice' seminar, Magdeburg, 5–7 October, available at www.sachsen-annalt.de.

Assouline, G. (1999), 'National report of Denmark', in C.M. Enzing, J.N. Benedictus, E. Engelen-Smeets et al. (eds), *Inventory of Public Biotechnology R&D Programmes in Europe*, Brussels: DG Research – European Commission: DK3-DK43.

BMBF (2001), 'Rahmenprogramm Biotechnologie: Chancen nutzen und gestalten', Bonn, Berlin: Federal Ministry of Education and Research (BMBF).

BMFT (1991), 'Biotechnologie 2000: Programm der Bundesregierung', Bonn: Federal Ministry for Research and Technology (BMFT).

Bundule, Maija (2006), 'Report on the national research system in Latvia', available at http://portal.unesco.org/education/en/files/55162/11979839605 Latvia_National_Research_System_Report_EN.pdf/Latvia_National_Research_System_Report_EN.pdf.

Critical I (2005), 'Biotechnology in Europe: Critical I comparative study', Brussels: EuropaBio.

Critical I (2006), 'Biotechnology in Europe: Critical I comparative study', Brussels: EuropaBio.
Danish National Research Foundation (2005), 'Research and development in Denmark' working paper, available at http://www.dg.dk/filer/publikationer/Research%20and%20Development%20in%20DK%202005.pdf.
Enzing, C. et al. (2007), 'Inventory and analysis of national public policies that stimulate biotechnology research, its exploitation and commercialisation by industry in Europe in the period 2002–2005', final report, Luxembourg: Office for official publications of the European Communities.
European Commission (2002), 'Life sciences and biotechnology: a strategy for Europe', COM (2002) 27, Brussels: European Commission.
European Commission (2005a), 'European trend chart on innovation: annual Innovation Policy Trends and Appraisal Report: Latvia', Brussels: European Commission.
European Commission (2005b), 'Special Eurobarometer 225: social values, science and technology', available at http://europa.eu.int/comm/public_opinion/archives/ebs/ebs_225_report_en.pdf, accessed 1 August 2006.
European Commission (2007), 'Communication from the Commission to the Council, the European Parliament, the European Economic and Social Committee and the Committee of the Regions on the mid-term review of the strategy on life sciences and biotechnology', COM (2007) 175 final, Brussels: European Commission.
Eurostat (2006), 'General economic background and innovation indicators', available at http://epp.eurostat.ec.europa.eu, accessed October 2006.
GAIN (2006), 'Denmark biotechnology annual 2006', GAIN Report DA6008, available at http://www.fas.usda.gov/gainfiles/200607/146208201.pdf, accessed 1 October 2006
Gaskell, George, Agnes Allansdottir, Nick Allum, Cristina Corchero, Claude Fischler, Jürgen Hampel et al. (2006), 'Europeans and biotechnology in 2005: Patterns and Trends, Eurobarometer 64.3.
Giessler, S. and T. Reiss (1999), 'National report of Germany', C.M. Enzing, J.N. Benedictus, E. Engelen-Smeets et al. (eds), *Inventory of Public Biotechnology R&D Programmes in Europe*, Luxembourg: DG Research – European Commission
Indicators of Science (2007), 'Statistical handbook', Moscow: SU-HSE.
Indicators of Science (2008), 'Statistical handbook', Moscow: SU-HSE.
Indicators of Innovation Activities (2008), 'Statistical handbook', Moscow: SU-HSE.
Knowles, L.P. (2004), 'A regulatory patchwork? Human ES cell research oversight', *Nature Biotechnology*, **22**(2), 157–63.
Marklund, G., R. Nillson, P. Sandgren, J.G. Thorslund and J. Ullstrom (2004), 'The Swedish National Innovation System 1970–2003', Stockholm: VINNOVA, available at http://publiceng.vinnova.se/main.aspx?ID=4566479F-996C-400C-AC57-0ADA07F0F1F7, accessed 30 October 2006.
Ministry of Economics of the Republic of Latvia (2006), 'The national economy of Latvia: a macroeconomic review', available at http://www.mfa.gov.lv/data/file/en/plitem_file_13777_no27.pdf.
Ministry of Industry, Employment and Communications and Ministry of Education and Science (2004), 'Innovative Sweden: a strategy for growth through renewal', Stockholm: Ministry of Industry, Employment and Communications.

Norden (2006), 'Legislation on biotechnology in the Nordic countries: an overview', TemaNord 2006:506, available at http://www.norden.org/pub/ovrigt/ovrigt/uk/TN2006506.pdf, accessed 1 October 2006.

OECD (2005), 'Main science and technology indicators', vol. 2005/1, Paris: OECD.

OECD (2008), 'Main science and technology indicators', vol. 2008/1, Paris: OECD.

Patel, P., A. Arundel and M. Hopkins (2008), Sector report 'Monitoring, analysing trends and identifying challenges in biotechnology', Brussels: Europe Innova.

Pīrāgs, Valdis and Elmars Grēns (2002), 'Latvian genome project: unified genome database of the Latvian population', available at http://bmc.biomed.lu.lv/gene/print/Latvian%20Genome%20Project-raksts%20Judith%20Sandor.doc.

Reiss, T. et al. (2007), 'Consequences, opportunities and challenges of modern biotechnology for Europe (Bio4EU): Task 1 and 2', Brussels: European Techno-Economic Policy Support Network (ETEPS Net).

Russian Statistical Yearbook (2007), Moscow: Rosstat.

Sandgren, P. (2005), 'Annual innovation policy trends and appraisal report, Sweden 2004–2005', European Trend Chart on Innovation, Brussels: European Commission.

Van Beuzekom, B. and A. Arundel (2006), 'OECD Biotechnology Statistics 2006', OECD, available at www.oecd.org/dataoecd/51/59/36760212.pdf.

VINNOVA (2002), *Effective Innovation Systems and Problem-oriented Research for Sustainable Growth: VINNOVA's Strategic Plan 2003–2007*, Stockholm: VINNOVA.

Zika, Eleni, Ilias Papatryfon, Oliver Wolf, Manuel Gómez-Barbero, Alexander J. Stein and Anne-Katrin Bock (2007), *Consequences, Opportunities and Challenges of Modern Biotechnology for Europe*, Luxembourg: Office for Official Publications of the European Communities.

# 10. Biotechnology in Denmark and Sweden

**Carl Magnus Pålsson and Birgitte Gregersen**

## INTRODUCTION

Being a 'basic technology' or 'general purpose technology' biotechnology (like ICT) introduces fundamentally new technical principles that over time are expected to influence a broad range of other technologies. As these basic technologies develop, mature and integrate with other technologies it subsequently becomes more and more difficult to classify firms and activities according to their primary technology input and output. Reflecting this trend, recent analyses often use the term 'life science' as a wider industrial classification for biotechnology-related products and processes. Following Gestrelius, Sandström and Dolk (2008) 'life science industry' in this chapter refers to biotech companies within the three sectors: biotechnology, pharmaceuticals and medical technology. Applying this broad definition, life science holds strong positions in Denmark and Sweden. If we take traditional performance measures such as patents, publications, new products, pipeline development and new companies, Danish and Swedish life science industry in most categories belongs to the top five in Europe – if measured per capita.

There are several interrelated factors explaining this. Some relate to the overall general characteristics of the Danish and Swedish National Innovation Systems linked to the welfare state, some are specific for the life science sector, and some are even specific for particular technology areas and sub-sectors within life science. Specific framework conditions include, for instance, dedicated research funds for life sciences, and European and national policies aiming at stimulating life science implementation. Technology-specific framework conditions include for instance regulations for testing drugs, or tax reductions for biofuels. As in most OECD countries, life science is a prioritized policy area in both Sweden and Denmark.

Looking across the various specificities between the two countries, and the different life science technologies especially, the following factors have played a key role as drivers for the industrial success of Danish and Swedish life science sectors:

- *Home market and path dependency in the knowledge base.* The origin of the Danish and Swedish life science industry has two main tracks back in time. One track relates to the Danish and Swedish welfare systems, especially within health care and the environment. Both countries have a long history of research and industrial development within pharmaceuticals. The other track relates to the national production structures: in Denmark especially a long tradition of agriculture and food production and in Sweden raw materials (wood, pulp and paper industries).
- *Education, research and access to qualified employees.* The high research intensity of life sciences makes companies in this field very dependent on access to university research and graduates. This is clearly reflected in the fact that most companies locate their activities in close connection to the dominating universities and research institutes within the field. This is especially visible in the Danish case, where more than 80 per cent of the companies and 90 per cent of the employment in life sciences are located in the bi-national cluster, Medicon Valley, covering the Copenhagen area and the Scania region of Southern Sweden. The rest of the Danish life science sector is located in close connection to the universities in Århus, Odense and Aalborg. In Sweden we see similar clustering of the life science industry around Stockholm/Uppsala, Gothenburg, Malmö/Lund and Umeå – with Stockholm/Uppsala hosting more than half of the Swedish employment in life science.
- *Network and triple helix collaboration.* A central feature of both the Danish and the Swedish national innovation systems is a relatively elaborated tradition regarding collaboration and networking between key actors, supported by public targeted funding for stimulating collaboration between universities, semi-public research institutions and industry.
- *Access to financial resources* is a key factor for any research-intensive industry and it has also been high on the policy agenda in both countries. In Denmark several of the larger life science companies (Novo Nordisk, Novozymes, Leo Pharma, Lundbeck, Carlsberg, Oticon) are controlled by specific foundations, with access to research funds high on the agenda.
- *Attractive environment for clinical trials.* Especially within life science the possibilities to use systematic population and health data registers, allowing for patients to be tracked over time, provides

an important research base and attractive environment for clinical trials. Both Denmark and Sweden provide such opportunities.

It is important to underline that it is not one single factor but the interdependency and co-evolvement of these various factors that have shaped the development of the Danish and Swedish life science sector. Within the broader context delineated above and with reference to the background information on biotechnology industry characteristics and policy environment in Denmark and Sweden as presented in Chapter 9, we take a closer look at two key aspects. First, the role of universities as one of the key actors, and second the role of public policies. Despite many similarities between the two Scandinavian neighbour countries, there are – as the following reveals – also interesting differences that are relevant seen from a policy learning perspective.

## BIOTECHNOLOGY IN DENMARK AND SWEDEN AT A GLANCE

### Sweden

Sweden historically has a record of strong medical research, forming a foundation for current activities and strengths in the biotech field. Two large international corporations have – part of – their roots in Sweden. Both have merged with international partners in recent years, and currently only one of them, AstraZeneca, remains relatively intact. With approximately 28 per cent of the total employment in Swedish life science, AstraZeneca is still dominating the life science, industry. The other, Pharmacia, has recently gone through a process of mergers and divisions, resulting, it has been argued, in a loss of competence that is problematic, not least for the general development of Swedish pharma and biotech.

In their study from 2007, Sandström, Bergqvist and Dolk identified approximately 617 companies with 34 400 employees in the life science industry in Sweden involved in manufacturing, consultancy, product development and/or research and development (R&D) (see Table 10.1).

#### Pharmaceuticals

Within pharmaceuticals AstraZeneca accounts for nearly half of the 19 500 employees in the sub-sector. Pfizer and Bivitrum are two additional important players. Within pharmaceuticals, drug discovery and drug development is by far the largest business segment (Sandström et al., 2007).

*Table 10.1 Number of companies and employment in Swedish life science industry, 2006*

| | Companies | Employment |
|---|---|---|
| Pharmaceuticals | 229 | 19 474 |
| Biotechnology | 251 | 8931 |
| Medical technology | 326 | 12 284 |
| Total[1] | 617 | 34 468 |

*Note:* 1. 'Total' adds up to less than the sum of the three sectors due to overlap between the sectors

*Source:* Sandström et al. (2007)

**Biotechnology**

The majority of the 251 companies in the biotechnology sector are active either in biotech tools and supplies, bioproduction or drug discovery. AstraZeneca is not active in this sub-sector, but a few other large companies, like Pfizer bioproduction and GE healthcare Biosciences, are included (ibid.).

**Medical technology**

The medical technology sector employs 12 280 people in 326 companies. Sixty per cent of the total employees in the sector work within the three dominating business segments: electromedical equipment, active and non-active implantable devices, and medical disposables (ibid.).

The real growth in life science companies has occurred since the end of the 1990s. Among companies in existence in 2004, 90 per cent were formed after 1990, and 63 per cent after 1999. Seen over the last decade (1997 until the present) there has been a modest positive trend in terms of relative results, but with a weakening towards the end. During the same period there has been an increase both in terms of biotech companies and employees in that sector. It is particularly in the pharmaceutical and in the equipment/instrument industries that we find biotech making contributions to R&D and to production. In both industries we find combinations of small innovative companies, linking academic research and corporate R&D to more well-established corporations (Norgren et al., 2007; Bergqvist, 2008).

For an extended period in the twentieth century and up until now, a rather limited set of large companies have been vital components of the Swedish economy. The dominance of MNCs in the Swedish economy applies also in the life sciences, where companies such as AstraZeneca and Pharmacia – as mentioned above – have merged into international conglomerates. As R&D and production are dominated by these actors and the priorities of the MNCs differ from the previously 'national' corporations, the sector as a whole has increasingly come to be appreciated as vulnerable, and in dire need of renewal. Still, however, the importance of AstraZeneca cannot be overestimated, and accumulated output within the sector to a considerable degree depends on the performance of that very corporation.

## Denmark

In Denmark industrial development of pharmaceuticals and biotech goes back to early industrialization and has its historical roots in food production. As Table 10.2 shows, also medical technology such as audiology, hearing aids and other medical devices builds on a long industrial history (Gestrelius, 2008).

The current turnover of the life science industry accounts for around 5 per cent of Danish GDP, making it a relatively important industrial player for the Danish economy (Gestrelius, 2008). Around 280 Danish life science companies have been identified, employing nearly 40 000 people (see Table 10.3). According to the study made by Gestrelius et al. (2008), 90 per cent of the companies are research-intensive companies and manufacturing companies. A few large companies (with over 500 employees) dominate the industry. This is especially the case within the business segment 'drug discovery and development', where Novo Nordisk alone employs around 10 000 people, corresponding to 25 per cent of the total employment of the Danish life science industry, or half the employment within pharmaceuticals.

### Pharmaceuticals

As indicated above, especially one company, Novo Nordisk, dominates the Danish pharmaceutical sector. Furthermore, a few other companies, like Leo Pharmaceuticals, H. Lundbeck and Nycomed, have more than 1000 employees each (Gestrelius et al., 2008). These companies also belong to the drug discovery and development segment.

### Biotechnology

In the Danish context there is a large overlap between the pharmaceutical and the biotechnology sector due to the fact that companies like Novo Nordisk, SSI, Ferring and around 50 young companies to a large extent base their products on biotech (Gestrelius et al., 2008). Novozymes, part of Novo A/S and owned by the Novo Nordisk Foundation, with its 5000

*Table 10.2 Selected Danish large and old companies within life science*

| Company | Established | Employees (FTE), Denmark 2006 |
|---|---|---|
| *Pharmaceuticals* | | |
| LEO Pharma | 1908 | 1200 |
| Lundbeck | 1915 | 2000 |
| Novo Nordisk A/S excl. NNIT & NNE Pharmaplan (1989) | 1920s | 10000 |
| Nycomed (as DAK) | 1922 | 700 |
| Ferring Pharmaceuticals | 1956/1999 | 400 |
| ALK-Abelló | 1923/1992 | 500 |
| *Biotech* | | |
| Carlsberg Research Centre | 1875 | 175 |
| Danisco (incorporated in Genencor in 2005) | 1872 | 1450 (excl. sugar division) |
| Novozymes (split from Novo Nordisk in 2000) | 1939 | 2200 |
| Chr. Hansen | 1874 | 850 |
| *Hearing aids and audiology measurements* | | |
| Oticon (W Demant) | 1904 | 1300 |
| GN Resound (GN Syore Nord) | 1943 | 400 |
| GN Otometrics (GN Store Nord) | 1960 | 200 |
| Widex | 1956 | 700 |
| Interacoustics (W Demant) | 1967 | 150 |
| Sonion | 1974 | 250 |
| *Medical devices and analysis instruments (excl. audiology)* | | |
| Radiometer | 1936 | 950 |
| Ambu | 1937 | 350 |
| Coloplast | 1957 | 2400 |
| Unomedical (first Pharma Plast, later Maersk Medical) | 1964 | 750 |
| William Cook Europe | 1969 | 600 |

*Source:* Gestrelius (2008).

employees is the largest company within the segment of industrial biotechnology. Novozymes, Chr. Hansen and Danisco specialize within the segment of enzymes, probiotics and cultures for food, health and industrial biotech. Furthermore, these large biotech companies are also active within energy and environmental biotechnology (for instance bioethanol) (ibid.).

*Table 10.3 Number of companies and employment in Danish life science industry, 2006*

| | Companies | Employment |
|---|---|---|
| Pharmaceuticals | 129 | 21 363 |
| Biotechnology | 142 | 22 308 |
| Medical technology | 89 | 11 797 |
| Total[1] | 280 | 39 375 |

*Note:* 1. 'Total' adds up to less than the sum of the three sectors due to overlap between the sectors

*Source:* Gestrelius et al. (2008)

**Medical technology**

The medical technology sector is dominated by three main business segments: medical disposables, electromedical and imaging equipment and audiological devices (ibid.). Within each of these three areas a few large international companies exist, for instance Coloplast, Unomedical and Dansac (medical disposables); Oticon, Widex and GN Resound (audiology/hearing aids) and Radiometer Medical (electromedical and imaging equipment).

## CLUSTERS AND TRIPLE HELIX RELATIONS

As life science represents an increasingly complex and widely distributed knowledge base, companies have become more and more dependent on collaboration with other companies, research institutes, hospitals and other actors (Cooke, 2005; Moodysson et al., 2008). This is reflected in the very clear tendency for the life science industry to cluster and localize around a few larger universities in order to have access to a broad palette of relevant collaboration partners. At the same time it is an industry with a high degree of international research collaboration and multinational companies, where global networks are crucial for innovation activities. Moodysson et al. (2008) seeks to explain this dichotomy around 'local node – global network' geography of innovation, or 'proximate' and 'distant' learning processes, by applying a distinction between 'analytical' and 'synthetic' modes of knowledge creation (Asheim and Gertler, 2005). In short, analytical knowledge creation mainly focuses on understanding and explaining based on discovery and application of science, while synthetic knowledge creation is mainly about engineering, and concrete, technical problem-solving. Both types of knowledge are present in most industries but to varying degrees, and dependent on the concrete innovation activity in focus. Based on detailed case studies of different innovation projects in the Swedish–Danish cross-border Medicon Valley cluster they found that 'Analytical knowledge creation tends to be less sensitive to distance decay facilitating global knowledge networks as well as dense local collaboration. Synthetic knowledge creation, on the other hand, has a tendency to be more sensitive to proximity effects between the actors involved, thus favouring local collaboration' (Moodysson et al., 2008, p. 1052). Bringing in such knowledge taxonomies that are more differentiated may help in explaining the clustering of the life science industry, and it might also be useful to include such considerations in (regional) policy-making aiming at attracting and maintaining various types of knowledge-intensive industries.

Figure A10.1 in the Appendix provides an overview of life science clusters in 31 regions (NUTS II level) of the Baltic Sea region. The region includes more than 102 000 employees in life science (2004). Out of these Copenhagen (Denmark) accounts for around 22 200 employees, Stockholm/Uppsala (Sweden) 16 900 employees, Malmö (Sweden) 5900 employees and Gothenburg (Sweden) 5400 employees. In other words, half of the employees are concentrated in these four clusters. The rest – except Kiel (Germany) with 14 700 employees and Helsinki (Finland), with 7000 employees – are rather small clusters (Graversen and Rosted, 2010).

Swedish biotech research, both academic and corporate, is located mainly in four regions: Uppsala/Stockholm, Gothenburg, Malmö/Lund, and Umeå.

*Uppsala/Stockholm* is the strongest cluster in Sweden, in which we find a focus on pharmaceuticals, bioproduction, biotech tools and functional food. The quality of research and institutions is high by international standards. Academic actors include the Karolinska Institute, Stockholm and Uppsala universities, the Swedish University of Agricultural Sciences, and university hospitals in both cities. Companies in the region include AstraZeneca, GE Healthcare, Biovitrum, Biacore and Pfizer. As a measure of the strength of a key actor in the biomed field, the Karolinska Institute in Stockholm is ranked below Harvard, Cambridge and Oxford, but ahead of Stanford, Imperial College and Johns Hopkins.

In *Gothenburg* the cluster strengths are found in biomaterials, stem cell research, medical technology, diagnostics and analytical tools, cardiovascular and metabolic diseases, and drug discovery. The regional academic infrastructure in the field is Gothenburg University, Chalmers University of Technology, and Sahlgrenska University Hospital. Among important corporations are found AstraZeneca, Nobel Biocare, NeuroSearch, and Cellartis.

In *Umeå* a focus on biomedicine, microbiology, protein chemistry, molecular genetics and plant biotechnology is found. The central academic actor is Umeå University, and we also find here the Swedish University of Agricultural Sciences (as in Uppsala), along with a range of research centres linked to the university.

In the *Malmö/Lund* region the main focus is on pharmaceuticals, agrobiotechnology, and environmental biotechnology. Lund and Malmö universities, the Ideon and Medeon Science Parks, and the university hospitals in the respective cities provide an infrastructure. Important companies include AstraZeneca, Active Biotech, Gambro, Bioinvent and Acadia. Of special interest in this study is the Malmö/Lund region, as it also forms part of the cross-national Øresund Region, together with Copenhagen/Zealand in Denmark (see below).

There are indications that the concentration to these regions has become even more pronounced in recent years (Sandström and Norgren, 2003; McKelvey et al., 2003; Ministry of Industry, Employment and Communications, 2005; Biotech Sweden, 2009).

## Medicon Valley

Under the 'Medicon Valley' initiative or 'brand', officially in place since 1997, the cross-national Øresund Region has positioned itself as one of the leading biotech regions in the world, alongside Cambridge, Boston and San Francisco. Obviously it builds on a longer history than that, going back even to the nineteenth century. It is one of the major life science clusters in Europe; the Swedish part of it constitutes 20 per cent of the national sector as a whole (Biotech Sweden, 2009). As mentioned in the introduction, in Denmark it accounts for nearly 80 per cent of the Danish biotech companies and for 90 per cent of the Danish employment in the sector.

The strength of the Øresund Region can be said to rest upon three pillars. The primary strength, according to a study by The Boston Consulting Group (BCG) (2002), is the presence of large and mature pharmaceutical companies (especially Novo Nordisk, AstraZeneca, H. Lundbeck, and LEO Pharma), an almost unique concentration; the only comparable concentration is found in Boston. These corporations provide a kind of basic structure for the region, for other actors, both academic and commercial, to build upon. The relative maturity of the corporations of the region may outweigh some of the generally problematic 'youthfulness of the Europe biotechnology sector' (Critical I, 2006), but this can be considered mere speculation. In addition to the mature companies is found a range of newcomers, prominent among which are: Acadia, Active Biotech, Alligator, BioInvent, Gambro, Biogaia, Probi and Camurus.

Add to this the broad range of academic research available in the region (primarily Copenhagen and Lund universities, and their associated university hospitals), and, finally, well established interaction of commercial actors and academic research around clinical research and trials. The actors in the region together contribute what is considered the three crucial elements in the success of biomedical R&D: basic academic research, corporate R&D, and clinical R&D. Relevant both to current conditions and to future development is the breadth of activities in the region. In BCG's evaluation, Medicon Valley comes out as one among four top biotech regions in the world; the others being Boston, San Francisco Bay Area, and Cambridge (UK). According to the Medicon Valley homepage, (www.mediconvalley.com), the cluster includes 477 companies, 10 universities and 33 hospitals (see Appendix Figure 10A.2).

Unlike some other rather specialized biotech regions in Europe the operations of corporations as well as the academic research performed in the Øresund region cover many areas, foremost among them diabetes, inflammatory diseases, neuroscience and cancer. Due to the presence and high profile of all relevant categories of actors, the commercial and appropriation potential within these areas are considered the highest. According to the study, in the Øresund region other research areas with a high reputation are also found, but with a lower commercial potential; those are haematology, infectious diseases, receptor studies, molecular biology, stem cell research and cardiology (The Boston Consulting Group, 2002).

As much of the evolving 'ecology' among mature Big Pharma corporations and smaller more flexible biotech companies concerns opportunities for interaction, the possibilities of the latter supplying the former with new drug candidates, it is crucial that conditions for such interaction do indeed emerge. The Ideon Science Park, in the immediate vicinity of Lund University can be appreciated as a facilitator to such ends. It was the first of its kind in Scandinavia, and has so far provided opportunities for such interaction. A broad range of new companies, as well as spin-outs from the former Pharmacia operation can be found in the science park.

However, it should be emphasized that the true and hoped-for synergies of R&D activities in 'Medicon Valley' as a whole, across national boundaries, remain to materialize. The quality of academic output in some studies is found to be high, but innovations in Denmark and Sweden are not on corresponding levels. Among weaknesses found is that the number of scientific discoveries leading to patents is high, but the commercial use of those patents is low. Patents can be appreciated either as actual innovation or as innovation potential. It seems that quite some effort is spent on securing patents, but not so much, in comparison, on developing research results into long-term business cases. Additionally, the relatively low-staffed Technology Transfer Offices (TTOs) have limited industry or commercial experience. These analyses point in the direction of focusing efforts on improving the innovation systems as the primary area of action, overcoming the often observed fragmentation of the regional innovation system (Medicon Valley Academy, 2004).

The various actors (academic research, corporations, and hospitals for clinical trials) still operate under the general conditions of the national systems (Denmark and Sweden, respectively). Differences apply, for example, for funding and for the handling of intellectual property rights, and it is unlikely that they will be harmonized in the near future. The Medicon Valley Alliance points to obstacles that impede integration and cooperation in the region, for instance how national boundaries in some cases prohibits cross-border investments. While earlier studies have underlined that though the general picture of the presence of relevant categories of actors applies, the actual interaction was limited (The Boston Consulting Group, 2002; Medicon Valley Academy, 2004), a recent study finds that cluster concentration actually matters for economic performance (Graversen and Rosted, 2010). For instance, 41 per cent of firms report a high level of collaboration on research. Graversen and Rosted (2010) conclude that 'cluster policies seems to matter for innovation and collaboration and through those channels leading to higher productivity' (p. 122).

## THE ROLE OF UNIVERSITIES IN BIOTECH RESEARCH

Research in life sciences accounts for 40 per cent of research performed at Swedish universities. This share is comparable to that found in other countries.[1] In the most recent years (2005–08) the funding has increased by approximately 20 per cent. In comparisons such as these it should, however, be noted that many countries of comparison are currently in a phase with even larger relative increases in funding (Norgren et al., 2007).

Access to qualified graduates and PhDs is especially important for a research-intensive industry such as life science, and one of the main explanations why life science industries cluster around the universities. Sweden and Denmark both have a relatively high share of graduates and PhDs within life science, and the 'production' of candidates is increasing. In Denmark around one third of the co-funded industrial PhDs are found within life science (period 2002–06).

Life science is a highly dynamic sector with a relatively large number of new small start-ups, and university spin-offs constitute an important part of the industrial dynamics. In Denmark, more than half of the start-ups have their roots in local Public Research Institutions and universities.

Both Sweden and Denmark belong to a group of countries (along with, for example, Switzerland, USA and Finland) that display a range of strengths and few weaknesses across R&D and commercialization. Measured in terms of patenting, Denmark performs exceptionally well in comparison with other EU countries, a position that it had already conquered in the mid-1990s. However, both Sweden and Denmark suffer from some weaknesses when it comes to publicly funded R&D, and Sweden, additionally, when it comes to general funding available for biotech research (Reiss et al., 2004; Patel et al., 2008).

From the perspective of companies, collaboration with academia is important for several reasons. This is especially pronounced for small and medium-sized companies. It is difficult, however, to obtain information on the exact volume of such collaboration. Seen across the entire population of companies (all industries), active in relevant fields, R&D collaborations primarily take place with national universities, but there are significant differences between different categories of companies. Collaboration comes in various forms: with R&D work together with research groups at universities, co-authoring of articles, and recruitment – not least in certain geographical locations. However, within biotech the overall collaboration for both Swedish and Danish firms is oriented towards international linkages, especially with US partners. Such cooperation is more important than that at a local or European level. Co-location in firm to firm collaboration is not found to be an important factor. In the Swedish case McKelvey et al. (2003) found that those intra-Swedish deals that do occur are mostly between the largest firms and SMEs, supporting the observation that the larger firms are indeed international in orientation, with relatively little connection to the national development. The increasing importance of tapping into strategic international collaboration is clearly reflected in The Life Science Ambassador Program launched by the Medicon Valley Alliance in 2008 (see Box 10.1).

When it comes to collaboration between universities and firms, the picture is slightly different. Here we find, regardless of the main trend, examples of local and national initiatives. For the main regions mentioned above it is possible to identify rather well-established and long-term forms for cooperation, involving firms and academic research. In Uppsala and Lund, for example, we find these built upon spin-offs from the former Pharmacia operation. Differentiating between three types of collaboration – firm–firm, firm–academia, and academia–firm – it is in the last category that we find most examples. McKelvey et al. (2003) found that Swedish

### BOX 10.1 THE LIFE SCIENCE AMBASSADOR PROGRAMME

The Life Science Ambassador Programme is a global network of innovative life science clusters around the world. The programme provides companies and research institutions within the life sciences industry with a unique opportunity to find partners, collaborators, investors and sponsors around the world.

The Life Science Ambassadors operate as full-time interlockers, identifying potential opportunities with strategic partners and facilitating communication between decision-makers so as to foster the development of global collaborations and partnerships. The Life Science Ambassadors are uniquely proficient in providing such connections because they are all multilingual and have extensive experience, knowledge and established networks in the life sciences industry from the cluster they represent.

The Life Science Ambassador Programme promotes innovation and increases the competitiveness of the participating clusters through the building of strategic partnerships and international alliances. Companies and academic institutions will have access to leading life science innovation environments and talents to further research and development, which may lead to commercial opportunities.

Currently, five regions are participating in the exchange of Life Science Ambassadors: Medicon Valley, Denmark and Sweden; Kobe-Kansai, Japan; British Columbia, Canada; Seoul, South Korea; and Boston, USA.

*Source:* http://www.ambassadorprogramme.com/, accessed September 2010.

biotech SMEs are more likely to engage in collaborations with universities, in forms that to some degree depend upon geographical proximity. In consequence, the biotech specializations at various universities have implications for what substance there might be in these collaborations.

## BIOTECHNOLOGY POLICY ENVIRONMENT

The debate on the future of life sciences and biotech in Sweden and Denmark takes place under conceptions similar to those found in the rest of Europe for the last couple of years. A recurrent theme is the development in Europe compared to the US. In these analyses Europe mostly comes out unfavourably, in the sense that it does not get enough productivity out of its R&D investments.

Since the late 1990s the notion of a 'Swedish paradox' has been put forward, claiming that the high input of R&D investments is not reflected in expected corresponding output in the form of innovation and growth (Bitard et al., 2008). The debate can be said to relate back to similar ideas in the 1980s on the prevalence of 'eurosclerosis', i.e. rigidities in markets resulting in the European nations lagging behind its main competitors globally. Both concepts have been criticized from theoretical as well as empirical points of departure (Granberg and Jacobsson, 2006; Ejermo and Kander, 2008), but they still frame much of the analyses of the current orientation of academic research.

A general funding issue, often brought up especially by actors directly involved in innovation activities, is that Sweden is one among only a few countries within the OECD not to have tax incentives for R&D, resulting, so the argument goes, in competitive disadvantages. According to the critics this disadvantage is both real and symbolic, in the latter case as it may send a signal to actors and observers that Sweden does not promote innovation (Royal Swedish Academy of Engineering Sciences, 2008).

In Denmark the discussion has focused more on the actual lack of public and private R&D investments (relative especially to Sweden and Finland) and the likelihood not to be able to fulfil the Bologna targets. However, especially in relation to biotech or life sciences, the access to funding has been relatively favourable. In Denmark specific foundations control several of the larger companies. This applies, for instance, to Novo Nordisk, H. Lundbeck and Leo Pharma. The same goes for some of the medical technology firms, for example the Oticon and Danfoss foundations. In addition, biotech has been a priority area for the Danish State investment fund ('Vaekstfonden'), which provides a broad spectrum of finance solutions for SMEs.

As in many other settings, in both countries there has been an increased focus on how to make academic research more responsive to the needs of the private sector. In Denmark this discussion has been linked to a policy with a strong focus on commercialization of university research. Due to certain idiosyncrasies of the Swedish research system the question of 'industry relevance' has some special features that concern the division of labour between various categories of research-performing actors. Over the second half of the twentieth century universities in Sweden came to take on the role that in comparable countries is performed by research institutes. In Sweden the industrial research institutes sector is small, and there are no institutes specialized in biotech (Sandström and Norgren, 2003). In a formulation widely in circulation the universities should be 'the research institutes of society'. This, in turn, has meant that the transformational pressure on universities, to link and to some degree adapt academic research to the demands of industry, arguably has been higher in Sweden. Obviously this is an issue of considerable importance to actors inside and outside academia, as it concerns fundamental ideas and ideals regarding the ethos and justification of universities.

Another recurrent theme in the Swedish debate is the dominance of a few major corporations in the national economy. Swedish industry in general is heavily export-oriented, and top-heavy in the sense that across sectors a few large corporations have come to dominate, being responsible for roughly 70 per cent of business R&D (Reiss et al., 2008). In terms of volume the most important sectors are non-electrical machinery, motor vehicles, pulp and paper, telecommunications and iron and steel (Marklund et al., 2004). Many of the companies in those sectors trace their roots back to the nineteenth or early twentieth century, and over the years the country as well as certain regions have come to depend considerably on the strength of those companies on the global market. In recent decades many of them have been absorbed into international MNC groups, thus losing their previous national focus. This, it has often been observed, makes Sweden vulnerable to the vagaries and priorities of these MNCs. Thus, renewal from below has emerged as a concern, as there seems to be a lack of expanding SMEs that can fill any void emerging when the MNCs change their focus.

In contradistinction to Sweden, the Danish 'mode of innovation' has historically been dominated by SMEs continuously making incremental innovations based on learning by doing, learning by using and learning by interacting, especially with customers and suppliers (Christensen et al., 2008). However, when it comes to private R&D expenditure, the distribution is much skewed. In 2001 2 per cent of all firms conducted nearly 40 per cent of the total private R&D. In particular the two major companies, Novo Nordisk and Sauer Danfoss, dominate the research scene. Although the main part of private R&D expenditure is in-house, there is an increasing tendency to establish laboratories outside Denmark. This also goes for Novo Nordisk. Around 5200 people work in research out of Novo Nordisk's 29000 employees; 55 per cent of the employees are located outside Denmark (see www.novonordisk.com, accessed September 2010). Although this trend is prevalent in most countries hosting large research-intensive international companies, it may be a cause of concern especially for a country with only a relatively few R&D-intensive firms.

## POLICIES AND AGENDA-SETTING AT NATIONAL LEVEL

Handling the boundaries for what to consider relevant to the development of 'biotech' proper is difficult, as much development of the field – both general and more narrow – is presumed to occur on the very boundaries between various disciplines, not only the life sciences but also, for example, information technologies (cf. Lacasa et al., 2004). In a sense, biotech can be appreciated as paradigmatically 'modern', as it is not so much a traditional 'discipline'; rather, it can be understood as a set of interrelated techniques and instruments, drawing on various bodies of knowledge. Such observations make it necessary also to consider the balance between vertical (sector-specific) policy instruments and horizontal (generic). In addition to this the application fields corresponding to the dynamics of different industrial sectors should be taken into account. The outcome of considerations such as the above has implications for the choice among policy measures; how much and what type of intervention, or absence of it, is adequate. Over time one may notice in Sweden and in Denmark a shift towards more horizontal/generic funding instruments (cf. Reiss et al., 2004; Lacasa et al., 2004).

The national policy on biotech is thus obviously linked to the more general research and industry policies. The 'toolbox' is similar, grounded in a rather common understanding of how companies, universities and research institutes may interact for production and welfare gains.

Table 10.4 uses a typology developed by Reiss et al. (2005) and Lacasa (2007) (see a similar typology in Chapter 12 on Germany, this volume). The four sub-areas (development of the knowledge base and human resources; knowledge transmission and application; market; and industrial development) provide a framework for the seven listed policy measures.

From an overall view the two countries look rather similar when it comes to policy implementation, although some differences exist if we take a closer look at some of the issues.

### Development of the Knowledge Base and Human Resources

*VINNOVA:* in Sweden a major overhaul of the research funding system was implemented in 2001, including, among other things, the establishment of The Swedish Governmental Agency for Innovation Systems (VINNOVA), with a strong mandate to influence the interaction among academic and commercial actors. Other agencies reorganized or were established at the same time, and the explicit purpose of the 2001 reform was threefold:

- to focus national efforts on a set of scientific areas appreciated to be of special importance;
- to promote collaboration between actors involved in fundamental research and those actors more oriented towards development activities; and finally
- to facilitate the so-called 'third task', i.e. society-wide information on and dissemination of research and results.

The reform, and especially the establishment of VINNOVA, has been appreciated as a move to initiate a national, comprehensive policy on innovation, involving also a higher degree of coordination among funding agencies (D'Este and Costa, 2007).

The reform, not least establishing an agency that in name and in deeds embodies the innovation system approach, can be understood as an instrument well anchored in a Swedish policy tradition of interventions compensating for market failures. Among the concrete initiatives taken we find several such instruments, such as linking actors to each other and thus creating or strengthening networks, also opening up cooperation across technological and scientific fields (cf. Lacasa et al., 2004).

*The Swedish Foundation for Strategic Research*: another important actor in financing biotech research has been The Swedish Foundation for Strategic Research (SSF), a semi-public body established in the mid-1990s. The mission of the foundation is to support 'both pure basic research and applied research, as well as research that bridges the gap between these extremes', a statement that might be understood as a tentative definition of 'strategic research', one that bears resemblance to VINNOVA's bridging role.

As Figure A10.3 in the Appendix shows, the Danish funding system for research and innovation consists of several bodies from which life science and biotech can apply for funding (in competition with other research areas): *The Danish Council for Independent Research* funds specific research activities within all scientific areas divided into five main areas, of which medical sciences is one. Another important body is *The Council for Strategic Research*, with the ambitious mission to ensure Denmark's position as a global frontrunner regarding welfare, wealth and science in the short and long term. Furthermore, The Danish Council for Strategic Research seeks to promote international cooperation in research, including cooperation with the new high-growth countries. *The Danish National Advanced Technology Foundation* is an independent body that offers grants in the form of co-funding for high-technology research and innovation initiatives and projects. The Danish National Research

*Table 10.4 Typology of biotechnology policies for Denmark and Sweden, 2004*

| Sub-areas of the Biotechnology Innovation System | Policy goals | Policy area | | | | | | |
|---|---|---|---|---|---|---|---|---|
| | | Education | Research | Exploitation | Industrial development | Regulation | Fiscal measures | Demand |
| Development of the knowledge base and human resources | 1. To promote high level of biotechnology basic research | | 5 5<br>4 3 | | | 5 5 | | |
| | 2. To promote high level of industry-oriented (and applied) research | | 5 5<br>4 3 | | | | | |
| | 3. To support knowledge flow between scientific disciplines | | 4 4<br>3 3 | | | | | |
| | 4. To assure availability of human resources | 3 4 | | | | | | |
| Knowledge transmission and application | 5. To facilitate transmission of knowledge from academia to the industry and its application for industrial purposes | | | 4 4<br>2 1 | | 4 5 | | |
| | 6. The adoption of biotechnology for new industrial applications | | | | | | 3 0 | 1 nd |
| | 7. To assist firm creation | 0 0 | | | 3 2<br>3 1 | 4 5 | 2 0 | |
| Market | 8. To monitor and improve the social acceptance of biotechnology | | | | | | | 5 4 |

| | | | | | | | | | |
|---|---|---|---|---|---|---|---|---|---|
| | 9. To facilitate the introduction of new products | | | | | 3 | 5 | | |
| | 10. To strengthen the economic sectors exploiting biotechnology | | | | | | | 3 | 0 |
| | 11. To keep/attract large firms (important market, important for firm development: tacit knowledge etc.) | | | | | | | 3 | 0 |
| Industry | 12. To encourage business investment in R&D | | | 3 | 3 | | | 4 | 0 |
| | 13. To improve firms' competitiveness | 0 | 0 | | | 4 | 5 | 4 | 0 |
| | 14. To exploit regional potentials | | | 1 | 1 | | | | |

*Notes:*
Light grey: generic policies; dark grey: biotechnology-specific policies
0: no policy in place; 1–5 scale for policy activity, where 5 is the highest level.
With pattern: DK; without pattern: Sweden

*Source:* Reiss et al. (2005).

Foundation has the objective to promote and stimulate basic research at the frontiers of all scientific fields. The main funding mechanism is the Center of Excellence (CoE). *The Danish Council for Technology and Innovation* administers a number of initiatives to promote innovation and dissemination of knowledge between knowledge institutions and enterprises:

- cooperation and interaction (innovation consortia, approved technological service, industrial PhD, knowledge pilot, networks of high technology);
- entrepreneurship and commercialization (technology transfer, business incubators);
- initiatives at a regional level (innovation environments, regional growth environments, regional ICT initiative);
- international innovation (pre-projects to the European Commission's 7th Framework Programme).

In short, the development of the Danish research funding system has followed the same development path as most European research funding systems in the sense that an increasing share of the funds are allocated to selected high-tech areas (biotech, nanotech, ICT, life science, renewable energy) at the same time as demands for consortia and excellence are put forward.

*Linkages to international research, and to international actors*. In research, as in the national economy in general, small economies are obviously highly dependent on international contacts. Furthering contacts with the internationally leading research environments poses special challenges for small countries. Thus, an important policy component has been to facilitate mobility for international researchers to Sweden and Denmark. Apart from the obvious issues related to research as such, reforming tax regulations for specialists in various fields, not only science, is also among the specifics discussed. Internationalization and international contacts are important in another dimension as well: new companies now finding themselves, ideally, in a growth phase, searching for partners and markets abroad are pressed to mobilize the resources necessary to link up to international actors. Supporting new companies, with limited resources, in establishing such contacts is an important general policy field. The instruments are organizations like, for instance, Invest in Sweden Agency, and its sister organization Invest in Denmark Agency. There are also regional initiatives for the same purposes, and Medicon Valley Alliance can be found among these.

**Knowledge Transmission and Application**

In both Sweden and Denmark various policy measures exist to stimulate collaboration between universities and firms. One example is the Danish 'Center Contracts', providing funding for research projects that include partners from universities, private companies and approved technological service institutes – GTS (Advanced Technology Group). The systemic view on innovation processes that informs much of the Swedish and Danish policy landscape, more or less prescribes an increased focus on network formation – academy, industry and public bodies on all levels. 'The triple helix' has entered as the general, rather uncontroversial heuristic within which policies are formulated and implemented. Much focus in the debate in many countries has so far been on strengthening the science base, appreciated as a prerequisite for the general system performance, and this is in line with much of the development in countries of comparison (Lacasa et al., 2004). This goes for Sweden and Denmark as well, but the orientation of funding agencies, with an explicit agenda of furthering interaction among academia and industry, may very well be seen as a development in step with more generally observed trends, more application-oriented initiatives increasing in relative weight. This is a shift that has brought considerable debate and occasional resistance, however strong the drift in line with the policy.

Among initiatives taken to further actual interaction between universities and companies, we find the formation of a large number of 'competence centres' and networks. The guiding idea for these is to promote problem-oriented research, and still with high scientific quality. Related to this, the ambition to promote and vitalize the – so far – rather insignificant research institute sector is raised. This is considered to be especially important for SMEs as they, normally, do not have enough research capacity in-house. The institutes may serve both as R&D performing actors in themselves, and as bridges between academia and companies (VINNOVA, 2002).

*Entrepreneurship and university–industry linkages*. As a response to and confirmation of policy trends throughout industrialized economies, in general in both Sweden and Denmark there has been an increased emphasis on academic entrepreneurship, and linking industry and universities. Several new companies stem from university research, and it is an obvious policy to discover and promote such opportunities. Academics often do not consider themselves entrepreneurs in a strictly commercial sense, and thus it becomes all the more important to implement support structures that allow a certain division of labour between different actors, such as the academics, TTOs at the universities, and companies in different phases of development. As the latest initiative (among many) to stimulate entrepreneurship activities the Danish government has set up a Foundation for Entrepreneurship in order to create a coherent national commitment to education and training in entrepreneurship.

Among the measures taken in Sweden to further interaction and mobility of competencies between academia and industry we find, for example, adjunct professors, usually persons with a predominantly industrial experience having 20 per cent of their position at a university, and the remainder in another – private or public – organization, and industry-based doctoral students. Both forms of interaction have increased in recent years. A measure that has been suggested is to systematically encourage senior university researchers to take temporary positions in industry.

## Industry

*Venture capital and financing R&D*: in their benchmarking study, Ernst & Young (2008) ranked Denmark among the European 'Top 5' in 2007 when it comes to the ability to raise venture capital. In 2006 ten Danish specialized biotech venture capital firms invested 9 billion DDK (1.2 billion EUR) in life science; of these around 1.5 billion DKK (200 million EUR) in new start-ups (Vaekstfonden, 2007). Although nearly half of the total Danish VC investments are in life science (2006), it may still be too small, considering that on average it requires around 400 million DDK for each start-up.

Though there has been considerable concern regarding the availability of venture capital, the statistics are mixed. In a recent OECD report, Sweden ranked number one in 2007 both if measured as life science venture capital investment as a percentage of GDP or as a share of all national venture capital investments (OECD, 2009). Denmark ranked second and tenth respectively. Despite these seemingly positive rankings, an important field of interest is the financing of early stage commercialization. Seed investments is one priority, tax incentives for young innovative companies another. The Swedish Fund for Industrial Development (Industrifonden) is the most important public, policy-directed actor, often aligning itself with private investors in providing funds for seed investment. VINNOVA has launched a 'concept testing programme' aimed at small biotech companies, through which they can obtain one year of proof of concept tests. In the last decade about 20 new biotech companies have started each year in Sweden. Most of those are spin-offs from university research, often located in proximity to universities. Supporting such processes, providing adequate economic and infrastructure conditions for starting new companies, in general, has evolved into a central concern for policy-makers and actors at all levels. Among instruments used we find, for example, seed financing, incubators, science parks and university spin-off organizations. The motives for public intervention are often considered to be, on the one hand, the market failure associated with the high risks involved, and, on the other, the potential public and societal benefits associated with the value added by new knowledge-intensive products and firms. In biotech these considerations are especially pertinent, as the period from discovery to – possible – application can be quite extended, involving high risk and high costs. The Swedish government has made some efforts to intervene and to amend this situation. In the case of biotech specifically, a handful of government agencies have been instrumental for early stage financing, establishing public venture funds, notably NUTEK (The Swedish Business Development Agency), and SIC (The Innovation Centre Foundation). Regionally the Foundations for Technology Transfer have contributed, as well as holding companies linked to the universities. Within the regional initiatives the ambition has been to enrol a range of actors, trying to further the coordination that has often been missing. One crucial aspect has proved to be the commercial viability of biotech projects, and the concern to ensure that the public funding agencies have enough commercial competencies to make proper evaluations of such (Sandström and Norgren, 2003; SwedenBio, 2005).

*Strengthening of regional cooperation:* strong research environments are, by nature and by necessity, geographically situated. Thus, strengthening the interaction in the regional settings, near the major universities, has become an important objective in both countries. The fragmentation among actors in these settings has also evolved into an issue that needs to be solved, the ambition being to coordinate research, industry activities and funding. The universities have increasingly taken on the role as regional facilitators of cooperation, providing organization and infrastructure. Judging from the current state of affairs there is still a lot to be done in this field, and it hardly comes as any surprise that regional actors, both inside and outside academia, are looking to the universities to take the lead, to provide local or regional platforms for cooperation.

The really strong links of cooperation are, expectedly, found among science and technology-intensive spin-off companies, located near the universities. Unfortunately they have not been growing strongly, and in some sense they do not compensate for the cuts resulting from structural reforms in the big, established life science companies. Here is found an absence of a certain type of actor, a middle level, something that is central to understanding recent policies. However, changing this situation is a very long-term effort, stretching over decades, and the scope for policies actually amending the situation is difficult to evaluate.

## SUMMARY AND CONCLUSIONS

Sweden and Denmark show very high performance within life science. They build on a strong knowledge base, knowledge transfer and application in combination with various private and public funding schemes in both countries, and they both have long experience in designing and implementing programmes to promote biotechnology. Despite many similarities there are also significant differences between the innovation systems of the two small Nordic welfare states.

Sweden has the highest R&D expenditures as a percentage of GDP in Europe (over 4 per cent) and the highest business expenditures on R&D in the OECD, which serves to highlight the importance of the corporations in the research portfolio. However, the relative role and weight of corporate R&D efforts is decreasing as a proportion of GDP, a serious issue in any country. Public investments in R&D are also decreasing, measured as a proportion of GDP. Public investments in the corporate sector are decreasing, mostly due to cuts in defence spending, historically in Sweden an important area for public investments, as an instrument for industry as well as R&D policies. In contradistinction to Sweden, R&D expenditure as a percentage of GDP has increased in Denmark – although at a slow speed. In 2008 Denmark spent 2.88 per cent of GDP on R&D, of which the private sector spent 2.01 per cent (Danish Agency for Science, Technology and Innovation, 2010).

According to the Biotechnology Innovation Scoreboard 2002 (European Commission, 2003) Denmark and Sweden came out as two of the leading countries for nine out of 12 indicators. Sweden had more biotechnology publications and dedicated biotech companies, while Denmark scores highly in terms of drug approvals and US patents. Looking at newer data, the two countries still perform well. Using the number of product candidates in the various phases of the pipeline in 2009, Denmark ranks third after the UK and Germany. Sweden comes out as number 6, after France (4) and Switzerland (5) (Ernst & Young, 2010).

Differences exist in the institutional framework for university IPR. In Sweden the individual professors enjoy full ownership to IP based on public research activities, while Denmark adopted a variation of the American Baye-Dole Act model in 2000 with the purpose of generating future revenues from patents taken out by universities and increasing the commercialization of public research. It is an important question whether and how the changed IPR regime in public research affects collaboration with external partners in the short and the long run. One effect might be that companies actually become less motivated in collaboration with universities if they have to share patent rights.

Foreign-owned firms have, due to a number of mergers in recent years, come to play a crucial role. The dependence on a few large firms, increasingly foreign-owned, exacerbates the vulnerability of the Swedish innovation system to decisions outside the purview of national actors (for further details see Reiss et al. (Chapter 9, this volume); D'Este and Costa, 2007). Mergers and acquisitions are also present in the Danish life science industry, but the specific ownership structure based on foundations has preserved several of the old, large Danish life science companies.

Both countries have a long tradition of implementing various policies to promote biotechnology, for example, specific research programmes, venture capital, public–private partnerships, and industrial PhDs. However, the establishment of VINNOVA reflects a more systemic and long-term perspective on the Swedish policy approach than what currently characterizes the Danish innovation policy. Following Swedish government directives, a comprehensive national strategy for biotech was introduced in 2005, covering the field in the narrow sense, as well as how they connect to application areas, and the potential renewal in a range of industry sectors. A number of primarily structural goals over the period 2005–15 are delineated in this strategy, for example:

- a 50 per cent increase in the number of employees in the life sciences sector;
- a doubling of Swedish net export from the biotech sector;
- a more diversified industry structure, with more SMEs;
- a broadened research base.

However, the current economic crisis has also hit the biotech industry. First of all, more difficult access to capital – which is a key for any research-intensive industry – has slowed down investments in new innovation activities. Lay-off of workers has followed mergers and acquisitions, and many VCs have given priority to existing portfolios (IRIS Group, 2009; Ernst & Young, 2010). This situation makes public policies that maintain and further stimulate investments in high-risk R&D important.

## NOTE:

1. Danish universities account for 40 per cent of the public R&D spending within health; 32 per cent of universities' R&D spending is within health.

# REFERENCES


Asheim, B.T. and M.S. Gertler (2005), 'The geography of innovation: regional innovation systems' in J. Fagerberg, D. Mowery and R. Nelson (eds), *The Oxford Handbook of Innovation*, New York: Oxford University Press.

Bergqvist, H. (2008), *A Benchmarking Study of the Swedish and British Life Science Innovation Systems: Comparisons of Policies and Funding*, VA 2008:12, Stockholm: VINNOVA.

Biotech Sweden (2009), *The Swedish Life Science Industry – Guide 2009*, Vol. 07, Stockholm: Biotech Sweden.

Bitard, P., C. Edquist, L. Hommen and A. Rickne (2008), 'Reconsidering the paradox of high R&D input and low innovation: Sweden', in C. Edquist and L. Homen (eds), *Small Country Innovation Systems: Globalization, Change and Policy in Asia and Europe*, Cheltenham, UK and Northampton, MA, USA: Edward Elgar.

Bloch, C. (2006), 'Biotechnology in Denmark 2995', *Working paper* 2006/4, The Danish Centre for Studies in Research and Research Policy.

Boston Consulting Group, The (2002), *Commercial Attractiveness of Biomedical R&D in Medicon Valley: The Role of R&D in Attracting Regional Investments*, Copenhagen: The Boston Consulting Group.

Christensen, J.L., B. Gregersen, B. Johnson, B.-Å. Lundvall and M. Tomlinson (2008), 'An NSI on transition? Denmark', in C. Edquist and L. Homen (eds.), *Small Country Innovation Systems: Globalization, Change and Policy in Asia and Europe*, Cheltenham, UK and Northampton, MA, USA: Edward Elgar.

Cooke, P. (2005), 'Rational drug design, the knowledge value chain and bioscience megacentres', *Cambridge Journal of Economics*, **29**, 325–41.

Critical I (2006), *Biotechnology in Europe: Comparative Study for EuropaBio*, Brussels: EuropaBio.

Danish Agency for Science, Technology and Innovation (Forsknings- og Innovationsstyrelsen) (2010), *Erhvervslivets forskning, udvikling og innovation i Danmark*, Copenhagen: Danish Agency for Science, Technology and Innovation.

D'Este, P. and J. Costa (2007), 'BioPolis: Inventory and analysis of national public policies that stimulate research in biotechnology, its exploitation and commercialisation by industry in Europe in the period 2002–2005; National Report of Sweden', SPRU – Science and Technology Policy Research, Brighton, UK, available at http://ec.europa.eu/research/biosociety/pdf/biopolis_sweden_en.pdf, accessed 2 September 2009.

Ejermo, O. and A. Kander (2008), 'The Swedish paradox revisited' in C. Karlsson, B. Johansson and R. Stough (eds), *Entrepreneurship and Innovation in Functional Regions*, Cheltenham, UK and Northampton, MA, USA: Edward Elgar.

Enzing, C. et al. (2007), 'BioPolis: Inventory and analysis of national public policies that stimulate biotechnology research, its exploitation and commercialisation by industry in Europe in the period 2002–2005', Final Report, March 2007, Luxembourg: Office for Official Publications of the European Communities.

Ernst & Young (2008), *Biotech in Denmark – Growing Stronger*, Ernst & Young.

Ernst & Young (2010), *Beyond Borders – Global Biotechnology Report 2010*, Ernst & Young.

European Commission (2003), '2002 European Innovation Scoreboard' Technical Paper No. 7, Biotechnology Innovation Scoreboard, European Commission.
Gestrelius, S. (2008), 'Why is Danish life science thriving? A case study of the life science industry in Denmark', VA 2008:09, Stockholm: VINNOVA.
Gestrelius, S., A. Sandström and T. Dolk (2008), 'Natinal and regional cluster profiles – companies in biotechnology, pharmaceuticals and medical technology in Denmark in comparison with Sweden', VA 2008:10, Stockholm: VINNOVA.
Granberg, A. and S. Jacobsson (2006), 'Myths or reality: a scrutiny of dominant beliefs in the Swedish science policy debate', *Science and Public Policy*, **33**(5), 321–40.
Graversen, A.B. and J. Rosted (2010), 'Klyngepolitik på et faktabaseret grundlag – Læring fra et pilotstudie af life science i Østersøregionen', *FORA Rapport*, Copenhagen.
IRIS Group (2009), 'Towards a strong biotech cluster in the Copenhagen region: An analysis of the key success actors for internationally leading biotech regions', prepared for: DANISH BIOTECH, The Capital Region of Denmark, and the Danish Association of the Pharmaceutical Industry (Lif), Copenhagen.
Lacasa, I.D. (2007), 'BioPolis – Inventory and analysis of national public policies that stimulate research in biotechnology, its exploitation and commercialization by industry in Europe in the period 2002–2005 – National Report of Denmark', Karlsruhe: Fraunhofer Institute for Systems and Innovation Research (Fraunhofer ISI).
Lacasa, I.D., T. Reiss and J. Senker (2004), 'Trends and gaps in biotechnology policies in European Member States since 1994', *Science and Public Policy*, **31**(5), 385–95.
Lindman, J., J. Timsjö and N. Özbek (2008), *Looking over the Shoulders of Giants: A Study of the Geography of Big Pharma R&D and Manufacturing Operations*, VA 2008:13, Stockholm: VINNOVA.
Marklund, G. et al. (2004), *The Swedish National Innovation System 1970–2003: A Quantitative International Benchmarking Analysis*, VA 2004:01, Stockholm: VINNOVA.
McKelvey, M., H. Alm and M. Riccaboni (2003), 'Does co-location matter for formal knowledge collaboration in the Swedish biotechnology-pharmaceutical sector?', *Research Policy*, **32**, 483–501.
Medicon Valley Academy (2004), 'From bioscience to new jobs in Medicon Valley: A Medicon Valley Academy strategic report', Copenhagen: Medicon Valley Academy.
Ministry of Industry, Employment and Communications (Näringsdepartementet) (2005), 'Läkemedel, bioteknik och medicinteknik: en del av Innovativa Sverige' (Pharmaceuticals, biotechnology and medical technology: a part of Innovative Sweden), December, Stockholm: Ministry of Industry, Employment and Communications.
Ministry of Science, Technology and Innovation (2010), 'Research barometer 2010" [in Danish], *Forskning: Analyse og Evaluering* 7/2010, Copenhagen: Ministry of Science, Technology and Innovation.
Moodysson, J., L. Coenen and B. Asheim (2008), 'Explaining spatial patterns of innovation: analytical and synthetic modes of knowledge creation in the Medicon Valley life-science cluster', *Environment and Planning A*, **40**, 1040–56.
Norgren, L., R. Nilsson, E. Perez, H. Pohl, A. Sandström and P. Sandgren (2007),

*Needs-driven R&D Programmes in Sectorial Innovation Systems*, VA 2007:15, Stockholm: VINNOVA.
OECD (2009), *OECD Biotechnology Statistics 2009*, Paris: OECD.
Patel, P., A. Arundel and M. Hopkins (2008), Sector report: 'Monitoring, analysing trends and identifying challenges in biotechnology', final report, May, Luxembourg: Europe Innova.
Reiss, T., S. Hinze and I.D. Lacasa (2004), 'Performance of European Member States in biotechnology' *Science and Public Policy*, **31**(5), 344–58.
Reiss, T., R. Lindner and U. Schmoch (2008), 'Developing universities: The evolving role of academic institutions in economic growth. The public and private appropriation of knowledge; Research Report No. 1 on Biotechnology in Europe: Background information on biotechnology industry characteristics and policy environment in Denmark, Germany, Latvia, Russia and Sweden' (unpublished).
Reiss, T., I.D. Lacasa, V. Mangematin, F. Corolleur, C. Enzing, C., A. van der Giessen, J. Senker and L. Nesta (2005), *Benchmarking of Public Biotechnology Policy. Final report*, Contract no. FIF.20030837, Brussels: Commission of the European Communities, Fraunhofer.
Rosiello, A. (2005), 'Comparing biotechnology innovation systems: The cases of Scotland, Sweden and Denmark', *Innogen Working Paper* 35, University of Edinburgh and the Open University.
Rosiello, A. (2007), 'The geography of knowledge transfer and innovation in biotechnology: The cases of Scotland, Sweden and Denmark', *European Planning Studies*, **15**(6), July.
Royal Swedish Academy of Engineering Sciences (2008), 'Nationell policy för forskning och innovation: En framtidsinriktad omvärldsanalys', IVA-M 381, Stockholm: Royal Swedish Academy of Engineering Sciences.
Sandström, A. and L. Norgren (2003), *Swedish Biotechnology: Scientific Publications, Patenting and Industrial Development*, VA 2003:2, Stockholm: VINNOVA.
Sandström, A., H. Bergqvist and T. Dolk (2007), *National and Regional Cluster Profiles: Companies in Biotechnology, Pharmaceuticals and Medical Technology in Sweden 2007*, VA 2007:16, Stockholm: VINNOVA.
SwedenBio (2005), *Biotechnology in Sweden: A National Biotech Agenda for Growth*, SwedenBio, The Swedish Biotechnology Industry Organization.
Teigland, R., D. Hallencreutz and P. Lundequist (2005), 'Uppsala BIO: the Life Science Initiative: experiences of and reflections on starting a regional competitiveness initiative', Research Paper 2005:3, Uppsala: Center for Research on Innovation and Industrial Dynamics (CIND).
Vaekstfonden (2007), 'Biotechnology and venture capital in Denmark – perspectives for the future', Copenhagen: Vaekstfonden.
Valentin, F., R. Lund Jensen and H. Dahlgren (2008), 'How venture capital shapes emerging bio-clusters: A cross-country comparison', *European Planning Studies*, **16**(3), April.
VINNOVA (2002), *Effective Innovation Systems and Problem-oriented Research for Sustainable Growth: VINNOVA's Strategic Plan 2003–2007*, VP 2002:04, Stockholm: VINNOVA.
VINNOVA (2005), *Strategi för Tillväxt: Bioteknik, en Livsviktig Industri i Sverige*, (*Strategy for Growth: Biotechnology, a Crucial Industry Sector in Sweden*), VP 2005:02, Stockholm: VINNOVA.
VINNOVA (2006), *End of an era? Governance of Swedish Innovation Policy*,

(authors: J.G. Thorslund, L. Elg and P. Sandgren), VA 2006:01, Stockholm: VINNOVA.

VINNOVA (2009), *Internationellt Jämförande Studie av Innovationssystem inom Läkemedel, Bioteknik och Medicinteknik*, VA 2009:18, Stockholm: VINNOVA.

# APPENDIX

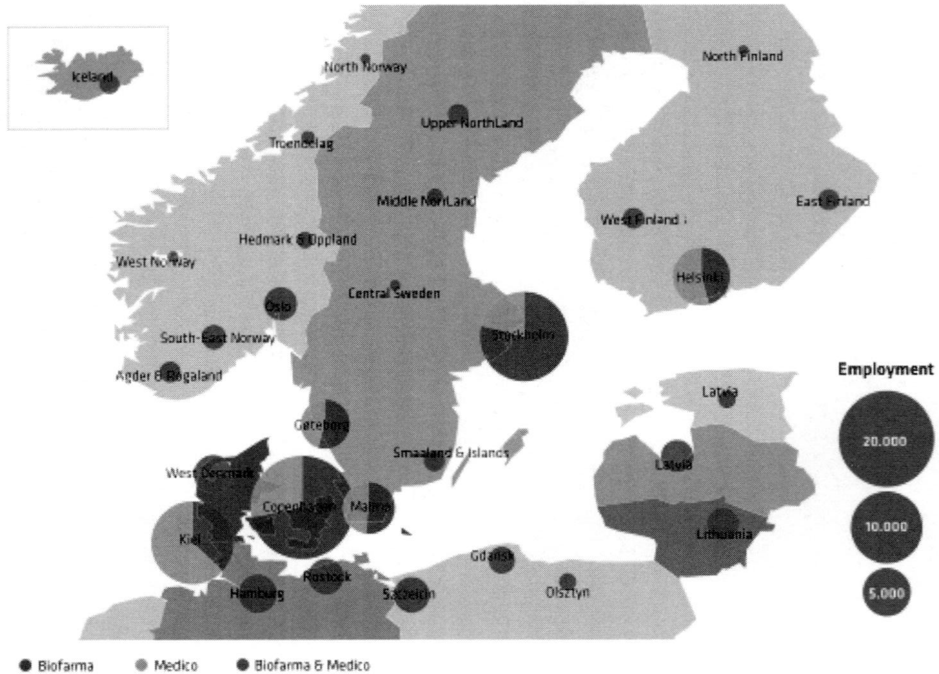


*Source:* Graversen and Rosted (2010, p. 23)

*Figure A10.1 Employment in life science in the Baltic Sea region*

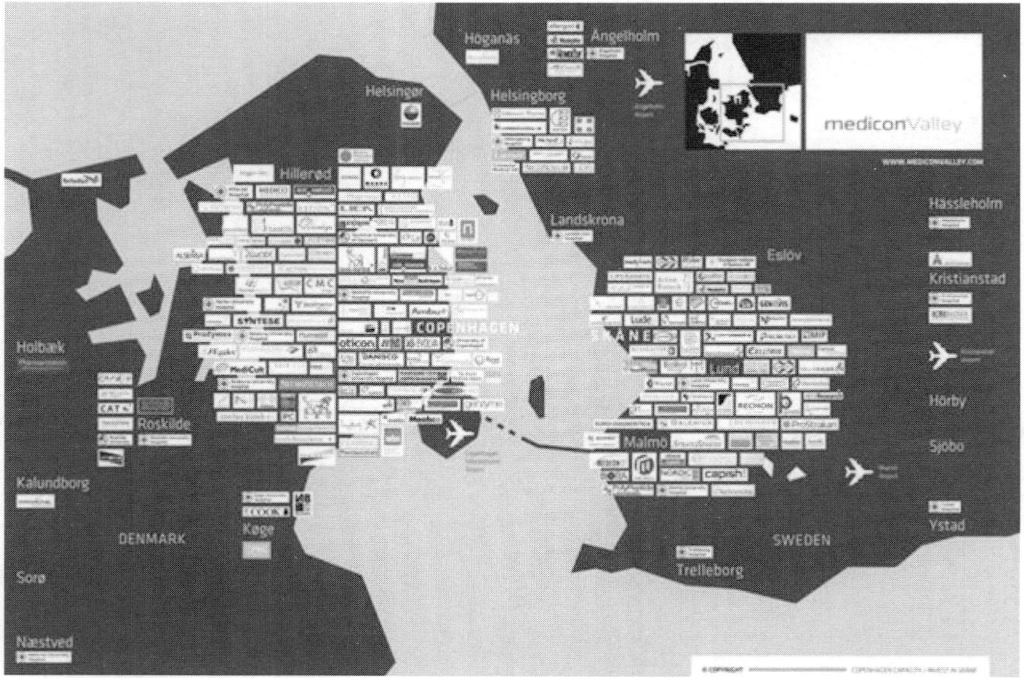


*Source:* www.mediconvalley.com

*Figure A10.2 Medicon Valley cluster map*

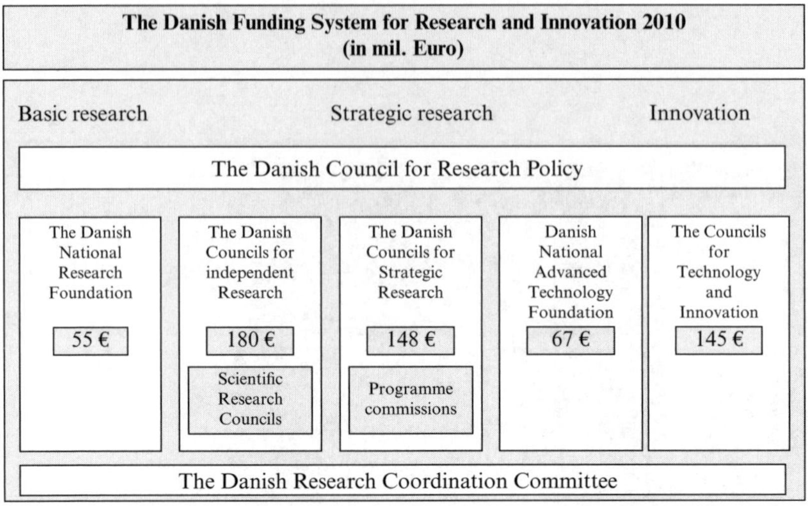


*Source:* Danish Agency for Science, Technology and Innovation (http://en.fi.dk/councils-commissions/the-advisory-and-funding-system-for-research-and-innovation)

*Figure A10.3 The Danish funding system for research and innovation, 2010*

# 11. Biotechnology appropriation in a small country: from historical legacies to contemporary challenges in Latvia

**Anda Adamsone-Fiskovica, Janis Kristapsons, Aija Lulle and Erika Tjunina**

## INTRODUCTION

Biotechnology is one of the keywords for the twenty-first century with research and applications of the results currently being promoted and advanced intensively all over the world. While the evolution of this field can be traced back to a more distant past, its upsurge has been more marked since the second half of the twentieth century, both in terms of research and policy developments, with the process frequently being referred to as a 'biotechnology revolution' (e.g. Russell, 1988).

Nowadays biotechnology is among the key priorities in the field of scientific and technological development worldwide, exerting an impact first and foremost on agriculture, food science, and modern medicine. The *Strategy for Europe on Life Sciences and Biotechnology 2002–2010* elaborated by the European Commission (EC) states that 'life sciences and biotechnology are widely recognised to be, after information technology, the next wave of the knowledge-based economy, creating new opportunities for our societies and economies' (EC, 2002, p. 7). Along with food and agriculture, biotechnology is defined as one among nine thematic priorities for collaborative research within the European Union (EU) 7th Framework Programme (FP) for 2007–2013 (EC, 2006, p. 4). However, prior to that it had been argued that Europe does not have a unified biotechnology policy in place yet, whereby policy-makers in individual European countries have developed a variety of different policy concepts and instruments with the goal of fostering innovation in biotechnology (Caracostas and Brichard, 2004, p. 342).

Though being a small EU country, Latvia is no exception in this respect, with biotechnology having occupied a significant role in the scientific and economic development of the country over the course of the last century. At the same time there is a range of differences in the way research and the corresponding policy domain have evolved in various national contexts. The main aim of this study is thus to trace the historical and current developments in the field of biotechnology research and policy in Latvia, by means of focusing on the following key aspects:

- first, it examines the impact and implications of the recent changes of economic policy on the development of biotechnology fields in Latvia;
- secondly, it aims to identify the agenda-setting agents in biotechnology research and policy, examining interaction mechanisms and power relations in the public and private sectors, as well as the roles of domestic and external actors;
- thirdly, it addresses the assessment of the current policy-making practices and outcomes by different stakeholders;
- fourthly, it detects the key factors hindering or facilitating the development of this sector and policy area, under the current conditions in Latvia.

## RESEARCH METHODOLOGY

This chapter is primarily based on a qualitative methodology (Denzin and Lincoln, 2000), encompassing triangulation of methods. References to quantitative data are only fragmentary due to their scarcity and/or doubts over their validity. The study is based on a review of secondary sources, available studies and analytical reviews so far carried out on the development of biotechnology in Latvia, as well as an analysis of related policy documents. Furthermore, the main body of the study is based on 19 expert interviews (13 face-to-face, six by phone) conducted from March to December 2008 with leading researchers (mainly heads of units), managers and policy-makers in the field of biotechnology in Latvia representing key research institutes (independent and those associated with universities), related ministries, professional associations as well as business enterprises. Semi-structured expert interviews allowed researchers both to probe respondents' answers and clarify them, and at the same time to take advantage of the flexibility of the interview format (Haralambos and Holborn, 2007, p. 826).

Many of the interviewed experts hold positions in both research and administrative/industry organizations and therefore were given not only the role of an evaluator of biotechnology policy and practice, as is most common in expert interviews, but were also seen as active agents involved in policy-making and evaluation of policy. Interviews were mainly conducted in a non-directive way to avoid offering opinions and expressions of approval or disapproval, while in some cases experts were actively encouraged to reflect more in detail upon policy and practice issues with direct questions and active tactics as advocated by some scholars (see, for example, Holstein and Gubrium, 1995) to ensure richer and fuller data.

The study employs elements of policy analysis as a research method and an analytical tool, with a primary focus on decision-making models and the policy process approach (Fischer et al., 2007). The emphasis is therefore laid on the political processes and involved stakeholders in terms of tracing the means being used, the role of various stakeholders and power relations as represented in the policy process. This meso-level approach is supplemented with a meta-policy approach that also looks into the wider contextual factors of the policy process in respect of both the supranational context and national social, economic and political structure.

It has to be noted that the existing national literature on the development and state-of-the-art of biotechnology in Latvia so far has mainly been produced (though not exclusively) by stakeholders closely involved in the respective realms of research and/or policy. Thereby this study aims to provide an additional external view on this realm with an emphasis on the complexities of advancing specific research and policy agendas. Though biotechnology as a multidisciplinary field represents a number of rather diverse subfields of research and industrial applications (for example, environmental, pharmaceutical, food biotechnology) with both the borderline of the field being a rather blurred one and the field internally featuring considerable differences in terms of performance, development paths, and key problems faced by different specialization fields, the study attempts to give an overall view of the national dynamics of this field as a whole.

The chapter has been structured so as to give first a condensed historical retrospect on the development of biotechnology research in Latvia over the last century, followed by a mapping of the current biotechnology sector in terms of its research and industrial components, and a general outline of the present policy landscape in the field. A further major thematic block is devoted to the analysis of the agenda-setting agents and mechanisms in both biotechnology research and policy, by means of focusing on the role of public vs. private, individual vs. collective, and domestic vs. external actors and influences. Interview excerpts and references to other data sources are made throughout the text to make it more illustrative and well-grounded. The report is concluded by a summary of the main findings, and a further reflection on the research results and their implication for the future development of the biotechnology field in Latvia.

## Historical Legacies

Biotechnology has established scientific traditions and a rather strong research potential in Latvia, developed already at the beginning of the twentieth century on the basis of practical applications of microbiology (Beķers and Jākobsons, 2001) thereby positioning Latvia as a respectable scientific centre in the East European region. The first Latvian microbiologists taught and conducted research at the Tartu University (Estonia) in the late nineteenth and early twentieth centuries. The years of the first independent Republic of Latvia (1918–40) witnessed the establishment of the Institute of Microbiology at the University of Latvia (UL) (1921), the development of general microbiology and its medical and veterinary branches as well as the advancement of industrial and agricultural microbiology (see Table 11.1). In the 1920s, different kinds of sera and vaccines were being produced at the Serum station founded by the institute with a range of pharmaceutical companies becoming involved in the production of medications in the 1930s and 1940s (Kalviņš, 2007, pp. 72–4).

After World War II and the incorporation of Latvia into the Soviet Union (USSR), the development of industrial microbiology and industrial biotechnology was promoted by leading researchers in biochemistry, physiology and biotechnology (Beķers and Jākobsons, 2001). In 1957, the Institute of Organic Synthesis, which is considered to be the cradle of molecular biology in Latvia, was founded, thereby setting the ground for research in the field of gene engineering and modern biotechnology undertaken at the institute since the 1980s (Grēns, 2006). Several new laboratories and research groups were formed at the institute, attracting a growing number of young researchers. In the late 1950s, the Vitamin and hormone factory established during WWII was incorporated in the Institute of Organic Synthesis thereby becoming its Experimental factory, which was subsequently also merged with the Penicillin factory (established in 1945) (Līdaka, 1996, pp. 277–9). The Experimental factory of the institute later formed the basis of the currently largest pharmaceutical company in Latvia: Grindex.

The period 1970–90 was marked by the development of lysine and its industrial application, both in Latvia and beyond (see 'Inventions and inventors of Latvia' 1999). According to some estimations, the Soviet Union accounted for a quarter of the world's lysine production, with some early work in scaling up lysine production for industry carried out at the Livani Experimental Biochemical Factory in Latvia (Rimmington and Greenshields, 1992, pp. 5, 21). Initially as the Riga Citric Acid Factory of the Latvian Academy of Sciences (LAS) (re-established as the Experimental Biochemical Factory of the LAS Institute of Microbiology in 1966) it was involved in the development and introduction of production technologies for citric and itaconic acids, with several similar factories being set up in the USSR and licences sold to foreign companies. Considerable efforts were made in the development of bioreactors and control instruments for implementation in biotechnological processes (Beķers and Jākobsons, 2001).

Notably, in 1987, the Interbranch scientific and technical complex 'Biotechnology of Latvia' was formed as one of six complexes of this kind (Beķers, 2009, pp. 68–9) to coordinate biotechnology R&D across Latvia and to facilitate communications with industry, altogether embracing more than 22 research institutes and production facilities (Rimmington and Greenshields, 1992, p. 90). These complexes were promoted as part of the efforts initially undertaken on the USSR level in trying to cope with the economic decline of the time, by means of taking advantage of scientific achievements. This centralized policy initiative facilitated the inclusion of biotechnology of plant and animal cells as well as genetic engineering in the strategic research plans along with classical biotechnology (Beķers, 2009, pp. 68–9). As also noted by Rimmington and Greenshields in their comprehensive analysis of the development of biotechnology in the successor states of the former Soviet Union, in the more traditional areas of industrial biotechnology, such as the production of bulk products of agriculture, the USSR achieved major successes, at the same time lagging behind world developments with regard to new technologies based on genetic engineering (1992, p. 27). However, from 1985 Soviet planners had switched the emphasis from biotechnology for low-grade products for agriculture to producing pharmaceuticals. This, in turn, led to the expansion of biotechnological research and development (R&D) that was undertaken in over 20 research and production organizations (five LAS institutes, two higher education institutions, nine companies, two design organizations) (Beķers and Jākobsons, 2001). There was considerable industrial potential in Latvia, with several biotechnological production units and a range of companies applying biotechnological methods operational at that time.

Mention should thus be made of yet another biotechnology-related Interbranch scientific and technical complex 'Biogen', which linked 30 of the key research, design and production facilities across the Soviet territory (Rimmington and Greenshields, 1992, p. 97). Among others it incorporated the science production association Biolar, headed by the Scientific Research Institute of Applied Biochemistry with its associated production

*Table 11.1 Main institutional actors and selected research themes in Latvia's biotechnology landscape: development chronology*

| | | | | | | | | |
|---|---|---|---|---|---|---|---|---|
| MAIN ACTORS | | | | | | | | |
| *UNIVERSITIES* | **INSTITUTE OF MICROBIOLOGY** | | | | INSTITUTE OF MICROBIOLOGY& BIOTECHNOLOGY | | | |
| | **KLEISTI SERUM STATION** | UL Faculty of Biology | | | | | **Kleisti Biotech-nology Park** | |
| *ACADEMY OF SCIENCES* | | | **INSTITUTE OF ORGANIC SYNTHESIS** | Labo-ratory of Molecular Biology | Institute of Molecular Biology | **BIO-MEDICAL RESEARCH AND STUDY CENTRE** | | Pharma-ceutical and biomedical research centre |
| | | INSTITUTE OF WOOD CHEMISTRY | | | | | | |
| *FACTORIES* | | Citric acid factory (closed down in 1992) Vitamin and hormone factory | | | **GRINDEX** | | | |
| | | | | **OLAIN-FARM** | | | | |
| | | | | **Livani Lysine factory** (closed down in 1992) | | | | Competence centre for Environ-ment, bio-energetics and biotechnology |

| | | | | | | | | | | |
|---|---|---|---|---|---|---|---|---|---|---|
| MAIN ACTORS | *SPIN-OFFs* | | | | | | | Biotechnical Centre<br>Biosan<br>Elmi<br>Asla Biotech<br>Silvanols<br>Biolat | | |
| | YEARS | 1921 | 1946 | 1957 | 1964 | 1970 | 1990 | 1993 | 2005 | 2010 |
| MAIN THEMES | *INDUSTRIAL* | | | | Lysine synthesis | **Lysine production** | | | | |
| | *HEALTH & GENERIC* | | | **Pharma-ceuticals** | | | | | **Latvian Genome database** | |
| | *AGROBIO-TECHNOLOGY & FOOD* | | | | | | | **Protein biotechnology** | | |
| | *ENVIRONMENT BIOFUEL* | | | | | | | Biofuel technology | | |
| | *BIO-ENGINEERING* | | | | | | | Laboratory equipment | | |

facilities based in the town of Olaine, as well as the Institute of Organic Synthesis in Riga (ibid., pp. 33, 124–5, 176).

Research and industry in biotechnology during the first independence period and the Soviet period have been positively assessed by all researchers interviewed, and described as a time when the biotechnology tradition was established, maintained and developed in Latvia. However, while intensive growth of industrial microbiology during the Soviet era in Latvia was ensured by the centralized financial support from Moscow, after the collapse of the Soviet Union and in the beginning of the renewed independent Republic of Latvia in 1991, financial support was discontinued, leading to a substantial decrease in science funding. Following the regaining of independence there was a reorganization of many research organizations and a closure of several industrial facilities, accompanied by a considerable reduction in the number of staff and overall activity in the field. This has been recognized as a huge mistake by researchers:

> The foundation of the experimental industry was the greatest achievement during the Soviet era; it enabled many further developments [. . .]. Regrettably, in most of the cases these factories failed to survive during the first shock after the state regained its independence. It was an example of a short-term vision; it seemed it might be easier to sell out everything or close down. But problems caused by hiking up energy prices were temporary; one could have survived through these and other hardships. (researcher)

There were too few resources for a country as small as Latvia to further support various biotechnology processes, including resource-intensive ones, for example, gene technology processes, on an industrial level (Grēns, 2006). As argued by Rimmington and Greenshields in the early 1990s, 'the collapse of Soviet power and subsequent suspension of payments and grants from Moscow could have potentially catastrophic consequences for Baltic biotech facilities' (1992, p. 87). Yet one can observe a certain revival of biotechnology R&D following the accession of Latvia to the EU in 2004 with a range of new projects and production units gradually emerging in the country.

## Mapping the Current Latvian Biotechnology Sector

The following mapping of the current biotechnology sector has been carried out in three areas: by means of characterizing the present research capacity; the industrial set-up; and the policy landscape.

### Research capacity

Nowadays, research in biotechnology is concentrated in several research institutes, centres and university units, most of which continue a line of work started as early as the Soviet period. The most visible among those are the Institute of Organic Synthesis (1957) and the Biomedical Research and Study Centre (1993). It can be noted that the latter originated from the former via an intermediary – the LAS Institute of Molecular Biology (later affiliated to UL) founded in 1990. In 1997, the International Centre of Biomedicine and Biotechnology, intended for ensuring the link with educational institutions and student training, was established within the Biomedical Research and Study Centre with support provided by UNESCO. It can be noted that an experimental biotechnological complex is being built at the centre. An initiative of a joint biopharmacy centre, involving the Institute of Organic Synthesis, the Biomedical Research and Study Centre and the University of Latvia, as well as an initiative of developing studies of pharmaceutical technologies, are also underway.

Recent years have witnessed a rapid growth of biomedicine and pharmaceutical research at the universities of Latvia. Organizations currently undertaking R&D in biotechnology and life sciences include the UL Institute of Microbiology and Biotechnology, the Department of Microbiology and Biotechnology of the UL Faculty of Biology, the A. Kirhensteins Institute of Microbiology and Virology of the Riga Stradins University and its clinical hospital, the Research Institute of Biotechnology and Veterinary Medicine, 'Sigra', at the Latvia University of Agriculture. The University of Latvia is also developing a centre of natural sciences (with a biotechnology focus) in a designated area (Kleisti) near the capital city. Last but not least, mention has to be made also of the Latvian State Institute of Wood Chemistry – an independent state research institute.

Many of these units have developed on the basis of individual scientific schools led by strong leaders who have also directly influenced the advancement of specific fields of specialization. Besides, the historical legacy plays a crucial role in the evolution of the specialization of individual organizations, as well as the national biotechnology profile as a whole, providing a strong path dependency. This is said to hold true also with regard to the overall organization and management of S&T in the USSR successor states, whereby 'continuity with the past is in fact much stronger than might at first appear'.

> However, the Baltic States (Latvia, along with Lithuania and Estonia) have been treated as comparatively more advanced in developing alternative social and economic structures in their transition towards market economy. (Rimmington and Greenshields, 1992, pp. 67–8) (ibid.)

In terms of specialization one can trace several major lines of research being pursued in Latvia, including plant biotechnology, molecular biotechnology, food biotechnology, and biotechnology equipment. Yet notwithstanding their qualifications and abilities, Latvian researchers are still in the quest of finding their unique niche on the international scene, with only some groups so far having made notable progress in this respect:

> In terms of development prospects it is important to find niches where we can be competitive and valuable on both national and international level. It is senseless to continue doing what all others do. Well, it is a way to demonstrate the ability to achieve something, and to some extent it also makes sense, yet nobody can rouse interest for others by this. (researcher)

Researchers also express scepticism about their achievements, for instance, in terms of scientific publications:

> The number of publications is insufficient, and we don't have much of those that we could say to be something unique [. . .]. There are certain developments and we have original achievements [. . .]; still, the publicity could have been higher [. . .] with more of our own articles [not only co-authored with foreign partners] needed to attract wider attention and encourage greater interest by the industry. (researcher)

According to the BioPolis study (Reiss, 2007), which provides internationally comparable data on publications and citations (Science Citation Index) as well as patents (European Patent Office), the distribution of publications in various sub-sectors of biotechnology in Latvia is generally comparable with ratios in the EU and the US. The biggest share of publications (2002–04) is in the so-called Red biotechnology, with half of all the articles (50 per cent) published in health and over a quarter (27 per cent) in generic fields. This trend can be explained by the influence that stronger countries have on the biotechnology research agenda.

When it comes to publications within the broader biotechnology sector, according to our own calculations that are largely in line with the data and conclusions provided by Reiss, the numbers have almost doubled during the last decade (1995–2004), reaching a level of 15 publications per 1 million inhabitants a year. But it is still a very moderate performance and reaches only 14 per cent of the EU25 level where, on average, 105 scientific articles are published annually per 1 million inhabitants. However, despite the smallness of the country and modest numbers of publications, the citation rate is significantly higher: 95 per cent of the EU25 average (Reiss, 2007, p. 30). It indicates that Latvia's biotechnology research is competitive on the international scene.

Yet again, only about between one and two European patents in biotechnology go to Latvia's scientists per year, thereby reaching just 10 per cent of the EU25 average (calculated per million inhabitants). As noted by

Reiss (2007, p. 17), between 1994 and 2003 only ten biotechnology patents in total were applied for from Latvia, which makes a discussion of performance in knowledge transmission based on patent indicators unfeasible. However, the overall picture might be more optimistic, since the given figures exclude patents in the field of pharmaceuticals, where Latvia has a better performance in terms of granted patents.

The low publication and patent rates can be related to the weak stimulus for publishing articles in highly prestigious scientific journals and the probably erroneous decision of the Latvian Council of Science made in the 1990s not to distinguish between national and international publications upon research assessment. In the case of European patents, the main obstacles have to do with insufficient financial resources as well as the lack of qualified assistance in the field of patenting. Simultaneously, there is also a common opinion that possible gains from patents are insignificant; it is difficult to put them into practice due to the limited market in Latvia, and ties with potential partners abroad are still weak.

An extended analysis, based on the above-mentioned BioPolis study of the policies and funding for biotechnology research and commercialization in the Central and East European (CEE) countries for the period 2002–05 (Senker et al., 2008), has rightly noted the low performance of Latvia based on the number of publications and patents. At the same time the authors acknowledge that this method for the assessment of performance of the CEE countries is not fully justifiable. The format for this practice is influenced not least by the legacy of the past on the publishing behaviour of CEE scientists, and the bias towards English language journals in publication databases. It also has to be noted that no account has been taken of the inventions made by the Latvian scientists that have become the basis for the patents filed by foreign companies. Besides, relative to the size of the Latvian economy, a rather notable pharmaceutical industry has been concentrated here, with a considerable share of its production (80 per cent) being exported (Kalviņš, 2009). Therefore biopharmacy or health biotechnology is among the few research fields in Latvia that has had and has retained rather strong positions.

Some additional illustration of the state-of-the-art is provided by the data on the participation of the Latvian R&D community in the EU FPs[1] that demonstrate the research capacity and cooperation with fellow researchers abroad. There were altogether 84 Latvian partners participating in 75 project proposals within the thematic priority of 'Life sciences, genomics and biotechnology for health' under FP6 (2002–06), out of which 10 were successful in attracting EU funding (Berķis, 2008, p. 11). The participation of Latvian small and medium-sized enterprises (SMEs) in this thematic priority was as high as 22 per cent, though mainly due to successful participation of a particular Latvian biotechnology company (Asla Biotech) (ibid, p. 12).

When it comes to academic education in the field, in the early 2000s there were more than 5000 students involved in the biomedicine-related study programmes offered by the University of Latvia, the Riga Stradins University, the Riga Technical University and the Latvia University of Agriculture (Grēns et al., 2007, p. 32) with a gradual increase discernible in subsequent years. Education in the field of biotechnology is currently provided also by the Rezekne Higher Education Institution and the Mechanics and Technology College of Olaine. Increasing efforts are being undertaken in improving the university infrastructure and boosting the number of graduates and qualified specialists in the area of life sciences, including biotechnology.

International cooperation is considered to be of vital importance for such a small country as Latvia, with many research institutes looking for collaboration partners abroad. Since 2002, Latvia has been involved in the international initiative ScanBalt BioRegion, launched under the EU FP6, which stands for a meta-region (region of regions) in life sciences and biotechnology in the Baltic Sea Region encompassing 11 countries and regions (Blank et al., 2003). The mission of ScanBalt is to promote the development of this BioRegion 'as a globally competitive meta-bioregion by forming a network of networks and promoting public dialogue on the opportunities and dilemmas of biotechnology and related disciplines' (ibid., p. 48).

A mapping analysis undertaken within the framework of this project, with respect to the competencies and capacities in life sciences and biotechnology in the individual bioregions and countries, resulted in a classification of bioregions into three groups (ScanBalt, 2007, pp. 21–2). Accordingly, the Latvian biotechnology cluster was enlisted among the 'Scientific fountains'. This denomination represents a geographical concentration of knowledge-producing activities characterized by a high R&D activity and a large pool of human capital. The other two groups cover 'Co-location Clusters' and 'Mode 3 clusters'. While the former is characterized by the creation of spin-offs and the attraction of companies who benefit from co-locating their own R&D activities in close proximity to university research institutes, the latter features a large pool of biotech-skilled workforce and supposedly high impact on local business. Regarding the relative regional competence indicator, the Latvian bioregion places emphasis on several biomedical research fields, with the regional pillars of this relatively young biotechnology scene represented by virology and cell biology (ibid., pp. 57–8).

The value of national human resources is being increasingly stressed, especially when looking from the point of view of potential cooperation partners. Nowadays it is the specific knowledge and experience possessed by Latvian researchers that is seen as the primary resource in the international market, rather than access to a cheap labour force as a motive for cooperation:

> We have people, experienced in particular fields of research, and they are eager to cooperate. Therefore it might actually be human resources that are essential for them. (researcher)

On the whole, the research capacity in the field of biotechnology in Latvia reflects its historical legacies and the turbulence during the fundamental changes of the country in the 1990s, with the dismantling of the previous system. Nevertheless, the country has demonstrated its ability for further development of the sector by establishing new laboratories, expanding academic activities and by demonstrating academic excellence in publications with high citation rates despite the relative disadvantage of a small country. The current research capacity in this area in Latvia can thereby be characterized as a generally notable and future-oriented one, with the biotech field being among the most advanced ones in terms of its potential for the future.

**Industrial component**

There are quite diverse estimations with regard to the number of companies that could be attributed to the industrial biotechnology sector in Latvia. But as noted by a foreign journalist 'the most remarkable thing about Latvia's biotech sector is that it has one', stating that Latvia's biotech business community is still in its infancy (Nimmo, 2005, p. 4). According to the Yellow Pages of the Baltic Sea network *ScanBalt Bioregion*, there are 26 biotechnology companies in Latvia, yet these cover a broad area of life sciences, thereby not specifically covering biotechnologies (ScanBalt, 2006). The Association of Biotechnology of Latvia (2009), established in 2006, has enlisted 15 most well-known Latvian SMEs applying biotechnology methods in their production, though at the end of 2010 it had only six companies among its members – a figure of dedicated biotechnology firms mentioned also by Malo and Norus in their study of the emerging biotechnology industry in Estonia, Latvia, Lithuania and Poland (2009, p. 487). The national experts interviewed for this study estimate the number of truly biotechnology-related companies to be no more than a dozen:

> What could be defined as a biotech company has always been a disputable issue. [. . .] One could include also all yogurt producers or breweries, despite the fact that they buy ready-made ingredients, since they anyway have a microbiologist there to manage the process. The question remains if a biotech company is one producing biotechnology itself or also those companies making use of biotech in their own production. We don't have more than 15 of those

> producing biotechnology [. . .] and in most cases they are small enterprises. (entrepreneur)

When reflecting on the small number and generally low capacity of companies involved in the biotechnology sector, experts draw parallels with booming sectors such as Internet technologies that used to have high profit expectations associated with them. Thereby experts bring forward an assumption that low research capacity and slow development of an industry might also be influenced by industry developers refraining from substantial investment in the area, where profits might come only after some uncertain time. Instead, they choose to invest in more common products, which have already secured their place in the consumer market:

> People understood that [biotechnology] will not be a guarantee of fast profit. [. . .] and that achievements will rather come in many years and they might not be commercially feasible at the very beginning. (civil servant)

But, of course, explanation for the limited scale of the industrial biotech sector in Latvia has to be sought for also in the more general historical legacies of the socialist command economy that was characterized by high-level R&D but an inability to translate R&D into industrial applications (Rimmington and Greenshields, 1992, p. 28). As in other branches of the economy, this barrier to innovation in the biotech field had been related to the lack of commercial incentives for the industrial managers at that time (ibid.), which became a brand new impetus under the market economy since the early 1990s. Though the Baltic States are said to have pioneered the setting up of small profit-oriented biotech companies and to have witnessed the emergence of many individual scientists possessing an entrepreneurial flair, not least in Latvia (ibid., pp. 84–7), in terms of international comparison this economic activity can be judged as rather modest.

Currently, the pharmaceutical sector is seen as the leading one in the biotechnology field, the pharmaceutical industry in Latvia having the most significant output. It is also one of the main beneficiaries of the growing interaction between industry and academia, while the highest value-added of all of Latvia's biotechnology and pharmaceutical sectors is generated by the production of biomedical equipment (LIDA, 2005). In 2002, the volume of industrial production related to the biomedical sector in Latvia made up EUR 70m or 0.81 per cent of GDP, yet the overall export volume of the sector is comparatively low (Grēns et al., 2007, pp. 29–30).

Unlike several larger pharmaceutical companies (Grindex, Olainfarm), those explicitly operating in the field of biotechnology are mainly small or medium-sized ones, the majority of which are concentrated in and around the capital of Latvia (Riga). Some of the strongest are spin-offs created during the 1990s, specializing in producing laboratory equipment. The motivation for the creation of spin-offs came as a breakthrough strategy out of the stagnation of research during the 1990s. As noted by Malo and Norus, following the transition from state ownership to market economies quite a few scientists in the countries of the Baltic Sea region, including Latvia, turned into entrepreneurs 'to capitalise on their expertise and the latest public research results' with many dedicated biotechnology firms having spun off from universities and public research organizations (2009, p. 487). Likewise, new research directions have been derived and built up based on the experience gained by Latvian scientists abroad.

As the Latvian biotechnology sector is a rather small one, most of the equipment produced by local companies is exported to the West, to other post-communist countries as well as to some other parts of the world. The main factors hampering the growth of the small companies generally include limited local opportunities for approbation of new technical solutions, difficulties to fulfil all technical demands in the export markets, the lack of trust of Western partners in companies from post-communist countries, as well as the lack of investments to expand their production and patent their products. However, in the future companies envisage further expansion of their activities, working not only as suppliers to other equipment producers but also working directly for end users.

Given the necessity for high production capacity of the internationally large branches of chemistry and biotechnology, the development prospects thereof in Latvia are assessed as rather problematic, though the national research potential in biomedicine and medical biotechnology is seen by science managers and researchers as internationally competitive and is characterized by good cooperation with local industrial companies (Grēns et al., 2007). Under these conditions the attraction of large foreign investors, who could be motivated by a skilled labour force and by comparatively lower labour and infrastructure costs, is considered to be one of the major opportunities for the further development of this field in Latvia. It has been argued that the collapse of the Soviet Union (along with its massive science budget) has opened the way for Western biotechnology and pharmaceutical companies seeking opportunities for cooperation and investment in the successor states of the USSR, with the Baltic R&D and production establishments becoming attractive not least due to their geographic position potentially serving as a bridge between the West and the former USSR (Rimmington and Greenshields, 1992, pp. 90–93).

Before the EU accession and even a couple of years after the country joined the EU, prospects of development were rather bleak. The main sectoral weakness at that time was related to the lack of marketing skills and options for local products and funding thereof; a limited local market;

an outdated research infrastructure and technical supply; as well as a slow inflow of young specialists and a continuous brain drain (Grēns et al., 2007). A considerable impediment for a substantial expansion on the industrial part was seen in the generally limited research funds; public funding for private companies was scarce and almost exclusively available through international cooperation (FMERG, 2007). This, in turn, has determined the specialization in services that has served as the source of regular income for the majority of biotechnology companies. While nowadays Latvia is witnessing the creation of a number of pharmaceutical production facilities and significant improvements in research infrastructure, along with the two other Baltic States, it is still seen as being in an earlier stage of development in terms of its biotechnology potential, largely depending on the availability of financial resources for the advancement of the field in the future (ibid.).

Researchers underline that Latvia does not yet take full advantage of international cooperation. In researchers' opinion these lapses are often related to the lack of strategic vision on behalf of governmental bodies responsible for the distribution of financial resources to launch or co-finance such collaborative projects:

> We are not participating in the EU technology platforms, and in this case Latvia, hoping to save a couple of thousands, peanuts actually, misses an opportunity to receive millions. (researcher)

International cooperation is perceived to be of paramount importance not only as a springboard to increasing visibility in the international research community or to be able to access other markets, but also for being competitive and to receive recognition at home:

> Cooperation with, let's say, Cambridge: what does it imply? Money, brains, a window to the world? Everything taken together, I would say. But the most important gain [from international cooperation] is the recognition at home. To be valued here, we have to be recognised there. (entrepreneur)

It should be mentioned that the imbalance between the public research sector and industry (both privately or publicly owned companies) has considerable implications for the overall sectoral structure of biotechnology in Latvia. The limited national absorptive capacity of the enterprise sector largely determines the current orientation of the Latvian research institutions primarily at collaboration partners abroad rather than locally:

> One has to look for a real partner who you can work with rather than picking around with those located closer to you. Because the whole science develops not according to the geographical proximity or principles of some kind of national affiliation but rather according to the scientific logic. (researcher)

Leading researchers emphasize that an intensive international collaboration is needed and it is very important to make contacts and collaborate with the best laboratories abroad, 'which would not allow [us] to fall into provincialism and self-satisfaction' (Grēns, 2006). It is being questioned whether a small country is actually able to maintain high-level science without foreign laboratories:

> I hardly think it is. The present industry of Latvia does not stimulate the development of high-tech, in most cases it is not capable of making use of such technologies either. (ibid.)

The weakness of the local industrial partners also underlies the expectations voiced in respect of the emergence of a major foreign investor that could provide much-needed impetus for boosting the developments in the local market. The issue has been addressed in most of the interviews, with respondents seeing international cooperation as an opportunity to increase competitiveness and sharpen the scientific perspective through what one entrepreneur denoted as a 'positive opposition'.

Yet, in experts' opinion, international cooperation cannot be seen as the only way of development, and should be critically reviewed in terms of the potential gains and losses, both immediate and long-term. The greatest concerns are related to the ability to ensure equal opportunities and equal share of future incomes to researchers from Latvia vis-à-vis more resource-intensive industries abroad:

> Latvia is a small country and our entrepreneurs are still financially rather weak, therefore we are forced to collaborate with foreign companies. It provides us with experience and allows us to earn money. [. . .] Yet, the problem is that if we continue to work mainly for foreigners, one should be aware that the main profit will go to them, but actually license fees of patents should go to Latvia's science, since our scientists have secured them. (Kalviņš, 2007, pp. 110–11)

To conclude, it should be noted that the current set-up of the national economy determines a rather marked distinction between the contribution made by the private and by the public sector to the advancement of biotechnology in Latvia. Namely, there is a considerable imbalance between the capacities of the public research sector and the local industrial actors. Nowadays the research capacity in biotechnology in Latvia is much stronger than the industrial component of the scientific-production complex, with the majority of domestic industrial actors having limited absorptive capacity for locally generated research results. As argued by Malo and Norus (2009), dedicated biotechnology firms in transition economies generally face a shortage of skilled labour, rather costly registration procedures as well as considerable financial constraints (for example, lack of venture capital) and a still underdeveloped regime for intellectual property rights, with all these factors taken together strongly impeding the development of strong biotech industry in these post-socialist countries.

While an urgent need to establish a national-level transfer centre of biotechnology in Latvia has been voiced by several stakeholders, international cooperation is also seen to be essential in boosting the research capacities, gaining international recognition and finding industry partners willing and able to launch production of biotechnology products.

**Policy landscape**

The policy landscape in the field of biotechnology in Latvia is outlined here by the identification of the respective public authorities, the evolution and scope of the national legislation governing the field as well as the policy measures and initiatives aimed at advancing the development of biotechnology on a national level. Following the systems perspective in the analysis of the biotechnology policy landscape put forward by Lacasa et al. (2004), we analyse here how vertical (biotechnology-specific) and horizontal (generic) policy instruments are implemented in Latvia and how they influence coordination and cooperation in networks involved in the biotechnology field.

Following the changes in the political and economic regime in the 1990s, there was a substantial slowdown both in research and in industry. However, in the last decade biotechnology has become one of the key national priorities in science defined by the Cabinet of Ministers, introducing both vertical and horizontal policy instruments to promote the advancement of the biotechnology field. The priority area of Organic synthesis and biomedicine was included in the list of research priorities back in 2001, yet a special four-year state research programme focusing on gene technologies and new synthesis technologies of biologically active substances was launched only as recently as 2005 (see Table 11.2). Additionally, in 2006, another four-year programme in the newly defined priority area of Agro-biotechnology was launched, with a focus on innovative and environment-friendly food production technologies (on all programmes for 2005–09 see Ministry of Education and Science, 2010, pp. 41–82).

On the whole, the governmental decision on the prioritization of selected branches of science has formed a substantial part of the policy-making process in Latvia, also having an impact on the field of biotechnology:

> Setting priorities in science development also might be seen as a way of policy-making. There is no biotechnology explicitly among these priorities, but it appears there as a subfield and as such is involved in at least two main

*Table 11.2 Major funding sources for the biotechnology sector in Latvia*

| | |
|---|---|
| STATE BUDGET | Basic and Applied Research Projects Programme (1994–)<br>Support for Joint Research Projects (1994–)<br>Support to market-oriented research (1994–)<br>State research programme 'Forestry and wood processing technology' (2005–09)<br>State research programme 'Material science' (2005–09)<br>State research programme 'Organic synthesis and biomedicine' (2005–09)<br>State research programme 'Medical science' (2006–09)<br>State research programme 'Agrobiotechnology' (2006–09)<br>State research programme 'Public health' (2010–13) |
| ABROAD | Research cooperation with foreign-based entities<br>Licensing<br>EU Framework programmes<br>EU Structural funds |
| NATIONAL COMPANIES | Olainfarm JSC<br>Grindex JSC<br>Elmi Ltd.<br>Biotehniskais centrs JSC<br>Asla Biotech Ltd.<br>Biosan Ltd. |

> directions: innovative food products and production technologies thereof and biomedicine, pharmacy, respectively. (civil servant)

However, the very process of priority setting has triggered discussions on the appropriate number and definition of priorities, as well as on the contents of the future programmes taking into account realistic prospects of a small country (Kristapsons et al., 2008, p. 21). While biotechnology is defined as a priority area in a range of policy documents, the actual prioritization of this field and the relative importance attributed to it can be questioned in the light of the scope and coverage of other priorities:

> All science sectors cannot be defined as priorities – that would undermine the whole concept of 'priority'. It seems to me that currently there is some fear in Latvia, anxiety, let's say – if I do not fall into the category of prioritised areas, I will be left outside and will not get any funding at all. That is totally wrong. Priority sectors are those where state invests proportionally more funding, presuming they would give higher output to the economy and welfare. [. . .] But we have to take into account that the whole process [of setting priorities] has just

started. In 1991–2005 we had five priority areas – it would have been enough, but there were efforts made to include more, and now we have nine already. (civil servant)

It should be noted that in August 2009, according to the Law on Research Activity (2005), a new reduced set of five priority fields for funding fundamental and applied research for the period of 2010–2013 was approved by the government (see Ministry of Education and Science, 2010, pp. 82–7). Among those a thematic block on public health includes biomedical technologies with biomedicine and pharmacy envisaged to form the basis of this research area.

Both horizontally and vertically, the advancement of biotechnology research has also been promoted through several state aid programmes funded from the EU structural funds (SFs) since 2005. Thus, for instance, several grants have been allocated for the modernization of applied research infrastructure at state research institutions dealing with biotechnology research.[2] State aid has also been envisaged for several activities funded in the framework of the EU SF planning period for 2007–13. At least three notable activities related to the biotech field launched in 2009–11 can be identified. Firstly, out of the total number of six competence centres to be set up by 2015 (around one third of funding to be additionally attracted from business companies) two – Competence Centre for Environment, Bioenergetics and Biotechnology and Competence Centre for Pharmaceutics and Chemistry – pertain to the biotech field. Another support measure is the programme for the establishment of research centres of state significance in the framework of which several of the nine centres, which are primarily aimed at establishing common research infrastructure, cover aspects of biotech (including the Pharmaceutical and Biomedical Research Centre). Thirdly, biotech has also been supported in the framework of yet another EU SF co-funded programme 'Support for science and research', which is mainly aimed at providing support for applied research and where quite a considerable share of the total number of 114 projects funded deal with biotechnology. Although against the backdrop of economic and financial crisis having evolved since 2008 there have been considerable cuts in the R&D funding from the state budget, with GERD having dropped from 0.61 per cent of GDP in 2008 to 0.45 per cent in 2009, to a certain extent this EU SF funding is envisaged to compensate for these cuts.

Policy initiatives can also be traced to several trade missions to other European countries organized by the Latvian Investment and Development Agency for researchers and entrepreneurs in the field of biotechnology in 2008. Finally, a notable bottom-up incentive in providing funding for national research in the field of medicine, including biotechnologies, is represented by the establishment of the TAIHO Latvian Foundation in

2004 by Japanese pharmaceutical company, Taiho Pharmaceutical Co. Ltd, Latvian Academy of Sciences, Riga Stradins University, University of Latvia, Grindex, and Institute of Organic Synthesis. Based on funding raised from donations by physical and legal entities, it aims at facilitating R&D of new medications in Latvia.

The main sectoral public authorities that could be considered as at least formally involved in the governance of the field of biotechnology in Latvia are:

- the Ministry of Agriculture;
- the Ministry of Environmental Protection and Regional Development;
- the Ministry of Health;
- the Food and Veterinary Service;
- the State Plant Protection Service.

It is hardly possible to identify a single public authority holding direct responsibility for the whole field of biotechnology and encouraging network activities among the actors involved in this sector. Generally the specific policy landscape in Latvia is rather fragmented, with various organizations holding responsibility for different aspects of biotechnology, but lacking an overall coordination and thereby a clear systemic approach to policy instruments enabling development of this sector. Each of the involved authorities covers only a minor or very specific aspect of the field. Thus, for instance, while there is a special Division of Biotechnology and new foodstuffs at the Ministry of Agriculture formed in 2006 as a result of the reorganization of the Ministry of Health, in practice it deals only with issues regarding genetic engineering and genetically modified organisms (GMOs). Thereby the primary function of the division is related to the elaboration of the legislative documents governing this specific area.

It is also illustrative that some of the interviewees involved in the field of biotechnology from the enterprise sector were not even aware of the existence of such a ministerial unit, not to speak of any kind of collaboration record. At the same time it has to be emphasized that the ministry has close collaboration with scientists upon drafting the laws, regulations and national position statements in the area of GMOs.

While the Ministry of Economics could be expected to cover at least the industrial part of biotechnology, there are no officers or divisions in the structure foreseen to be in charge of specific branches in general, except for the Energy Department, with its Division of fuel policy dealing with issues particularly related to biofuel (Concept on biofuel (1995); Programme on biofuel 2003–10 (2003); Law on biofuel (2005)). Also, there is no designated unit covering biotechnology research at the Ministry of Education and Science, thereby limiting the basis for collaboration between the ministry, related research units and companies in advancing this area:

> There are no particular officials responsible [for biotechnology] at the Ministry. The capacity is too limited. I know that in larger countries there is at least one person particularly responsible for an individual branch at the ministries. (civil servant)

When it comes to national administrative bodies involved in a horizontal research policy, one may mention the Latvian Council of Science in charge of the distribution of research project funding allocated by the state, which has a special branch commission in molecular biology, microbiology and biotechnology. This body could be seen as the first one in the domain of public administration during the period of regained independence of Latvia to encompass biotechnology as a designated field. It also encompasses the scientific expertise based on the academic record of all its members that provides it with a special say in this domain vis-à-vis the civil servants at the ministries. At the same time the Council is limited in its capacities regarding influencing the policy agenda. Almost all its members hold full-time positions at universities and research institutions and therefore can devote relatively little time for work at the Council. This, in turn, puts limits on the efforts for standing up for a thematic block that covers rather diverse and sometimes also potentially conflicting interests.

One of the first major developments in the field of biotechnology policy of the post-independence period was marked by the launch of the Latvian Population Genome Project in 2001. It can be noted that ambitious gene bank projects have been proposed in numerous countries, with the intention of mapping genes for common diseases, and an aspiration to improve the health of populations (see, for example, Austin et al., 2003; Gottweis and Petersen, 2008). Initially, the Latvian project was planned for a ten-year period and envisaged the creation of a unified national network of genetic information and data processing (Pīrāgs and Grēns, 2003). Since then the project has evolved into a long-term state-funded genetic research programme. On 13 June 2002, the Human Genome Research Act was passed governing the development of the national genome database and regulating genetic research in Latvia; several amendments have been added to it in subsequent years. It has been argued that the Act has stimulated the modernization of Latvia's pharmaceutical and biotechnology industries, and it is expected to attract foreign capital to research into genome and data usage (LIDA, 2005, p. 6).

On 20 March 2006, the Ministry of Agriculture issued a decree on the formation of a task force for drafting the Biotechnology Law; later it was elaborated and approved as the 'Law on circulation of genetically modified organisms' (2007). So far other legislative acts at national and international level (those Latvia have signed and/or ratified), pertaining to the field of biotechnology, include regulations for educational and research organizations, supervision authorities for inspecting and registering medicines, controlling food safety, and so on. Civil servants at the governmental bodies responsible for drafting and implementing legislative acts have expressed doubts if a special law on biotechnology is needed. It has been advocated that regulations required have been settled through other legislative acts and that a stricter regulation framework is not in the interests of the industry and scientists themselves:

> You see, nobody really needs a legislative regulation [a special law on biotechnology], because it would restrict everybody within a certain framework, but nobody wants to be restricted. Also scientists themselves, they would like to work freely. (civil servant)

International funding, and the EU funds in particular, is seen as a very important horizontal policy instrument for the development of the biotechnology field and also as a structural element for defining development directions. While the state does not fund any specific sector directly, it provides support through EU funds and the national budget allocated to science, within which a certain share goes to priority areas for both research and industry.

Stakeholders have differing opinions regarding what counts as a national policy in biotechnology. While there are selected policy instruments in place, they are not always seen as constituting a policy. Besides, due to limited administrative capacity the current policy in the field of biotechnology is rather fragmented and one cannot really speak of a unified biotechnology policy with the sector being generically supported by science, entrepreneurship and other policies. Moreover, one cannot speak of a specific biotechnology policy outside the context of the overall science policy in Latvia:

> There is a running state research programme, plans for the establishment of a competence centre [. . .], but since there is no common strategy or policy for the development of science in Latvia, I do not see a separate strategy and policy in the field of biotechnology either. (researcher)

> Biotechnology policy should be treated in the wider context of science development in Latvia. We are not talking about research policies in specific fields (e.g. IT, biotechnology). For example, we have drafted a concept paper on science and technology development in Latvia and now [April 2008] it is at the Cabinet of Ministers waiting for the approval [. . .] We are not focusing on specific sectors in the document; we are talking about science as a whole. (civil servant)

On a policy-making level, there is a persistent dilemma of either ensuring only horizontal policy measures or defining specific policies for particular branches (including biotechnology) in Latvia. In respect to specific branches, for the time being it is seen as more sensible for the state to simply provide a policy framework (i.e. generic instruments) to be filled with particular contents by the research community itself:

> The only science policy is to pick the right people or rather the right institutions to put the money into. The whole science policy should be targeted at a reinforced maintenance of objectively strong research institutes. (researcher)

Here it can be noted that according to the findings of a study on public policies promoting commercialization of biotechnology in 14 EU member states (Enzing et al., 2004) there was no justification found for the common proposition that countries with dedicated biotech instruments show better commercialization performance in biotechnology than countries without such instruments. Instead, it was shown that success is more likely to be determined by the systemic character of public policies that address all elements of the innovation system, including instruments that stimulate the life sciences knowledge base and its commercialization. As described above, the policy landscape regarding biotechnology in Latvia can be understood only in the wider context of science and research development in the country (including priority setting for particular sectors). The fragmented picture of the national policy landscape can be said to emerge as a result of an incremental policy approach, as opposed, in principle, to a proactive setting of long-term goals with a corresponding mix of legislative, policy and funding actions.

## AGENDA-SETTING IN BIOTECHNOLOGY RESEARCH AND POLICY

The following section focuses more specifically on the discussion of the actors determining the biotechnology research and policy agenda in Latvia, by looking at the policy-making process and policy assessments by stakeholders.

### Top-down vs. Bottom-up

The biotechnology policy in Latvia represents a mixture of top-down and bottom-up policy-making elements. Some researchers even declare that 'there is not any biotechnology policy after 1991'. This and similar assessments were found in most of the interviews with researchers, and highlight that there is more room for manoeuvre for a bottom-up approach. The lack of a strict top-down policy approach dominant in the Soviet era is sometimes, in the current policy climate, generalized to the position that no biotechnology policy exists at all. Today the national policy in the field of biotechnology is mainly induced by strong lobbying activities of individuals from the research community:

> One cannot say that science and technology policy and its formulation are guided only from above. There are sectoral ministries that set their own priorities. There is also formation of initiative task forces – these, in turn, stemming from science – providing recommendations followed by their discussion and only then moving further to the government for approval. The end result is a governmental document, a policy, but cooperation and joint elaboration thereof overlap. A civil servant cannot assign that – he does not or might not know, he has different functions. The initiative comes from the scientists of the specific branch. (civil servant)

> It is a difficult question [whether there is any state policy in the field of biotechnology currently in Latvia]. I think that much depends on the initiatives of scientists and entrepreneurs. What was it like in the Soviet times? Yes, then we could definitely speak of a state policy in the field of biotechnology on the scale of the whole USSR, and Latvia was not at all the worst one and therefore it received quite a lot from Moscow. (researcher)

Stakeholders perceive the centralized biotechnology policy pursued during the Soviet times as having been more structured and comprehensive, which in turn was conditioned by a generally stronger support for science and its application in the economy. Yet while under the socialist regime all policies were strictly defined by the central governing bodies on an all-USSR level, in practice such a regime does not represent a sharp contrast to the current capitalist system in respect of the biotechnology policy, due to a strong, though less apparent, presence of bottom-up lobbyism in the latter system. It has been recognized that the right channels for lobbying have to be chosen:

> If I go alone to the ministry, nobody really listens to me. Maybe it was different, when [names a person] was still there, then it was possible to talk about issues, since we knew each other. It is a bit different now. The lobbying should go through the university and the university should talk to the ministry. (researcher)

The decisive role of the research community in determining the research agenda is strongly related to their specific competences, which due to the high level of complexity of this field provide them with exclusive expertise and decision-making power. In practical terms researchers and key persons from the industry have power over the civil servants and policy-makers in governmental bodies, due to the possession of expert knowledge. In some cases the lack of specific knowledge at the legislative and policy-making level may even serve as a facilitator and accelerator to adopt regulations:

> The law [on GMO] was prepared in one year. Probably it was so fast thanks to the fact that people in the parliament and the Cabinet of Ministers had limited knowledge about these issues, and, taking into consideration the high complexity, nobody has really anything to add. (civil servant)

While a smooth adoption of a law can be seen as a positive factor, it does not remove doubts about the possible hidden lapses in the process or outcome, and it does not assure stakeholders that it is the best possible legislative framework for the country. At the same time, as noted above, not only ministries, but also the research community, lack human resources able and willing to devote enough of their time and energy to the legislative process:

> The capacity of the ministry [of Education and Science] is weak, the capacity of the Latvian Council of Science is limited, because almost all people there work also here [at universities]. There is a shortage of human resources, people with specific knowledge and, what's most important, having time to do it. (researcher)

To conclude, civil servants, researchers and industry players display ambivalent attitudes towards whether a top-down, a bottom-up or an integrated approach should be favoured, and which approach could be expected to be the most effective in Latvia. But they are rather united in their scepticism that authorities responsible for the sector lack a strategic vision, though admitting that Latvia's situation is not unique in this respect:

> Is there a strategic vision of Latvia's government or the Ministry of Education and Science? [. . .] Have they analysed in detail what we want to achieve with that in ten, twenty or thirty years? I doubt it. Yes, there are priorities, of course, and biotechnology is among them in a way, but is it grounded in research and assessment for the future development? But I can't say Latvia is unique in this case – I think it has drifted along in quite many EU countries. (civil servant)

At the meso level the university role is appreciated as crucial to a qualitative and sustainable development of biotechnology research and to the setting of a biotechnology agenda for the future. Obviously, the university is an integral partner in knowledge creation when it comes to broadening biotechnology research. This involves interdisciplinary cooperation with physics, medicine, chemistry, food technology and computer sciences.

Last but not least, the university is seen as the primary supplier of human resources required for the development of the sector.

## External Influences

The specific national biotechnology policy of a small country is only one factor influencing the orientation and development of the field. It is also strongly defined by the international setting, both in terms of research, business and policy trends. The EU policy becomes constitutive, for instance, in respect to the adoption of regulations for biotechnology:

> We are trying to find the golden mean [national vs. international], since we cannot avoid it simply because of the fact that we are part of the EU and we are bound by international agreements, e.g. World Trade Organisation [. . .] and therefore we draft these policy and legislative documents. (civil servant)

From the viewpoint of actors involved in the implementation of a policy and legislative framework, Latvia's priorities are closely linked to the developments worldwide and, as a civil servant interviewed puts it, 'we are trying to follow these trends and [. . .] to adopt them', since science is never purely national. At the same time it has been recognized that 'we have to consider our national interests and national capacity'. Being part of larger alliances of countries, as is the case with the EU, has been assessed as an opportunity, which is still flexible enough for national governments to set their own priorities:

> I don't feel that Brussels [the European Commission] is putting a pressure on us to set priority directions; we have set them in our laws, these are our rules, we could also choose not to do it. Each country sets its own priorities. (civil servant)

In practical terms, the EU factor is considered part of an ever more complicated procedure for funding or production. It is admitted by researchers that the whole sector on both the EU and the national level demands a high level and complexity of bureaucratic procedures. However, it cannot be explained away only in reference to EU requirements, since the biotechnology sector as a whole is regulated on several levels, *inter alia* pertaining to issues regarding health and safety, biological diversity and ethics.

A considerable external impact in Latvia can be traced with regard to a growing coordination of the sector. This has been demonstrated by the comparatively recent developments in respect to clustering, the intentions of establishing sectoral competence centres as well as, most importantly, the establishment of a sectoral association. As outlined in its mission statement (2009), the Association of Biotechnology has been established to promote the development of the sector in Latvia with an aim of embracing as wide as possible an application of biotechnology methods potentially useful for the development of the national economy and facilitating international cooperation. Among the tasks set for the Association, the consolidation of the Latvian companies interested in prospective biotech projects is seen as among the most important ones. As noted by the participants, this process has largely been facilitated also by the practical considerations faced in the international market:

> [The establishment of the Association of Biotechnology of Latvia] came out of a sheer daily necessity when making different contacts abroad. The existence of a national association was always the first question being asked at various international exhibitions and annual biotechnology forums. (project manager)

Furthermore, strong external influences can be traced to research agendas being pursued by different units and institutions on the international scene. Under existing cooperation schemes with foreign companies, the thematic orientation of Latvian researchers is largely conditioned by projects and interests defined by contract partners abroad. Concerns are voiced by researchers about their operation 'in a discriminative environment and under discriminative rules', stressing that it is of great importance to stand up for their rights and, prior to that, to learn how to protect those.

## Public Engagement

Biotechnology research and applications is increasingly acknowledged to be among the most controversial areas of scientific innovation and technology (Bauer and Gaskell, 2002). Several strands of development in the field give ground for heated public debate on both the national and the international level, with concerns on a range of accompanying social, ethical and legal matters. Besides, as argued by Rimmington and Greenshields, 'as with the nuclear industry in the West, the biotech industry became synonymous with hazardous pollution in the Soviet public consciousness' (1992, p. 49), which was largely due to the almost complete disregard of the Soviet biotech industry for rules regulating release of materials (microorganisms) into the workplace and the environment, causing serious health problems to the local population (ibid., pp. 52–65). While such breaches of safety rules that were seldom reported publicly were more common in several other countries of the USSR, cases of exposure to excessive concentration of microbiological material had also been documented in Latvia as in the case with the Tukums Enzymes Factory in the late 1970s (closed down in 1996).

While the change of the regime since the 1990s has opened far more opportunities for democratic discussion of any matters of public concern, so far active involvement of the general public in the debate on biotechnology research and applications as well as a more general definition of research priorities is still limited in Latvia. For instance, according to a study on the social aspects of the Latvian genome research project, the interest and views expressed by the public, via the rather limited discussion platforms available for debating biotechnologies, has not exerted an influence on the policy level and has not contributed to the development of a comprehensive regulatory mechanism of biotechnologies (Putniņa, 2008). The study points to the lack of forums for public engagement in the assessment and development of technologies in Latvia with scientific issues being predominantly treated as pertaining exclusively to the scientists themselves. The public is not given a voice on these matters, having only a passive role in this kind of assessment. In terms of the broader decision-making processes it also has been noted that the Act on Human genome research was passed rather rapidly and without much public involvement. The original framing of the project has limited the scope of public debate to only two sets of actors, namely politicians and scientists (Putniņa, 2003, pp. 233, 241–2). This exemplifies the unwillingness of some actors involved in the process to discuss the ethical and socio-economic dimensions related to biotechnology research in a wider society.

According to a European thematic survey carried out in 2005 Latvia has been ranked tenth on an index measuring the optimism for biotechnology among the 25 EU countries (Gaskell et al., 2006, p. 13). At the same time another survey reveals that Latvians generally are among the least optimistic in regard to the effect of biotechnology and genetic engineering on our way of life in the next 20 years, with only 54 per cent of respondents voicing belief in their positive effect (compared to the EU25 average of 65 per cent) (Eurobarometer, 2005, p. 74). Accordingly, Latvians also feature higher rejection rates of different applications of biotechnology and genetic engineering. At the same time, account has to be taken of the comparatively large share of people (27 per cent) providing an indefinite response ('don't know') to the general question on the anticipated effect, as well as the comparatively limited familiarity with a range of particular technologies in this field (gene therapy, pharmacogenetics, genetically modified food, nanotechnology, stem cell research) covered by the other survey. The latter factor might therefore imply limited public communication of information on the subject on which informed public views could be based.

The former thematic survey on public views on biotechnology also maps the segmentation of the European public on the general principles of governance of science and technology based on a choice between, first, decision-making based on scientific evidence or on moral and ethical criteria, and, secondly, decisions made on expert evidence or reflecting the views of the public (Gaskell et al., 2006, pp. 43–5). As in the majority of European countries, the Latvian respondents primarily (64 per cent) give preference to the so-called principle of 'scientific delegation', which implies opting for decisions based on expert advice and on the grounds of scientific evidence. Only 16 per cent opted for 'moral delegation' (based mainly on the advice of experts and on the moral and ethical issues involved) and 13 per cent for 'moral deliberation' (based mainly on the general public's view and on the moral and ethical issues involved). Finally, 8 per cent of respondents chose 'scientific deliberation' as the desirable mode of governance, thereby covering those who want decisions to be based on scientific evidence and to reflect the views of average citizens. While these results show that there is a comparatively high level of confidence in the scientific enterprise in Latvia, there is also an emerging articulation of the need for bringing 'non-specialists' and moral considerations into decision-making on research agendas, including biotechnology.

So far there has been a rather reserved attitude towards public engagement in the field of biotechnology research and policy-making as well as science in general. In some cases a sharp boundary between researchers and the general public is being drawn due to the perceived nature and mission of fundamental research:

> A fundamental scientist is absolutely disinterested in these public interests. One couldn't care less since otherwise he is no fundamental scientist if he is being guided by some kind of demand factors. (researcher)

It should be pointed out that such attitudes are not founded on ignorance, nor do they discredit people's rights to be informed about biotechnology. Such a stand rather embodies an attempt to distance fundamental researchers from external influences and to secure their scientific freedom from short-term considerations and, sometimes, biased vision or market demands.

Also, researchers flag the issues of how well informed citizens really are and who are the opinion leaders; in researchers' own words, 'who formulate those interests'. Some of them strongly oppose such forms of discussion, where qualitative arguments and well grounded discussions are disregarded per se:

> As more science gets involved in some kind of discussions with the so-called society at the level of Delfi[3] or TV interviews, the more I think it loses. [. . .] People don't understand that if you have managed to develop even a single successful medicine, that's great. But they want the show to go on and each other evening there must be a new one. [. . .] I'm not sure, but I think that such

> a publicity does not help to raise awareness about science, rather discredits it [. . .] because one can't hope for applauses from a non-qualified audience [. . .]. What you would get most often is a mere misunderstanding and blaming of scientists for 'our money spent for science which we don't understand and don't want, so let's close down all those institutes and save money'. (researcher)

Even if the necessity of public engagement is acknowledged, related initiatives are usually rather distanced from the general public. In some contexts, public engagement has been portrayed as a problem: for example, with regard to GMOs, people are said to be generally inactive and/or relying upon myths. As stated by a civil servant, 'there are mythical assumptions about what could happen with us and with the next generation'. Failures to inform the society are not seen as a possible inefficiency of informers themselves or their chosen methods ('We are trying to publish leaflets, to inform, to place some kind of information in the press'), but with stereotypical assumptions on the very nature of a human being ('. . .but people probably presume a scandalous, negative side of the story'). Yet policy-makers and public authorities also see their role in enabling public engagement in the legislative process:

> GMO is a very sensitive issue (. . .) therefore it is important to establish a procedure, which requires civil engagement, so the government is not the only one to make a decision. (civil servant)

Researchers, too, see that raising the public awareness on science policy issues not only needs to be carried out through ad hoc opinions in mass media, but also that it should constitute an integral part of all large-scale research projects. This is largely based on the good practice cases in other countries, from which Latvia still has much to learn:

> As far as I know, the US research projects specifically require so-called outreach, meaning that a research component has to be accompanied by envisaged means of promoting the research in society. It can include traditional approaches – leaflets, knowledge dissemination through schools, lectures. But it would be good if we could do even more; it's just a question of who has time and money for that. It is not really required here. EU-funded projects include an information component, but often it does not go beyond simple activities of printing posters. (researcher)

To conclude: public engagement in debating biotechnology and in shaping its future directions is still relatively low in Latvia and lacks profundity and comprehensiveness. Also there are very few powerful players to influence the general opinion; the primary actors are researchers and representatives of industry. Civil society is rather passive in discussing scientific issues and thereby raising the level of knowledge and engagement of the general public. While recently a couple of encouraging initiatives have emerged in Latvia[4], so far these can be treated as an exception rather than a rule in the general practice in Latvia. In light of this, researchers, industry people and policy-makers at the governmental level may need to pay more attention to and provide greater effort for public engagement in a constructive dialogue by treating it as part and parcel of the whole success of the biotechnology domain steered into the direction of a socially accountable mode of development.

## CONCLUSIONS

The development of S&T policy in contemporary Latvia cannot be understood without considering the various overlapping socio-historical contexts and rapid changes the country has gone through after the collapse of the Soviet system. While Latvia used to be in the forefront of the field of biotechnology research in the Soviet era, nowadays it has lost its former position. This can be traced not only in relative terms, due to the change of the point and scale of reference (USSR vs. global market), but also in absolute terms vis-à-vis its own performance in the light of the experienced stagnation of the 1990s. Yet, it would be an oversimplification to conclude that the country should just follow the best practices in adopting policy instruments successful in other European countries, since it is a complex interplay among the structural, cultural and historical legacies shaping the national innovation systems.

In the case of Latvia the effects of the former Soviet regime have turned out to be both a strength and a weakness. While the Soviet market provided financial and human resources, research infrastructure and support for industry, the decoupling of Latvia from scale and resources of this former platform left it in a rather deplorable situation, requiring major adjustments of the whole market structure of the country. Following the restoration of independence, Latvia as a small and catching-up country had to face global competition, forcing it to look for a particular niche and for comparative advantages in the field of biotechnology. Nowadays, the latter is primarily associated with the quality (rather than quantity and/or costs) of human resources.

The impact of the 'small-country' factor is evident with respect to the limited researcher community, the lack of marked (sub-sectoral) policies and governing bodies holding direct responsibility for the development of individual branches of the national economy. On the other hand, it has not precluded Latvia from the formation of several separate research schools in the field of biotechnology, for example, leading to strong developments in health and generic biotechnology, biomedicine and others. Thus, one has to be cautious about making this 'country-size' factor into something too deterministic with regard to the specific development of the biotech field in Latvia. Rather, it can be coupled with other phenomena, and with the 'path dependence' that has manifested itself in Latvia, in terms of historically defined national research trends. Besides, it should be noted that nowadays the policy processes continue to develop under sometimes conflicting but obviously interrelated influences of national and global objectives given the fact that some biotechnology research and industry areas are linked to the EU policies and supranational legislation.

The biotechnology sector of Latvia is rather diverse, as the involved institutions represent various disciplines. While there are many informal contacts in some branches between the players in the Latvian biotechnology sector – research organizations (institutes and universities), enterprises and public authorities – formal coordination mechanisms are poorly developed in Latvia. The establishment of the Association of Biotechnology of Latvia in 2006 is one of the attempts to improve coordination, by bringing together representatives from research, education, business and government organizations. However, some effort is still needed for this mechanism to become fully operational and effective. Moreover, the biotechnology sector is characterized by an asymmetry between well-developed research capacity in some branches of biotechnology, on the one hand, and limited demand from the local enterprise sector on the other. Industry in Latvia has to be strengthened to enable transfer of research-based knowledge to production of biotechnology products for both national and international markets.

For the time being rather diverse perceptions can be observed on the role of the state and the degree of its interference in the biotechnology sector. The current policy landscape in Latvia is an interplay of ramifications between liberal political rationality in the sense of welcoming of the free market and a state-planned top-down approach by setting research priorities. On the one hand, there are policy-makers, civil servants directly involved in the policy implementation as well as some researchers and industry representatives that are of the opinion that the state should pursue a laissez-faire policy. On the other hand, there are also those that urge more intense involvement of the state in terms of defining strategic directions for the development of research, and the economy as a whole. The currently pursued policy of the state demonstrates a compromise between the two positions, since research priorities are being set, yet this is done in a rather open-ended manner with a strong involvement of the representatives of the research community.

Given the current fragmented policy landscape, characterized by the lack of a key public body strategically responsible for the whole biotechnology sector, researchers play a key role of providing expertise in policy-making practices. Aside from research priority-setting they are also invited to assist in drafting legal acts governing the sector, to provide expertise for other policy planning documents, and so on. Thus, researchers may largely define the future directions and contribute to the 'bottom-up' process of priority-setting. Additionally, the universities are providers of general expertise, thereby influencing the entire sector on a policy level. Along with these processes, it should be emphasized that universities play a comprehensive role in enhancing the entire biotechnology sector, consistently looking for new niches and new interdisciplinary research, as well as improving infrastructure to attract human resources once lost to better equipped universities or to research institutions abroad. They also have an important role to play in informing and engaging the public and in taking account of the societal needs with regard to the development and application of biotechnologies in the national context. While over recent years there have been certain implications of the development of biotechnology field for the general practices of research and policy-making in Latvia in terms of public engagement/citizen activism (for example, involvement of NGOs in the decision-making process via their participation in advisory boards) and addressing social implications of this sensitive S&T domain (for example, biosafety, risk assessment), there is still quite considerable room for facilitating public engagement in decision-making.

Fundamental science and links with higher education currently represent the main stimulus for the development of biotechnology in Latvia. High-level science cannot be maintained in Latvia without further intense collaboration with laboratories and institutions abroad. A major factor determining the importance of cooperation with the top laboratories of other countries is represented by the current industry set-up of Latvia, which is largely incapable of contributing substantially to high-tech development. In the current situation the country does not possess the resources for developing a competitive biotechnology industry. Nevertheless, it has several high-class research institutions that are searching for industrial partners abroad and developing other cooperation models, in order to be able to commercialize their research results.

While there were rather stable and positive developments in the field of the overall research funding and policy landscape in 2005–07, the deep economic and financial crisis witnessed in Latvia in 2008/2009 has led to rather drastic cuts in the public R&D budget, problems that are very likely to continue in the coming years. This, of course, has an effect on the public funding for biotechnology research as well, inevitably leading to a recurrent brain drain of talented scientists of the younger and middle-aged generation. Likewise an efficient and prospective functioning of the respective industrial sectors can be inhibited. Therefore the development of the biotechnology sector in Latvia at this point – following the initial major restructuring in the early 1990s – can be said to be once again going through a critical moment; the prospects being largely determined by the capacity of the economy to make the utmost use of scarce resources.

## ACKNOWLEDGEMENTS

We would like to express our gratitude to all the interviewees for their time and responsiveness. Special thanks go to Inga Ulnicane-Ozolina for her contribution in conducting the expert interviews and sharing initial ideas on the report outline.

## NOTES

1. These are EU-level framework programmes for research and technological development. They represent the EU's main instrument for funding research in Europe.
2. 'Modernisation of research infrastructure in the fields of biotechnology, biomedicine, organic synthesis, environmental science and ecology at the University of Latvia' (2005–06); 'First round of the centre of recombinant biotechnology' (2006–07).
3. News and entertainment Internet site with active anonymous commentators, often making negative and groundless statements.
4. (1) Since 2010, the GMO-free movement is undertaking activities aimed at facilitating policy-making in the field of genetically modified organisms based on the principles of sustainability and public interests; (2) A research project 'Capacity building for interdisciplinary biosafety research' (2010–12) has been launched by the University of Latvia with an aim of forming a new, interdisciplinary research team in the field of biosafety and biotechnology management bringing together researchers from the fields of biology, medicine, social sciences and law. The project includes a special work package on increasing public participation in assessment of biosafety risks, which thereby represents an integral part of the project and its communication strategy (see Adamsone-Fiskovica and Bundule, 2011).

## REFERENCES

Adamsone-Fiskovica, A. and M. Bundule (2011), *Monitoring Policy and Research Activities on Science in Society in Europe (MASIS): National report: Latvia,* European Commission, DG Research, available at: http://www.masis.eu/english/storage/publications/nationalreports/masisnationalreportlatvia/, accessed 2 March 2011.

Association of Biotechnology of Latvia (2009), Brochure of the Association,

available at: http://www.latbiotech.lv/download/Prospekts_Latbiotech_2009_LV.pdf, [in Latvian], accessed 16 September 2009.
Austin, M.A., S. Harding and C. McElroy (2003), 'Genebanks: a comparison of eight proposed international genetic databases', *Community Genetics*, **6**(1), 37–45.
Bauer, M.W. and G. Gaskell (eds) (2002), *Biotechnology: The Making of a Global Controversy*, Cambridge: Cambridge University Press.
Beķers, M. (2009), 'Mārtiņš Beķers', in *Zinātne un mana Dzīve* [Science and my Life] [in Latvian], Vol. 1. Riga: University of Latvia, pp. 63–72.
Beķers, M. and J. Jākobsons (2001), 'Tehniskās mikrobioloģijas un biotehnoloģijas attīstība Latvijā' [Development of Industrial Microbiology and Industrial Biotechnology in Latvia], [in Latvian], *Latvijas Zinātņu Akadēmijas Vēstis.* [*Proceedings of the Latvian Academy of Sciences*, Section A, **55**(3/4), 14–27.]
Berķis, U. (2008), 'Life sciences, genomics and biotechnology for health', in D. Bērziņa and A. Zīlāns (eds), *Latvia in EC 6th Framework Programme 2002–2006*, Riga: Ministry of Education and Sciences, pp. 11–15.
Blank, W., B. Samuelsson and P. Frank (2003), 'ScanBalt BioRegion: A model case for Europe', *Journal of Commercial Biotechnology*, **10**(2), 147–53.
Caracostas, P. and M.-C. Brichard (2004), 'Introduction to special issue on innovation policies for biotechnology in Europe', *Science and Public Policy*, **31**(5), 342–3.
Denzin, N.K. and Y.S. Lincoln (eds) (2000), *The Handbook of Qualitative Research*, Thousand Oaks, CA. Sage Publications.
Enzing, C., A. van der Giessen and S. Kern (2004), 'Commercialisation of biotechnology: do dedicated public policies matter?', *Science and Public Policy*, **31**(5), 371–83.
Eurobarometer (2005), *Social Values, Science and Technology*, Special Eurobarometer 225/Wave 63.1, European Commission, available at http://ec.europa.eu/public_opinion/archives/ebs/ebs_225_report_en.pdf, accessed 14 December 2008.
European Commission (2002), 'Strategy for Europe on life sciences and biotechnology 2002–2010', Communication from the Commission to the European Parliament, the Council, the Economic and Social Committee and the Committee of the Regions, available at http://ec.europa.eu/biotechnology/pdf/com2002-27_en.pdf, accessed 14 December 2008.
European Commission (2006), 'Decision No. 1982/2006/EC of the European Parliament and of the Council of 18 December 2006 Concerning the Seventh Framework Programme of the European Community for Research, Technological Development and Demonstration Activities (2007-2013)', *Official Journal of the European Union*, 30 December 2006, available at http://eur-lex.europa.eu/LexUriServ/LexUriServ.do?uri=OJ:L:2006:412:0001:0041:EN:PDF, accessed 14 December 2008.
Fischer, F., G.J. Miller and M.S. Sidney (eds) (2007), *Handbook of Public Policy Analysis Theory, Politics, and Methods*, Boca Raton, FL, London, New York: CRC Press.
FMERG: Federal Ministry of Education and Research of Germany (2007), 'Biotechnology in the Baltic States', available at http://www.biotechnologie.de/bio/generator/Navigation/English/Facts-and-Figures/countries-in-focus, did=65814.html?List Blld=4423
Gaskell, G. et al. (2006), *Europeans and Biotechnology in 2005: Patterns and*

*Trends*, Eurobarometer 64.3, A report to the European Commission's Directorate-General for Research, available at: http://www.gmo-compass.org/pdf/documents/eurobarometer2006.pdf, accessed 14 December 2008.

Gottweis, H. and A. Petersen (eds) (2008), *Biobanks: Governance in Comparative Perspective*, Abingdon, UK and New York, USA: Routledge.

Grēns, E. (2006), 'Pirmie soļi molekulārajā bioloģijā: cilvēki, idejas, risinājumi, perspektīvas' [First steps in the molecular biology: people, ideas, solutions, perspectives] [in Latvian], *Zinātnes Vēstnesis* [Science Herald], 20 March, 6, (p. 319), available at http://ww3.lza.lv/ZV/zv060600.htm.

Grēns, E., I. Kalviņš, and A. Deniņš (2007), 'Biomedicīna un biofarmācija: gēnu tehnoloģijas un jaunu bioloģiski aktīvu vielu sintēzes teholoģijas', in E. Grēns (ed.), *Zinātne, Pētniecība un Inovācija Latvijas Izaugsmei*, ['Biomedicine and biopharmacy: gene technologies and synthesis technologies of new biologically active substances', in *Science, Research and Innovation: Advancing Latvia's Development*] [in Latvian], Rīga: Zinātne, available at http://www.president.lv/images/modules/items/PDF/item_1588_ZI_labots.pdf.

Haralambos, M. and M. Holborn (2007), *Sociology, Themes and Perspectives*, 7th edn, London: Collins.

Holstein, J. and J.F. Gubrium (1995), *The Active Interview*, Thousand Oaks, CA: Sage Publications.

Inventions and Inventors of Latvia (1999), website http://inventions.lza.lv, accessed 10 September 2009.

Kalviņš, I. (2007), 'Interviews', in B. Zeltkalne, *Lūgšana Liktenim. Stāsts par Akadēmiķi Ivaru Kalviņu* [Prayer to the destiny. A Story about the Academician Ivars Kalviņš], [in Latvian], Rīga: Alvīne un Alberts.

Kalviņš, I. (2009), 'Pētniecība, attīstība un inovācija: mūsu izaicinājums krīzes pārvarēšanai!' [Research, development and innovation: our challenge for managing crisis!] [in Latvian], *Daily Diena*, 12 June.

Kristapsons, J., A. Adamsone-Fiskovica and I. Ulnicane-Ozolina (2008), *Erawatch Analytical Country Report 2008: Latvia*, available at http://cordis.europa.eu/erawatch/index.cfm?fuseaction=home.downloadFile&fileID=894, accessed 14 December 2008.

Lacasa, I.D., T. Reiss and J. Senker (2004), 'Trends and gaps in biotechnology policies in European Member States since 1994', *Science and Public Policy*, **31**(5), 385–95.

LIDA (Latvian Investment and Development Agency) (2005), *Biotechnology and Pharmaceuticals in Latvia*, available at http://www.liaa.gov.lv/get_file.php?file=uploaded_files/publication_files/2293.pdf, accessed 14 December 2008.

Līdaka, M. (1996), 'Medicīnas preparātu ražošana Latvijā' [Production of medical preparations in Latvia], in *No Grindeļa Līdz Mūsdienām*, [in Latvian], Rīga: Nordik, pp. 275–81.

Malo, S. and J. Norus (2009), 'Growth dynamics of dedicated biotechnology firms in transition economies. Evidence from the Baltic countries and Poland', *Entrepreneurship & Regional Development*, **21**(5–6), 481–502.

Ministry of Education and Science (2010), *Science in Latvia: National Research Programmes*, available at http://izm.izm.gov.lv/upload_file/Zinatne/vpp/Zinatne_Latvija_EN.pdf, accessed 10 March 2011.

Nimmo, B. (2005), 'Bio-industries: the shape of things to come?', *British Latvian Trade* (The magazine of the British Chamber of Commerce in Latvia),

September, available at http://www.bccl.lv/pdf/BLT_september_2005.pdf, accessed 10 March 2011.

Pīrāgs, V. and E. Grēns (2003), 'The Latvian Genome Project', in J. Sándor (ed.), *Society and Genetic Information: Codes and Laws in the Genetic Era*, Budapest, New York: Central European University Press, pp. 225–31.

Putniņa, A. (2003), 'Population Genome Project in Latvia: Exploring the articulation of agency', in J. Sándor (ed.), *Society and Genetic Information: Codes and Laws in the Genetic Era*, Budapest, New York: Central European University Press, pp. 233–45.

Putniņa, A. (2008), 'Genoma izpētes projekta Latvijā sabiedriskie aspekti' [Social aspects of the Latvian genome research project], [in Latvian], study commissioned by the Latvian National Commission for UNESCO. Bio-ethics and bio-safety centre of the University of Latvia, available at http://www.policy.hu/putnina/genoms_atskaite.pdf, accessed 14 December 2008.

Reiss, T. (2007), 'BioPolis: Inventory and analysis of national public policies that stimulate research in biotechnology, its exploitation and commercialisation by industry in Europe in the period 2002–2005, National Report of Latvia', Fraunhofer Institute for Systems and Innovation Research, available at http://ec.europa.eu/research/biosociety/pdf/biopolis_latvia_en.pdf, accessed 1 April 2011.

Rimmington, A. and R. Greenshields (1992), *Technology and Transition: A Survey of Biotechnology in Russia, Ukraine and the Baltic States*, London: Quorum Books.

Russell, A.M. (1988), *The Biotechnology Revolution: An International Perspective*, Brighton: Wheatsheaf.

ScanBalt (2006), Yellow pages of ScanBalt Bioregion, available at http://www.scanbalt.org/yellowpages, accessed 10 October 2007.

ScanBalt (2007), 'ScanBalt: Top of Europe: Competencies in Life Sciences and Biotechnology in the Baltic Sea Region', final report of the Competence Region project, available at http://www.bcv.org/hosting/bcv/webres.nsf/url-names/file_downloads/$file/scanbalt_competenceregion_abstract.pdf, accessed 14 December 2008.

Senker, J., C. Enzing and T. Reiss (2008), 'Biotechnology policies and performance in Central and Eastern Europe', *International Journal of Biotechnology*, **10**(4), 341–62.

# 12. Biotechnology in Germany

**Thomas Reiss, Ralf Lindner and Ulrich Schmoch**

## INTRODUCTION

Biotechnology is one of the key enabling technologies and has become a driving force of dramatic changes of the innovation process in various sectors. Not surprisingly, the promotion of biotechnology has been on the agenda of nearly every European country since the 1990s (Reiss and Dominguez Lacasa, 2007). Also, many non-European countries, with the USA the most prominent example, have been investing heavily in the promotion of biotechnology (e.g. Reiss et al., 2007). More than 35 years have passed now since the emergence of modern biotechnology based on genetic engineering technologies, monoclonal antibody approaches and other innovative technologies. Accordingly the question of the status of biotechnology appropriation in different countries is not only interesting from the academic point of view, where diffusion patterns and conditions would be a main interest. Rather, exploring the appropriation of biotechnology can also contribute to the understanding of the effectiveness of various policy approaches aiming at the promotion of biotechnology.

Current research shows that the development and implementation of biotechnology is rather diverse across countries and also within countries (Reiss and Dominguez Lacasa, 2008; Gaisser and Reiss, 2008; Enzing and Reiss, 2008; Senker et al. 2008). This could be explained partly by historical, geographical, economic or demographic factors. However, governmental policy approaches are an important key to understanding why biotechnology shows such large differences in growth patterns between countries. In a recent EU-funded research project – BIOPOLIS – a detailed analysis has been elaborated on the promotion of biotechnology via public funding activities (Enzing et al., 2007). This inventory of public policy approaches was complemented by a broad set of biotechnology performance data in order to put the national promotion activities into perspective. Based on these two data sets – policy input and output measures

– general conclusions on the effectiveness of various policy approaches have been drawn within the BIOPOLIS project.

Starting from this research, this report will provide an overview on biotechnology research performance and related public policies in Germany. In particular, we will explore how policies targeting public research institutions such as universities and non-university organizations contribute to the appropriation of biotechnology by national innovation systems.

In a first step the level of biotechnology appropriation in Germany compared to Europe as a whole will be explored in the next section. As a proxy for the degree of appropriation, performance data as elaborated in the BIOPOLIS project will be used (Enzing et al., 2007). This analysis will be complemented by an investigation of the main research actors in the German biotechnology innovation system in the next section. Special attention will be given to the role of universities. In the fourth section we will elaborate on the biotechnology policy environment in Germany, followed by a discussion of the way in which research agendas are developed. Based on this information a discussion of policy effectiveness will be provided in the sixth section, followed by summarizing conclusions.

## LEVEL OF BIOTECHNOLOGY APPROPRIATION

As a first measure for the level of biotechnology appropriation in Germany the number of graduates in life sciences related to the size of the country in terms of population is presented in Table 12.1. Compared to the European average we observe a rather low number of students graduating in life sciences in Germany. Even considering methodological limitations of this indicator, in the sense that it does not indicate specifically graduates in biotechnology, but rather covers life sciences as a whole, this empirical evidence suggests a critical situation concerning the availability of skilled human resources for biotechnology in Germany. It should be noted, however, that this challenge is not specific for biotechnology, but rather for Germany as an innovation location as a whole, something which has been pointed out, for example, in the latest report of the independent Commission of Experts of Research and Innovation from 2008. The report notes that the German educational system has fallen behind in international comparisons, and accordingly a shortage of skilled labour turns out to be a problem already in many sectors and will represent a long-term constraint for the innovation system unless sustainable counter measures are adopted quickly (Commission of Experts for Research and Innovation, EFI, 2008).

With respect to the role of universities in the appropriation of

*Table 12.1 Indicators for the appropriation of biotechnology in Germany*

| No. | Indicator | Unit | Period | Germany | EU | Number of countries included in the EU data |
|---|---|---|---|---|---|---|
| 1 | Graduates in life sciences | number per million capita | 2002 | 82 | 189 | 17 |
| 2 | Share of biotech publications in all publications | % | 2002–2004 | 14% | 13% | 25 |
| 3 | Biotech publications per population | number per million capita | 2002–2004 | 403 | 319 | 25 |
| 4 | Citations rates | number | 2000–2004 | 7.10 | 7.28 | 25 |
| 5a | Share of health publications | % | 2002–2004 | 59% | 58% | 25 |
| 5b | Share of generic publications | % | 2002–2004 | 26% | 25% | 25 |
| 5c | Share of plant publications | % | 2002–2004 | 7% | 7% | 25 |
| 5d | Share of animal publications | % | 2002–2004 | 4% | 5% | 25 |
| 5e | Share of food publications | % | 2002–2004 | 2% | 4% | 25 |
| 6 | Ratio biotech patents applications over biotech publications | number | 2001–2003 | 0.11 | 0.07 | 25 |
| 7 | Biotech patent applications per population | number per million capita | 2001–2003 | 44 | 22 | 25 |
| 8 | Number of biotech firms per population | number per million capita | 2001–2004 | 4.31 | 5.25 | 15 |
| 9 | VC invested in biotech per population | € per capita | 2002–2004 | 2.6 | 3.0 | 13 |

*Source:* BIOPOLIS research (Enzing et al., 2007)

biotechnology in Germany the problems related to the availability of skilled human resources seem to indicate that one important contribution of universities to supporting the appropriation process, namely the provision of qualified scientists, is inadequately developed.

The significance of biotechnology research in Germany is indicated by the share of biotechnology publications in comparison to the sum total of publications. With a ratio of 14 per cent over the period 2002 to 2004 the significance of biotechnology in Germany is slightly above the European average. We also observe a slight growth of this indicator since the mid-1990s (Enzing et al., 2007), pointing to an increasing significance of biotechnology research activities in the German research system. As regards the productivity of the knowledge production process in terms of biotechnology publications per population, Germany performs well above the European average, indicating a rather high level of biotechnology appropriation in the research community. On the other hand, citation rates in Germany are below the European average and we even found a decrease of this indicator between 1994/98 and 2000/04 (Reiss and Dominguez Lacasa, 2008). Even though citation data need to be evaluated with some caution, this observation points to the need for further in-depth analyses of the status of biotechnology research in Germany in terms of international attention and quality.

Analysing the thematic orientation of biotechnology research in Germany measured as shares of publications in specific fields reveals two main focuses: health biotechnology and generic biotechnology. This profile is in line with the situation in Europe as a whole.

In the following some indicators will be discussed which provide information on the application of biotechnology knowledge. Thereby some insight into the appropriation dynamics of biotechnology in the business sector will be possible. The ratio of biotechnology patent applications over biotechnology publications gives a first indication of the significance of application-oriented activities related to research activities. Accordingly, Germany is performing way above the European average, indicating a very well-developed ability in knowledge transfer (Table 12.1). This observation is supported by the high level of patent applications in biotechnology. The well-developed ability to transfer knowledge seems not to translate automatically into industrial development in biotechnology. This is indicated, for example, by the below-average number of biotech firms in relation to population in Germany. And in particular the low level of venture capital invested in biotechnology points to a critical problem of the innovation system. Since the period under investigation (2002–04), the availability of venture capital for biotechnology slowed down even further, so that nowadays financing business activities in biotechnology has become one of the most severe problems.

In summary, this analysis indicates that the appropriation of biotechnology in Germany has proceeded well in the science sector and that there are also promising knowledge transfer activities. However, appropriation of biotechnology in the industrial sector has not commenced satisfactorily. These conclusions concern only the biotechnology industry in a narrow sense and do not consider the diffusion of biotechnology into other business areas.

## BIOTECHNOLOGY RESEARCH ACTORS

In this section we will first provide an overview of the main public research actors which are at least partially active in biotechnology. Secondly, we will concentrate on an analysis of the role of universities in this context.

### Overview

The gross expenditure for R&D in Germany amounted to 55 739 million euros in 2005 (BMBF, 2008). About two-thirds of this total budget (39 596 million euros) are financed by the industry, the remaining third by public bodies. Similarly about two-thirds of total research activities in Germany are carried out in industrial R&D facilities. Concerning the public research landscape, Germany is characterized by a broad diversity of different actors. These include universities, the Max Planck Society, the Fraunhofer Society, the Helmholtz Association and the Leibniz Association.

German universities are responsible for both teaching and for conducting research activities. It is difficult to determine the share of resources designated to each task. By means of the so-called 'R&D coefficients', the official publications of the Ministry of Education and Research (BMBF) estimate that 9222.1 million euros were invested in research activities in the year 2005 (BMBF, 2008).

According to this source, the budget for research at universities increased by 17 per cent between 1998 and 2005. Unfortunately, there are no data available on the share of this budget directed to biotechnology research. Among the different disciplines considered in the official statistics, in 2005 natural sciences absorbed the largest volume of university research resources, with 2700 million euros (29 per cent of all resources), followed by medical sciences with 2307 million euros (25 per cent). We estimate that almost all universities in Germany with medical or natural science research facilities carry out biotechnology-related research. However, to provide accurate figures would require specific additional field surveys which would go far beyond the scope of this analysis.

*Table 12.2 Non-university research organizations*

| Institution | Annual budget (2005)* | Activities |
|---|---|---|
| Max Planck Society | €1200 million | Fundamental interdisciplinary research |
| Fraunhofer Society | €1200 million | Industry-oriented research |
| Helmholtz Association | €2700 million | Fundamental research and research services |
| Leibniz Association | €1100 million | Fundamental and industry-oriented research, research services |

*Source:* *BMBF (2008)

As regards the non-university research organizations, the most important institutions are given in Table 12.2.

The Max Planck Society (MPG) runs 79 institutes in Germany which are engaged in fundamental interdisciplinary research in three areas: biology-medicine, chemical-physics and social sciences. In 2005, the budget of the Max Planck Society amounted to 1200 million euros; the German federal and state governments financed 82 per cent of the Society's budget, around 13 per cent of the budget comes from competitive funding and about 5 per cent from donations from supporting members, evaluations, royalties and membership.

The Fraunhofer Society coordinates 56 research institutes engaged in industry-oriented research. The Society has an annual research budget of about 1200 million euros. The institutes have a financing scheme combining institutional grants and external funding (also industrial): up to 30 per cent of the budget is public institutional funding from the federal government and the *Länder*; the remainder comes from contract research. Due to the contract research activities of the institutes, the scientific and technological research focuses are mainly demand-driven. Biotechnology is, hence, one of the Society's many research fields.

The 15 German research centres of the Helmholtz Association carry out fundamental research in projects requiring special manpower, funding and equipment. With an annual budget of approximately 2700 million euros, the Helmholtz Association is Germany's largest research institution. The research orientation is structured in six research areas, which are distributed across all 15 centres: energy, earth and environment; health (with a 346 million euro budget in 2004); key technologies (encompassing scientific computation; information technology with nanoelectronic systems; nano- and microsystems and advanced engineering materials); structure of matter; transport; and space (with a 180 million euro budget in 2004).

Finally, the Leibniz Association (WGL) is the umbrella organization of 82 institutes all over Germany with an annual budget of approximately 1100 million euros. An important characteristic of this society is the diversity of the institutes in terms of organizations and research focus. The institutes are grouped into five sections: humanities and education, economic and social sciences, life sciences, physical sciences, and environmental research. In 2005 the section 'life sciences' had a total budget of 290.59 million euros, of which 74 per cent (215.66 million euros) was institutional funding from the federal government and the *Länder*.

## The Role of Universities in Biotech Research

The role of universities in biotech research in Germany will be discussed based on the following indicators:

- human resources involved as indicated by the approximate number of researchers;
- publications and patents;
- specialization as indicated by the focus of their research activities.

### Human Resources

Currently there are no statistics available with information on the number of personnel involved in biotechnology research activities at universities in Germany. The only exception is a survey of the Federal Statistical Office from 1992 where employment in biotechnology at universities and other research institutions was analysed (Hetmeier et al., 1995). Based on this survey the share of personnel concerned with biotechnology within all university personnel could be determined. Using these ratios and the available numbers on total personnel at universities (BMBF, 2008), rough estimates for biotechnology personnel can be given. Accordingly, in the year 1992 in total 12 200 people were employed in biotechnology at German universities. This number increased to 13 300 people in the year 2003 (Nusser et al., 2007). This estimate assumes that the share of biotechnology employees in all university employees did not change since 1992, which seems to be a very conservative assumption. If we consider, for example, that the share of biotechnology publications in all publications almost doubled between 1992 and 2004 (Reiss et al., 2003), and that in addition the research expenditures of the BMBF for biotechnology increased by 85 per cent between 1996 and 2004 (Nusser et al., 2007), it seems reasonable that also the share of biotechnology employees at German universities increased considerably. Accordingly, a recent estimate assumes that in 2004 between 25 000 and 29 000 employees were concerned with biotechnology research at German universities (Nusser et al., 2007).

**Publications and patents**

As an indication of the role of universities in fundamental biotechnology research publication activities can be used. For methodological reasons it is rather demanding to carry out a precise analysis of the share of publications generated in universities as it requires a precise identification of the institutional affiliation of the authors. Since it was not possible to carry out such an analysis within the framework of this project, we refer to a preceding study by Reiss and Hinze (2004) focusing on publications in biopharmaceuticals, which cover roughly 60 per cent of all biotechnology publications (see Table 12.1). According to this analysis, by far the most publications in the 1990s came from universities, with a share of 73 per cent of all publications in 1994 and a ratio of more than 77 per cent in 1999. Public sector research organizations such as the Max Planck Society were the second most important contributors to the scientific output. However, their contribution decreased slightly from about 14 per cent to 11 per cent between the two periods under investigation.

The important role of universities in generating biotechnology-related knowledge is also reflected in the scientific partnerships of pharmaceutical firms, as measured by co-authorship in scientific publications, assuming that joint publications reflect results of common research activities. Reiss and Hinze (2002) performed such an analysis and could show that during the 1990s universities had been by far the most important scientific cooperation partners for the German biopharmaceutical industry, being co-authors in 65 per cent of all co-authored papers. Non-university public research organizations followed in second place, with a share of about 13 per cent at the end of the 1990s.

The direct contribution of universities to technology generation as reflected in patent activities does not cover all technology areas, but is focused on knowledge-intensive ones. Most of the patent applications have a disciplinary origin in chemistry and mechanical engineering. The institutional basis of this focus is a specific strength of German industry in these sectors. In addition, universities provide patent applications in electrical engineering and in medicine at medium level, and physics at a low level. On average, the contribution of universities to the German domestic patent applications is at a level of about 4.5 per cent (Schmoch, 2004). This share is not very high, but with regard to the primary orientation of universities to basic research, still substantial. When the focus is directed to specific knowledge-intensive technologies such as biotechnology, semiconductors, organic chemistry, materials, control, surface technology, medical technology or polymers, the share is substantially higher, with a level between 10 and even 30 per cent, whereas in low-tech fields, the contribution of universities is negligible. So universities fulfil their primary mission to engage in complex research, including biotechnology.

A more detailed analysis of the contribution of universities to patenting in biotechnology has been performed by Reiss and Hinze (2004), focusing on biopharmaceuticals (see above). They could show that in the 1990s the most important patent applicants in this field were pharmaceutical firms, followed by biotech firms and public sector research organizations including universities. However, they could also observe an interesting dynamic change in this configuration, in the sense that pharmaceutical firms lost their impact considerably, as indicated by a change of their contribution to patenting: from 43 per cent in 1994 to 34 per cent in 1999. At the same time biotech firms and in particular universities were gaining importance. In 1994 the share of universities in all patent applications in biopharmaceuticals was about 16 per cent. By 1999 this share increased to roughly 23 per cent.

In sum, the analysis of patent applications and publications indicates that in the German biotechnology innovation system universities are playing an increasingly important role in the generation of knowledge and technology. Obviously, this contribution of universities seems to be acknowledged by industry as indicated by an increasing number of common research activities.

**Specialization**

In order to get an overview of the specialization of German universities in biotechnology, we analysed the descriptions of scientific activities of a total of 412 institutes affiliated with 65 universities, which are active in biotechnology. Data were retrieved from the database www.biotechnologie.de.

Almost two-thirds of the analysed university institutes have a clear focus on health-related biotechnology research activities. Basic biotechnology research such as developing general methods, bio-computing, bioinformatics or analytical methods are the second most important field of activities. About 25 per cent of all institutes are active in this area. Plant biotechnology, environmental biotechnology and industrial biotechnology are other important fields, where between 12 per cent and 14 per cent of the institutes indicate activities. Less important are animal biotechnology, food biotechnology, marine biotechnology and ethical, legal and social aspects of biotechnology where less than 2 per cent of the institutes seem to be active.

We also analysed the specialization of the universities for applied research in Germany. Data was based on information provided by 17 of these institutions. Interestingly, their specialization pattern is quite different compared to universities. The main focus is on environmental biotechnology where almost 70 per cent of the analysed institutes indicate activities, followed by basic biotechnology research activities mainly related to developing analytical and other methods and industrial biotechnology, both with shares between 50 per cent and 60 per cent. Health-related biotechnology activities make up only about 35 per cent of all activities, which is a sharp contrast to the research focus of universities.

In summary, the specialization analysis indicates that, on the one hand, the specialization patterns of universities well reflect the specialization patterns already obtained by analysing publications with health-related biotechnology as by far the most important field of activities of German universities. On the other hand, the analysis also indicates that there exists a division of labour between universities and universities for applied research in a sense that the latter have a much stronger focus on those fields of biotechnology which are closer to direct applications in industry and in the environment. In addition, universities of applied research are also very active in developing basic methodologies needed for biotechnology.

## BIOTECHNOLOGY POLICY ENVIRONMENT

To characterize the biotechnology policy environment a typology of policies will be used which had been developed by Reiss et al. (2005) and Reiss and Dominguez Lacasa (2007). Accordingly, the total portfolio of national biotechnology policies is divided into four broad sub-areas which are relevant for potential policy interventions: (1) the development of the knowledge base in human resources; (2) knowledge transmission and application; (3) the market; (4) industrial development.

These four sub-areas provide the framework for key processes of the innovation system. In order to support these processes, specific policy goals can be formulated for each sub-area. These policy goals first can be assigned to certain policy areas which basically cover the whole range of possible policy portfolios within the policy-making system: (1) education policies; (2) research policies; (3) exploitation policies; (4) policies related to industrial development; (5) fiscal policies; (6) regulation; (7) demand-oriented policies. Table 12.3 summarizes the described policy typology.

Using this taxonomy the biotechnology policy environment in Germany in the mid-2000s can be characterized as follows. In general terms the knowledge basis was the most promoted area of the innovation system. The market for biotechnology products on the other hand seemed to be

*Table 12.3 Typology of biotechnology policies*

| Sub-areas of the Biotechnology Innovation System | Policy goals | Policy area | | | | | | |
|---|---|---|---|---|---|---|---|---|
| | | Education | Research | Exploitation | Industrial development | Regulation | Fiscal measures | Demand |
| Development of the knowledge base and human resources | 1. To promote high level of biotechnology basic research | | X X | | | X | | |
| | 2. To promote high level of industry-oriented (and applied) research | | X X | | | | | |
| | 3. To support knowledge flow between scientific disciplines | | X X | | | | | |
| | 4. To assure availability of human resources | X | | | | | | |
| Knowledge transmission and application | 5. To facilitate transmission of knowledge from academia to the industry and its application for industrial purposes | | | X X | | X | | |
| | 6. The adoption of biotechnology for new industrial applications | | | | | | X | X |
| | 7. To assist firm creation | X | | | X X | X | X | |

*Table 12.3* (continued)

| Sub-areas of the Biotechnology Innovation System | Policy goals | Policy area | | | | | | |
|---|---|---|---|---|---|---|---|---|
| | | Educa-tion | Research | Exploi-tation | Industrial develop-ment | Regula-tion | Fiscal measures | Demand |
| Market | 8. To monitor and improve the social acceptance of biotechnology | | | | | | | X |
| | 9. To facilitate the introduction of new products | | | | | X | | |
| | 10. To strengthen the economic sectors exploiting biotechnology | | | | | X | X | |
| | 11. To keep/attract large firms (important market, important for firm development: tacit knowledge etc.) | | | | | X | X | |
| Industry | 12. To encourage business investment in R&D | | | | X | | X | |
| | 13. To improve firms' competitiveness | X | | | | X | X | |
| | 14. To exploit regional potentials | | | | X | | | |

*Note:* For illustration crosses indicate where specific policy instruments could be localized in the total policy portfolio. In addition, a differentiation can be made between generic policies and biotechnology-specific policies as indicated by light grey or dark grey fields respectively.

the neglected area of the German biotechnology innovation system in 2004 in terms of policy attention. Regarding the different policy areas active in promoting biotechnology, research policy has focused on supporting basic and applied research. The intensity of research policy was very high compared to the other policy areas. However, research policies seemed to be neglecting the promotion of knowledge flow between scientific disciplines. Another strong policy area was fostering industrial development. With a lower level of policy engagement than research policy, industrial policy especially targeted firm creation and the exploitation of regional developments. Germany had a very strong focus on cluster initiatives to promote innovation in biotechnology. On the other hand, policy engagement on exploitation was quite low.

Other policy gaps appear in the areas of education, regulation, fiscal policy and demand. Two policy gaps are particularly worrisome, in particular in combination with the performance indicators described in the previous sections. These are the low support for the education of qualified human capital, and the disregard for creating a suitable regulatory framework for innovation considering research, market and industrial issues. Only the issues of patenting biotechnology inventions and the special intellectual property regime for public research organizations have been tackled. Furthermore, Germany so far does not implement any kind of fiscal measures to create incentives in the innovation system. In addition, the adoption of biotechnology by companies in established industries was not an important issue in the policy profile of the mid-2000s.

## DETERMINATION OF BIOTECHNOLOGY RESEARCH AGENDAS

Biotechnology research agendas in the public domain are determined by a combination of top-down and bottom-up approaches. Main actors in the top-down scheme are two federal ministries, the Ministry of Education and Research and the Ministry of Economic Affairs and Technologies. In addition, the 16 '*Länder*-Governments' also develop their own biotechnology research agendas mainly driven by initiatives of two *Länder* ministries. Accordingly, there are between 30 and 35 institutions at the ministerial level involved in the determination of biotechnology policies with the underlying research agendas.

An important role is performed by intermediaries in Germany; these include the German Research Foundation and project management agencies. In the field of biotechnology there are four such agencies involved. These intermediaries are on the one hand responsible for the implementation of policies; on the other hand they support the ministerial levels in strategic considerations. Since all applications for research projects funded by the federal ministries and also by many *Länder* ministries are administrated by these intermediary organizations, broad knowledge on current research ideas and research activities is available at these institutions. This intelligence can be brought into discussions with the federal ministry level on future research agendas.

In a bottom-up way the main research organizations in Germany, in particular the Max Planck Society, the Fraunhofer Society and the Helmholtz Society, determine their own research agendas by a combination of strategic top-down planning from central strategic departments, and bottom-up idea development from individual researchers or research groups. In consultation with the federal ministerial level these research organizations contribute to the general agenda-setting in biotechnology research. In addition to these organizations there are advisory and coordination bodies involved in biotech-related policy-making. These include the Science Council, the Joint Commission on Education Planning and Research Promotion and the Planning Committee for University Building.

The ways in which new research programmes are developed can be very diverse. They may range from rather simple expert consultations to medium-term strategic processes involving a large number of stakeholders. Examples for the latter are various foresight activities of the German Government which are aimed, among other things, at the identification of future research topics including biotechnology. For example, since September 2007 the German Ministry of Education and Research has been running a broad foresight process in order to safeguard Germany's long-term innovative capacity as a centre for research and education. The process aims at identifying new key areas in research and technology, naming and determining areas for cross-cutting activities in the field of research and innovation, analysing fields of technology and innovation with regard to their potential for strategic partnerships and determining priority fields of action for research and development. Biotechnology and life sciences play an important role in this process (see www.bmbf.de/en/12673.php).

## POLICY EFFECTIVENESS

In general, there is great difficulty in identifying direct links between specific policy activities and any outcomes. One of the problems relates to the fact that there is a time lag between the date at which a policy is introduced and its results. Secondly, national economic, institutional, cultural or legal conditions may interfere intensively with policy activities and could be supportive or prohibitive to the fulfilment of policy objectives. Finally, policies are not active in isolation; rather, the achievement of desired outcomes may only be possible in conjunction with other complementary activities. These limitations should be kept in mind when considering the following observations, which are based mainly on the results of the EPOHITE project (Reiss et al., 2003).

In order to manage analytically the described time lag between policy activity and desired outcomes, we compare biotechnology policies in the 1990s with information on the level of biotechnology appropriation as presented in section 2 above. The German policy profile of the mid-1990s indicates a rather broad approach towards supporting biotechnology (Reiss et al., 2005). In most areas policy instruments had been implemented. Nevertheless, policies aiming at promoting basic and applied biotechnology research had been the most intensive ones during that period. The rather broad policy approach implies that for most policy goals at least some policy activity had been implemented. However, the support of knowledge flow between scientific disciplines and the assurance of the availability of human resources and market-oriented policy goals seemed to have gained less support by the various policy areas. A particularly high level of policy activity aimed at exploiting regional potentials in biotechnology.

Comparing these policy activities with the situation of biotechnology appropriation in the mid-1990s seems to indicate that the broad approach towards supporting biotechnology research in public research organizations has been quite successful, as illustrated by the quite high level of appropriation of biotechnology in the science sector. The rather low policy activities, aiming at supporting the development of human resources as observed already in the 1990s, obviously go in parallel with current problems in Germany in terms of securing the appropriate level of qualified human resources. Another field where policy activities so far have not been as successful is the interlinkage between different scientific disciplines. Even though there had been a number of policy approaches supporting scientific networks, interdisciplinarity in the science domain is still not very well developed. Obviously, there is still a need for appropriate policy incentives for improving interdisciplinary research and education at all levels of the science system.

## SUMMARY AND CONCLUSIONS

Our analysis indicates that the appropriation of biotechnology in Germany has proceeded well in the science sector and there are also promising knowledge transfer activities. However, appropriation of biotechnology in the industrial sector has not yet commenced satisfactorily. Concerning the role and contribution of universities to this condition, it is necessary to differentiate between the contributions of universities to education on the one hand and to research on the other hand. In the educational domain we observe problems in the availability of qualified human resources, in quantitative as well as in qualitative terms. The available data indicates that so far universities have not been able to provide a sufficient number of highly qualified scientists for biotechnology research. In addition, there is still a problem in Germany in terms of interdisciplinary education and qualification. In particular from the perspective of industry, interdisciplinary skills are not developed well enough among German graduates. This indicates that there remains a need to modify and improve university education aiming at teaching in a more interdisciplinary way. In addition, for the German university system in general it seems necessary to rethink the established career paths and incentives which still rely to a large extent on highly disciplinary skills, publications and communities.

Concerning research activities, our analysis indicates that German universities are playing an increasing role in the generation of knowledge and technology. And we also observe that this contribution of universities is acknowledged by industry, as indicated, for example, by an increasing number of common research activities.

Research agendas in the public domain of the German biotechnology innovation system are determined by a combination of top-down and bottom-up approaches. Two federal ministries mainly operate in a top-down manner: the Ministry of Education and Research and the Ministry of Economic Affairs and Technology. In addition, ministries at the *Länder* level are key actors. Concerning the bottom-up approach, the large German research organizations such as the Max Planck Society, the Fraunhofer Society or the Helmholtz Association are playing an important role. Universities are represented in bottom-up agenda-setting mainly by their participation in some of the important commissions such as, for example, the Science Council.

Finally, we could identify the following main challenges for policy in order to support the appropriation of biotechnology in Germany. It seems to be crucial that policy plays a supportive role in the further development and implementation of interdisciplinary research and education at all levels of the science and research systems and at universities in particular. In addition, it is important that policy activities uphold their supportive role for universities in terms of knowledge transfer to industrial actors. There has been a remarkable change in the role of universities in this respect during the last decades in Germany. Now it seems important to continue to further such a change.

## REFERENCES

BMBF (2008), *Bundesbericht Forschung und Innovation*, Bonn: Berlin Federal Ministry of Education and Research (BMBF).

Commission of Experts for Research and Innovation (EFI) (ed.) (2008), *Research, Innovation and Technological Performance in Germany: EFI Report 2008*, Berlin: EFI.

Enzing, C. et al. (2007), 'Inventory and analysis of national public policies that stimulate biotechnology research, its exploitation and commercialisation by industry in Europe in the period 2002–2005', final report, Luxembourg: Office for Official Publications of the European Communities.

Enzing, Christien and Thomas Reiss (2008), 'The effectiveness of biotechnology policies in Europe', *International Journal of Biotechnology*, **10**(4), 327–40.

Gaisser, Sibylle and Thomas Reiss (2008), 'Biopharmaceutical innovation capacities: benchmarking Europe and implications for CEE', *Studies in Ethics, Law and Technology*, **2**(2), Article 6.

Hetmeier, H.-W., W. Göbel and P. Brugger (1995), 'Ausgaben für biotechnologische Forschung', Wiesbaden: Statistisches Bundesamt.

Nusser, M., B. Soete and S. Wydra (2007), *Wettbewerbsfähigkeit und Beschäftigungspotenziale der Biotechnologie in Deutschland*, Düsseldorf: Hans-Böckler-Stiftung.

Reiss, T. and I. Dominguez Lacasa (2007), 'Benchmarking national biotechnology policy across Europe: A systems approach using quantitative and qualitative indicators', *Research Evaluation*, **16**(4), 331–9

Reiss, T. and I. Dominguez Lacasa (2008), 'The performance of European countries in biotechnology: How does Europe compare to the USA?', *International Journal of Biotechnology*, **10**(4), 303–26.

Reiss, T. and S. Hinze (2002), 'Diffusion of biotechnology into the pharmaceutical industry during the 1990s', paper presented at the tenth Triple Helix Conference 2002, Copenhagen.

Reiss, T. and S. Hinze (2004), *The Biopharmaceutical Innovation System in Germany*, OECD case study on structure, performance, innovation barriers and rivers, Stuttgart: Fraunhofer IRB Verlag.

Reiss, T., V. Mangematin, C. Enzing, J. Senker, et al. (2003), 'Efficiency of innovation policies in high technology sectors in Europe (EPOHITE)', Contract No. HPVI-CT-2001-000005, Final Report, European Commission, DG RTD, Brussels.

Reiss, T., I. Dominguez Lacasa, V. Mangematin, F. Corolleur, C. Enzing, A. van der Giessen, J. Senker and L. Nesta (2005), 'Benchmarking of public biotechnology policy', Final report to the European Commission Enterprise Directorate General, Karlsruhe: Fraunhofer – Institut für System- und Innovationsforschung, ISI.

Reiss, T. et al. (2007), Consequences, opportunities and challenges of modern biotechnology for Europe (Bio4EU): Task 1 and 2', Brussels: European Techno-Economic Policy Support Network (ETEPS Net).

Sandgren, P. (2005), 'Annual Innovation Policy Trends and Appraisal Report Sweden 2004–5', European Trend Chart on Innovation, Brussels: European Commission.

Schmoch, U. (2004), 'The technological output of scientific institutions', in W. Glänzel, H. Moed and U. Schmoch (eds), *Handbook of Quantitative Science and Technology Research: The Use of Publication and Patent Statistics in Studies on R&D Systems*, Dordrecht, Norwell, New York, London: Kluwer Academic Publishers, pp. 717–31.

Senker, Jacqueline, Christien, Enzing and Thomas, Reiss (2008), 'Biotechnology policies and performance in Central and Eastern Europe', *International Journal of Biotechnology*, **10**(4), 341–62.

# 13. Biotechnology: national policy and development priorities in Russia

**Galina Sagieva**

## INTRODUCTION

Biotechnology belongs to the group of breakthrough, rapidly-developing technologies. It allows society to deal with major socio-economic challenges such as increasing crop productivity; production of bioactive substances, vaccines; biosynthesis of antibiotics, hormones, interferon; early diagnostics and treatment of various diseases, and so on. Many experts believe that only bioprocesses have the potential to provide radical, global solutions to such challenges as production of food and environment protection. The development of biotechnologies is expected to create opportunities to greatly increase food reserves; develop new sustainable energy sources; prevent and efficiently treat severe diseases; further develop waste-free production technologies; and reduce adverse impact on the environment.

In Russia, biotechnology is seen as a major priority of S&T development; in policy statements it is designated 'live systems', and as such enjoys active government support through various targeted federal programmes. However, the country still lacks a system of statistical monitoring in the biotech sphere (data on internal R&D expenditures in this priority area has been collected since 2008 – but it is clearly not enough to measure overall activities in the biotechnology field). The statistical classification of products and economic activities (OKPD) does not have designated classes and groups for biotech products.

Modern biotechnologies based on advanced R&D results have high potential and are becoming increasingly more efficient, which strengthens and diversifies their effect not just on the human environment but directly on the human organism and its components, including genetic structures. Creation of genetic databases, introduction of passports with biometric data, identification of genetic anomalies, genetic therapy, development of cloning techniques and so on create new, ever more serious challenges to modern society. This development puts increasing demands on responsibility, and stresses the need for government regulation – and, therefore, for adequate information support. The development of S&T and innovation policies, decision-making and evaluation of results, is impossible without comprehensive and reliable information, based on statistical data.

## METHODOLOGICAL APPROACHES

This study is based on data collected by interviewing R&D organizations' representatives and the analysis of laws and bylaws regulating the science and innovation sphere, including biotechnology.

Twenty respondents were interviewed for this study: ten representatives of organizations specializing mainly in biotechnology R&D and ten representatives of companies manufacturing biotech products.

The study uses estimates obtained via polling of experts, to select priority development areas and critical technologies; these allow identification of the most promising areas of biotechnology development in Russia.

The priority area which covers biotechnology also includes bioengineering, cellular technology, biomedical and veterinary technology for human and animal life support and protection, genome and post-genome drug development technologies, biocatalytic, biosynthetic and biosensor technologies, and bio-information technology.

Requirements regarding content, selection and implementation of the Russian national S&T priorities are set in several conceptual and legal documents. These include Main Areas of the RF Government Activities until 2012; Concept of Long-Term Socio-Economic Development of the Russian Federation (RF) until 2020, developed by the RF Ministry of Economic Development and Trade; the RF National Security Concept; the federal law 'On Science and State S&T Policy'; the federal law 'On State Forecasting and Socio-Economic Development Programmes of the Russian Federation'; the RF President's messages to the RF National Assembly; the RF Government Regulation No. 340 of 22 April, 2009[1].

The latter regulation approved the 'Rules for Defining, Adjusting and Implementing Priority Development Areas for Science and Technology in the Russian Federation' and the 'List of Critical Technologies for the Russian Federation'. In particular, the regulation specified that 'Priority Development Areas' and 'Lists of Critical Technologies' must be adjusted at least once every four years, taking into account global technological development trends and the country's socio-economic priorities. The above documents are approved by the RF President. S&T development priorities and critical technologies are implemented via the framework of federal and departmental targeted programmes, other programmes and projects to promote high-tech spheres of the economy, financed with public and non-budget funds.

The ongoing work on selecting priority development areas and critical technologies in Russia started in the 1990s. According to the federal law 'On Science and State S&T Policy' the responsibility for setting priorities lies with the RF public authorities. The most important source of information for selecting and adjusting critical technologies is the long-term forecast of Russia's S&T development. Lists of critical technologies are compiled by the RF Ministry of Science and Education, jointly with other leading ministries, major S&T centres and leading scientists and professionals engaged as experts.

The main purpose of these lists is utilization of the innovation potential through application of critical technologies. Currently the lists compiled by the RF Ministry of Science and Education in 2004–06 are valid, while the next round of selection (adjustment) of priority development areas and critical technologies is under way; the latest draft includes biotechnologies.

## DEVELOPMENT OF BIOTECHNOLOGY IN RUSSIA

In the situation of financial crisis and an urgent need to overcome it as soon as possible, the efficiency of S&T and innovation policy becomes especially important. A major component is full utilization of the technological potential, while channelling available resources to promising, high-priority areas of the country's innovative development. Accordingly, R&D in the biotechnology field is extremely relevant and important, both in terms of its current state of development, and in shaping appropriate government policies.

### Biotechnology in Russia: the Current State and Development Trends

Innovative development and modernization of the Russian economy imply an active use of advanced technologies. Biotechnology as a priority development area for the Russian S&T field is included in a number of special government support programmes. For instance, the federal targeted S&T programme 'R&D in priority S&T development areas in 2002–2006' provided funding for research in such areas as human genome, biodiversity, genetic diagnostics, genetic therapy and so on. The federal programme 'National technological basis in 2002–2006' included a section to support biotechnology. Financing of biotech R&D is also envisaged in

*Table 13.1 Funding of biotechnology R&D through the federal programme 'R&D in priority areas of Russia's S&T complex development in 2007–2012' out of the RF federal budget for 2007*

| | Total | Capital investments | R&D |
|---|---|---|---|
| Live systems (€ millions, calculated via purchasing power parity of national currencies) | 81 | 4.4 | 76.6 |
| Share of 'Live systems' in the total amount of funds allocated through the federal targeted programme (%) | 24.1 | 25 | 75 |

the federal programme 'R&D in priority areas of Russia's S&T complex development in 2007–2012' (see Table 13.1).

Total expenditures on priority biotechnology R&D in 2007 amounted to €200 million, twice the level of the previous year.

According to the collected data, R&D organizations[2] usually conduct research in several priority areas at once. For example 20 per cent of the surveyed organizations conducted R&D work in four priority areas; 40 per cent of the organizations in three, 20 per cent in two, and the last 20 per cent conducted R&D only in one priority area.

Most of the funds were allocated to finance research in bioengineering technologies: 60 per cent of the surveyed organizations reported they spent between 40 and 60 per cent of their total internal R&D expenditures on research in this area. Next come biomedical and veterinary R&D (human and animal life support and protection technologies) (40 per cent of the organizations reported their appropriate expenditures amounted to 40 to 60 per cent of the total, while in 20 per cent of the organizations this figure was between 20 and 40 per cent).

According to the interviews, all surveyed organizations planned to increase their biotechnology R&D expenditures in the medium term. An absolute majority of the surveyed organizations (about 70 per cent) expected that average annual growth of their internal expenditures will be between 5 and 15 per cent, while the rest expected to increase their spending even more.

The overall growth of biotechnology R&D was taking place against a background of a changing structure of expenditures on priority R&D. Plans were under way to increase expenditures on bioengineering, biomedical and veterinary technologies for human and animal life support and protection, and on cellular technologies. About a third of the surveyed organizations intended to start work on genome and post-genome drug development technologies in the next one or two years.

To summarize: the surveyed organizations were especially active in such areas of live systems R&D as bioengineering technologies; biomedical and veterinary human and animal life support and protection technologies; biocatalytic, biosynthetic and biosensor technologies. Also, in the next few years they expected to significantly increase their R&D activities in cellular technologies and genome and post-genome drug development technologies.

In the course of the survey leading Russian researchers were asked about the most important areas in which Russian R&D organizations conduct research. Experts were also asked to compare Russian research results in these areas with the best international achievements; see Table 13.2 for processed results of the survey. Note that weighted average values were calculated for the following scale: 1 – significant lag behind the top world level; 2 – small lag; 3 – on a par with the top world level; 4 – on a par with the top world level plus some unique results in the area.

The collected data show that the surveyed organizations quite actively conduct R&D on development of drugs to prevent malignant tumours; increase stability, work capacity and life span of human organisms in normal and extreme conditions; highly selective (highly efficient) medicines; biocompatible and biodegradable materials and so on.

Russian research results received the highest marks in the following areas: tools to increase stability, capacity for work and life span of human organisms in normal and extreme environmental conditions, under effect of stress factors etc.; tools for medical diagnostics and assessment of food stuffs quality based on use of atomic force microscopy, mass spectrometry and other new principles; biocompatible and biodegradable materials for use in medicine, agriculture and industrial production; and new biomaterials.

According to the findings in the interviews, the level of Russian biotechnology research in most cases is on a par with the international level. Theoretical research in genetic, cellular and protein engineering technologies and structural biology, is mostly on a par with top international level. Basic research in biomedical and veterinary life support technologies, protection of humans and animals in general, match the best international examples, and in certain areas exceed them. Bio-information technologies, genome and post-genome technologies are approximately at the same level as in the leading Western countries.

Russia's leading positions in biotechnology R&D have been noted by the respondents in the interviews, mainly regarding some rather specific,

*Table 13.2 Major research areas in which Russian R&D organizations are active, and how they compare with top international results*

| Products, services, technologies and techniques | Share of organizations which conduct (Yes) or do not conduct (No) such research (per cent of their total number) | | Average estimate of research level (points) |
|---|---|---|---|
| | Yes | No | |
| Molecular diagnostics of most common human diseases with the help of biological microchips | 25 | 75 | 1.8 |
| Molecular diagnostics of infectious agents and their life products in food and environment with the help of biological microchips | 0 | 100 | 1.8 |
| Creation of new materials for treatment and diagnostics, based on nanoparticles containing biomolecules | 50 | 50 | 2 |
| Tools for medical diagnostics and assessment of food quality based on use of atomic force microscopy, mass spectrometry and other new principles | 38 | 62 | 2.5 |
| Highly selective (highly efficient) medicines | 62 | 38 | 1.8 |
| Treatment techniques and products based on stem cells | 14 | 86 | 1.8 |
| Individual treatment techniques and prophylactics | 17 | 83 | 1.5 |
| Highly sensitive and selective biosensors | 38 | 62 | 1.8 |
| Biocompatible and biodegradable materials for use in medicine, agriculture and industrial production | 38 | 62 | 2.3 |
| New biomaterials | 63 | 17 | 2.3 |
| Tools to increase stability, work capacity and life span of human organism in normal and extreme environmental conditions, under effect of stress factors etc. | 89 | 11 | 2.5 |
| Transgenic highly resistant and high-yield crops, producers of drugs, vaccines and food substances, materials for fine chemical synthesis, construction materials and raw materials for light industry | – | 100 | 1.3 |

*Table 13.2* (continued)

| Products, services, technologies and techniques | Share of organizations which conduct (Yes) or do not conduct (No) such research (per cent of their total number) | | Average estimate of research level (points) |
|---|---|---|---|
| | Yes | No | |
| Transgenic and cloned agricultural animals with higher productivity, producers of physiologically active substances, cells, tissues and organs | 14 | 86 | 1.2 |
| Genetically engineered micro-organisms for producing biologically active compounds, remediation of environment, enriching soil and mining | 43 | 57 | 2.2 |
| Deep processing of waste (agricultural, consumer, food industry) to make useful products | 63 | 17 | 2.3 |
| Biological pesticides, biological techniques to protect crops from pests | 50 | 50 | 2.2 |
| Medicines based on humanized monoclonal antibodies | 14 | 86 | 1.4 |
| Drugs capable of preventing development of certain kinds of malignant tumours | 78 | 22 | 1.9 |
| Techniques for long-term cultivation and preservation of organs | – | 100 | 1.7 |
| Cancer prevention techniques based on genetic diagnostics | 14 | 86 | 1.5 |
| 'Restoration' treatment of damaged organs using embryo stem cells | 14 | 86 | 1.3 |
| Techniques for selecting stem cells to grow specific organs in vitro | – | 100 | 1.3 |
| Artificial organs (pancreas, kidneys, liver etc.) using human cells and tissues | 25 | 75 | 1.3 |
| Use of biodegradable materials in agriculture, forestry and fisheries, in particular to grow crops on open soil under film | 13 | 87 | 1.4 |
| Biogas obtained from cattle's waste and food waste, for use as fuel | 14 | 86 | 1.7 |
| Generating biomass energy out of agricultural products and by-products | 12 | 88 | 1.8 |

*Table 13.2* (continued)

| Products, services, technologies and techniques | Share of organizations which conduct (Yes) or do not conduct (No) such research (per cent of their total number) | | Average estimate of research level (points) |
|---|---|---|---|
| | Yes | No | |
| Direct biological absorption of carbon dioxide produced by thermal power stations | – | 100 | 1.2 |
| Transformed cattle capable of extracting physiologically active substances such as antimicrobe proteins, embryo stem cells etc. | 13 | 87 | 1.2 |

occasionally quite narrow fields (in particular bioengineering, nanobiotechnology for creation of membranes and catalytic systems, biocatalytic, biosynthetic and biosensor technologies).

In certain areas (for example some aspects of cellular technologies) the lag behind the world leaders according to the experts is growing, due to exhaustion of the previous research's potential and a lack of favourable conditions for further research in new areas. This lag, along with traditionally underdeveloped mechanisms for commercialization of technologies, is getting in the way of making breakthrough discoveries in important development areas. At the same time theoretical research in cellular engineering technologies is on a par with the international level; in particular, Russian scientists conduct stem cell research, work on deciphering molecular mechanisms of hereditary and cancerous diseases, mechanisms for subcellular data transfer; and biological safety problems.

Respondents noted certain improvements in the S&T sphere in the recent period, due to increased government support and an overall increase of R&D funding. The increased level of R&D activities in Russia created conditions for a more rapid development of major technological areas, and their application via production of high-tech marketable products, ideally competitive on domestic and international markets.

These products included new-generation, highly efficient biocompatible materials for medical purposes; medicines targeting membrane proteins and receptors; biochip-based tools for medical diagnostics and assessment of foodstuffs quality; means to minimize adverse impact of anthropogenic and natural disasters on human health and environment; systems for utilization and disposal of highly toxic waste; water treatment and restoration of surface water areas; treatment of waste and drainage water from industrial enterprises; treatment of sewage in settlements; processing waste from medical and prophylactic institutions; processing biological waste of food production and agricultural enterprises.

The polling of experts to select critical technologies also had a more specific objective: to obtain a more detailed assessment of R&D results in certain areas of biotechnology, compared with the top international level. Accordingly, lists of the most promising basic R&D results in biotechnology were composed from literature analysis, and lists of breakthrough technologies being developed in Russia. In each of the selected R&D areas domestic and international results were compared with each other. Out of 20 basic R&D areas under consideration, for approximately 50 per cent of them the experts agreed on approximate parity or a slight lag, while for the other 50 per cent a significant lag was noted.

Parity or a slight lag was noted in such R&D areas as understanding the mechanisms of factors affecting reproduction of stem cells, and wide distribution of techniques for their *in vitro* reproduction for use in medical practice; understanding of bipolar disorder aetiology on molecular level; understanding the relationship between synaptic and memory/cogitative organism adaptation to external environment; understanding molecular mechanisms of apoptosis, which would allow extraction and transfer of special cells into live organisms; understanding the whole mechanism of neural net creation on molecular level; understanding mechanisms of vasoconstriction in atherosclerosis; understanding 'agreement' mechanisms in Alzheimer's disease; understanding mechanisms of carcinogenic mutation.

Significant lag was noted in areas such as understanding the mechanism of chemical toxicity destroying the endocrine system and such chemicals' effect on reproductive functions, behaviour, brain activity, immunity etc.; mechanisms of environmental and immune nature of allergic diseases; understanding of the aetiology of schizophrenia on the molecular level; understanding of human ageing mechanisms in each individual case; full understanding of molecular mechanisms of development and differentiation; understanding of mechanisms of cancer metastases dissemination.

Even more pessimistic was the evaluation of R&D in the area of breakthrough technologies. Out of 25 fields under consideration, the experts noted only two where the results obtained matched the level of the leading institutions in the world. These were development of techniques to forecast biological activity and functional properties of proteins of high order structures, and development of technology which would allow accurate calculation of original higher order protein structures. In eight other fields there was noted a slight lag, and in 15 more areas significant lag behind the top world level, or even total lack of relevant research.

A slight lag was noted in such R&D fields as efficient production of medicines including drugs with cultivated animal cells; creation of an artificial retina to restore eyesight of patients with degenerate retinas; development of antibacterial drugs without side-effects; development of techniques to identify most immunologically functional molecules responsible for rejection of transplanted organs; development of cellular treatment techniques to treat cardiac infarction; development of techniques to identify and classify genes related to diabetes, hypertension and atherosclerosis; development of techniques which would allow discoveries of new functions of proteins on the basis of their DNA parameters.

Significant lag was noted in such areas as development of technologies for quick genome analysis; creation of a full genome map for at least 50 major animal and plant species; growing specific organs out of specially selected animal cells; fully implantable artificial kidneys; development of treatment techniques to completely cure atopic dermatitis allergies; treatment techniques for various diseases based on controlling 'self-cure' system; developing technologies for separating stem cells into specific nerve and glia cells; development of light energy elements to stimulate photosynthetic response; development of techniques to check safety of genetically modified products' nutritional and environmental properties; safety of genetically modified foodstuffs including 'functional' ingredients capable of preventing hypercholesterolaemia, high blood pressure, hay fever etc.

Thus from the point of view of both respondents and experts, although in some areas Russian R&D results match the top international level, the vast majority of such results in the biotechnology sphere were not on a par with the leading world level; it might even be stated that they were significantly below it.

The respondents representing organizations engaged in development of new biotechnologies noted the following major barriers hindering their work:

- insufficient funding of research personnel (20 per cent of responding organizations);
- low demand for R&D results (20 per cent);
- lack of equipment (12 per cent);
- lack of financial resources (10 per cent);
- uncertainty about practical implementation of R&D results (8 per cent);

- faulty government S&T policy and inadequate legislation (30 per cent of responding organizations).

International patent statistics also confirm a significant number of unsolved problems in the support structure for Russian R&D in the biotechnology field. This can be seen, for example, in the number of patent applications for biotechnological inventions submitted by Russian institutions to the European Patent Office: in 2004 they were disastrously low – only 10, or less than 4 per cent of all Russian applications submitted to this office. The US were ahead of Russia almost 95 times over, Germany 38.4 times, Japan 35.6 times, France 14.6 times and the UK 12.4 times.

## Innovative Potential of Biotechnologies

The estimated volume of products manufactured with the help of biotechnology and consumed in Russia in 2005 amounted to about 60 billion roubles (about 1.7 billion Euros). Approximately 30 per cent of them were manufactured by domestic producers. The majority of the Russian market's demand was met by imported goods. Russian market capacity is estimated at about 100 billion roubles (about 2.8 billion Euros); that is, demand for biotech products remains significantly unmet.

Findings from interviews show that the surveyed companies[3] manufactured a sufficiently wide range of products based on biotechnologies (see Table 13.3 for a breakdown of enterprises manufacturing various product groups).

Larger enterprises usually have more than five different biotech products in their product range. Smaller companies typically specializse in a rather narrow range of products. Note that the bulk of the output is made up of biotech pharmaceuticals and microbiological plant protection products. Next come food additives; drugs and therapeutic cosmetics made of natural raw materials; and other products. The smallest share has biotech products for power generation and distribution; and green chemistry products.

The interviews show that biotech products made by the surveyed companies generally are not very competitive: according to their manufacturers, only 5 per cent of products can be called radically new, without an equivalent on international markets. About 30 per cent were considered competitive on international markets. The most competitive products, according to the respondents, included enzymes; drugs and therapeutic cosmetics made of natural raw materials; next came food and forage additives. The bulk of pharmaceuticals and microbiological plant protection products were among the less competitive products.

*Table 13.3 Breakdown of enterprises manufacturing products with the help of biotechnologies (per cent of the total number)*

| Product groups | Share of enterprises manufacturing products with the help of biotechnologies |
|---|---|
| Biotech pharmaceuticals | 80 |
| Food and forage additives | 60 |
| Enzymes | 60 |
| Transgenic plants and animals | 40 |
| Drugs and therapeutic cosmetics made of natural raw materials | 60 |
| Microbiological plant protection products | 60 |
| Biogeotechnologies | 10 |
| Biotechnologies used for environment protection | 40 |
| Green chemistry | 20 |
| Biotechnologies for power generation and transformation | 10 |
| Other products and services | 50 |

It is a known fact that the Russian biotech industry mostly produces generic products, amounting to 90 per cent of the total output, and, additionally, most of them are obsolete. Advanced high-tech products make up just 10 per cent of the overall output.

The critical technologies selection methodology adopted in Russia is based on identification of products and services which can contribute to a radical improvement of the GDP's qualitative structure and help achieve sustainable growth. Accordingly, one component in the critical technologies selection process involves compiling lists of major products and services which on the one hand would be competitive, with good prospects on internal and external markets, and on the other hand may have significant innovative components based on Russian R&D results. Manufacturing those kind of products affects the innovative potential of critical technologies, their ability to create new markets and to extend existing ones.

The expert polling exercises included adjusting lists of major innovative products made with various critical technologies in the priority R&D area under consideration.

According to the experts, major innovative products with the biggest potential sales, made with the help of bioengineering technologies included the following:

- immunobiological preparations based on genetically engineered humanized antibodies; new generation vaccines;
- transgenic highly resistant and high-yielding crops;
- genetically engineered micro-organisms for producing biologically active compounds, remediation of environment, enriching soil and minerals;
- individual diagnostics, prophylactics and therapeutic services.

Major innovative products made with the help of biocatalytic, biosynthetic and biosensor technologies included the following:

- key compounds (chiral syntons) for production of drugs;
- biocompatible and biodegradable materials;
- enzymes for food and pulp-and-paper industries;
- enzymes for agriculture;
- biocatalysts for chemical industry;
- enzymes for molecular biology, genetic engineering, biosensor systems;
- molecular diagnostics of most common human diseases, infectious agents and their life products in foodstuffs and environment; individual diagnostic tools;
- environment monitoring and protection systems to deal with consequences of anthropogenic and natural disasters.

The following product categories are thought to have the biggest potential sales of products that are primarily manufactured with the help of biomedical and veterinary technologies for human and animal life support and protection:

- tools to increase the human organism's stability, work capacity and active life span, in normal and extreme environments;
- technologies and techniques for physical adaptation of the human organism to changing environment conditions, and subsequent rehabilitation;
- diagnostics and treatment techniques and equipment.

Cellular technologies allow for making the following major innovative products:

- foetal tissues and stem cells for medical purposes;
- stem cell-based products;
- producers of cells, tissues and organs for transplantation surgery;
- stem cell banks.

According to the experts, the most competitive products with biggest potential sales, manufactured with the help of genome and post-genome technologies, included:

- highly selective drugs targeting cell membranes;
- subcellular drugs;
- drugs based on modified DNA fragments, for genetic therapy;
- drugs based on humanized monoclonal antibodies.

Innovative competitive products made with an extensive use of bio-information technologies included the following:

- new highly efficient models, computing techniques and algorithms for solving various problems in the live systems technologies area, and relevant software;
- databases and knowledge bases in the live systems technologies area, and high-speed access systems for them;
- provision of computer and information resources for solving various bio-information problems.

The following product groups have the highest export potential:

- genetically engineered micro-organisms for production of biologically active compounds, remediation of environment, enriching soil and minerals;
- new biomaterials;
- tools to increase stability, work capacity and active life span, in normal and extreme environments, under influence of stress factors etc.;
- molecular diagnostics of most common human diseases, infectious agents and their life products in foodstuffs and environment with the help of biological microchips.

The respondents noted the following factors hindering implementation of biotechnologies in production:

- underdeveloped cooperation links (57 per cent of all surveyed enterprises);
- insufficiently developed legislation regulating and promoting use of biotechnologies (97 per cent);
- insufficient financial support by the government (42 per cent);
- insufficient information about markets (42 per cent);

- high costs of biotechnologies (29 per cent);
- high commercial risks (29 per cent);
- lack of own financial resources (28 per cent).

Practically all respondents agreed that government support of innovation activities was too low.

The following measures to increase competitiveness of biotech products were named as the most important ones:

- increase (co)investments in industrial production and R&D (43 per cent of the surveyed enterprises);
- reduce taxes and eliminate cost-related economic barriers (43 per cent);
- increase the share of products purchased by the government (27 per cent);
- ease access to markets, eliminate bureaucratic barriers (27 per cent);
- speed up and simplify patent registration procedures (14 per cent);
- facilitate integration of science and industrial production (39 per cent of the surveyed enterprises).

## Socio-economic Effect of Biotechnologies

In the interviews attempts were made to estimate the potential effects of biotechnology implementation by asking appropriate questions to respondents representing both industrial and R&D organizations. The results describing effects of biotechnology implementation are summarized in Table 13.4.

Apparently, for the first group of the above organizations the most important effects included higher product quality, reduced energy costs, reduced environment pollution, increased output, creation of new markets; increased productivity and reduced production costs were seen as less important.

Enterprises using advanced biotechnologies saw creation of new markets as the most important effect. Other important aspects included higher product quality, reduced energy costs and reduced environment pollution.

Experts selecting priority R&D areas were asked to assess potential effects of biotechnology on various Russian industries, including energy saving, materials consumption and productivity.

According to them, the highest effect in terms of energy-saving biotechnologies is to be found in agriculture, food and consumer goods industries. According to the expert evaluation, energy savings in these industries may

*Table 13.4 Effects of biotechnology implementation (per cent of all respondent organizations and enterprises)**

| Effect estimate | Organizations conducting biotech R&D | Enterprises using biotechnologies in production |
|---|---|---|
| Increased output | 67 | 34 |
| Higher product quality | 89 | 50 |
| Creation of new markets | 56 | 86 |
| Increased productivity | 44 | 17 |
| Reduced production costs | 44 | 16 |
| Reduced energy costs | 78 | 34 |
| Reduced environment pollution | 78 | 34 |
| Other | – | 22 |

*Note:* * Respondents were allowed to select several answers

exceed 5 per cent. Somewhat smaller savings (between 1 and 3 per cent) can be achieved in the pharmaceutical industry.

In terms of materials consumption, the biggest effect was expected in the pharmaceutical industry (over 5 per cent). A smaller effect was estimated for agriculture (more than 3 per cent) and for the chemical industry (1–2 per cent).

The biggest growth of productivity due to use of biotechnology was expected in such industries as pharmaceuticals and animal breeding (over 5 per cent), chemical, light and food industries (3 per cent or more).

Experts were also asked their opinion on critical biotechnologies' effect on the rate of economic growth, social development and improved environment. The breakdown of their replies is summarized in Table 13.5.

According to the experts, the biggest effect in all of the above areas will come from biomedical and veterinary technologies for human and animal life support and protection, and biocatalytic, biosynthetic and biosensor technologies. Estimates of the effects of other critical technologies were rather varied.

Thus, according to both the experts and the respondents (developers and manufacturers), the Russian biotech sphere has both advantages and weaknesses hindering its development and negatively affecting the competitiveness of biotechnology. The main problems are connected with the legislation and regulation framework, and with the commercialization and efficient application of biotechnologies in industrial production. To improve the current situation, many respondents pointed to the need for an active government policy to promote development of this sector,

*Table 13.5 Socio-economic effects of critical technologies (per cent of the total number of experts assessing each critical technology's effect)*

| | Effect | | |
|---|---|---|---|
| | Significant | Medium | Insignificant |
| **Higher economic growth** | | | |
| Bioengineering technologies | 43 | 57 | – |
| Cellular technologies | 40 | 40 | 20 |
| Biomedical and veterinary technologies for human and animal life support and protection | 57 | 29 | 14 |
| Genome and post-genome drug creation technologies | 44 | 28 | 28 |
| Biocatalytic, biosynthetic and biosensor technologies | 62 | 38 | – |
| Bioinformatics | 38 | 38 | 24 |
| **Social development** | | | |
| Bioengineering technologies | 17 | 51 | 32 |
| Cellular technologies | 43 | 14 | 43 |
| Biomedical and veterinary technologies for human and animal life support and protection | 75 | 25 | – |
| Genome and post-genome drug creation technologies | 62 | 12 | 26 |
| Bio-catalytic, biosynthetic and biosensor technologies | 43 | 57 | – |
| Bioinformatics | 29 | 42 | 29 |
| **Improved environment** | | | |
| Bioengineering technologies | 38 | 50 | 12 |
| Cellular technologies | 16 | 33 | 51 |
| Biomedical and veterinary technologies for human and animal life support and protection | 44 | 34 | 22 |
| Genome and post-genome drug creation technologies | 29 | 29 | 42 |
| Bio-catalytic, biosynthetic and biosensor technologies | 50 | 37 | 13 |
| Bioinformatics | 29 | 42 | 29 |

by which is meant not merely direct financing, but the forming of an improved innovation climate, activation of processes of integration and cooperation between science (state research institutes) and business.

## NATIONAL BIOTECHNOLOGY POLICY: LEGAL FRAMEWORK FOR THE BIOTECH SPHERE

According to the interviews, Russian laws and the national policy in the biotech field are mostly concerned with safety regarding industrial, agricultural and environmental uses of DNA1 compounds; quality assurance in molecular genetic testing; general aspects of legal protection and licensing of intellectual property and problems of commercialization of technologies generated by national budgetary R&D and higher educational institutions within R&D funded by budget.

### Use of DNA1 Compounds

A whole host of legislation developed by the Russian Ministry of Education and Science jointly with the Russian Academy of Science and other stakeholders, and examined by international experts, includes norms and provisions regulating use of DNA1 compounds. In particular, the Federal Law 'On State Control of Genetic Engineering' mentions among major goals in this area: improvement of living standards and health care; environment protection and restoration; maintaining biodiversity; increasing efficiency of agriculture, mining and processing industry; maintaining and developing human resources, including vocational training (article 5 of the Federal Law).

Also, the law establishes principles for genetic engineering, based on public safety (biological and physical protection of personnel involved in genetic engineering and of the general public) and environmental protection, as well as clinical testing of genetic diagnostic and genetic therapy techniques on somatic cells level (in open and closed systems); open access to data on genetic engineering safety (risks and the steps taken to ensure safety); certification of products (ensuring they match environmental protection and sanitary standards, pharmacopoeia articles requirements and RF national standards); full disclosure of information on production technologies and products' properties (articles 5, 7, 10 and 11).

The law also stresses the need to develop and support international cooperation in genetic engineering by entering into appropriate international agreements (article 13).

According to the Federal Law 'On Sanitary and Epidemiologic Welfare', public safety and welfare are supposed to be ensured by government control and supervision; certification of potentially dangerous products and services; licensing potentially unsafe activities; official registration of chemical and biological substances, certain categories of products, industrial and domestic waste, products never previously imported to the Russian Federation; by socio-hygienic monitoring; conducting sanitary and epidemiologic welfare research; informing the public about outbreaks of infectious diseases, environmental conditions, sanitary and counter-epidemic (prophylactic) measures taken by appropriate agencies, etc. (Article 5).

Procedures for assuring quality and safety of food, materials and products (including biologically active additions, raw materials for food industry etc.), as well as for government supervision and control in this area are described in the Federal Law 'On Quality and Safety of Food Stuffs' and the RF government regulation 'On Government Supervision and Control in Assuring Quality and Safety of Food Stuffs'.

Since there are several government agencies responsible for regulating and supervising activities in the genetics field, the Inter-Departmental Commission on Genetic Engineering was established in 2005 to coordinate the work of relevant federal executive bodies (Ministry of Health and Social Development, Ministry of Agriculture, Ministry of Natural Resources and Ecology) on developing genetic engineering and biotechnology and ensuring adequate safety of these activities. The Commission is supported by the Federal Agency for Science and Innovations, according to the directive issued by the Ministry of Education and Science.

The Commission has the following main areas of responsibility:

- participation in developing government policy initiatives in the area of genetic engineering and biotechnology, and in their implementation;
- cooperation with federal and regional executive agencies, R&D, educational and other organizations to ensure coordination of government control in the field of genetic engineering and biotechnology;
- increasing efficiency and further development of licensing and notification systems for genetic engineering, including further improvement of Russian laws regulating biotechnology and development of methodologies for assessment and management of potential risks in this field;
- contributing to the development of international cooperation in genetic engineering;
- raising public awareness of prospects opened up by the use of modern biotechnological products;

- providing annual reports on the Commission's activities to the RF Government (para. 3).

**Quality Assurance in Molecular Genetic Testing**

Another important area of legal regulation in the biotechnology field is ensuring adherence to legal, ethical and professional principles and quality assurance of molecular genetic testing – both overall and during clinical testing; monitoring the quality of laboratory research; reporting results; and ensuring that laboratory personnel match necessary educational standards.

In Russian legislation, clinical testing is regulated by various laws, including:

- the Constitution of the Russian Federation;
- Federal Law 'Bases of Legislation of the Russian Federation on Public Health Care';
- Federal Law 'On Medications' (Chapter IX 'Development, Pre-Clinical and Clinical Research of Drugs').

Article 21 of the RF Constitution establishes the supremacy of individuals' well-being over scientific interests and the society's benefits as a whole; according to this principle no person can be subjected to medical, scientific or any other experiments or testing without their voluntary consent.

Additionally, fundamental legal norms include Article 43 of the Basic RF Legislation on Public Health Care, which describes the procedure for applying new prophylactic, diagnostic and treatment techniques, use of medications, immunobiological and disinfectant drugs and biomedical research. According to this, biomedical research can only be conducted in institutions which are members of the national or municipal health care system, and must be based on previous laboratory experiments. Biomedical research involving human subjects can only be conducted after obtaining their written consent. People cannot be forced to take part in biomedical research.

The Federal Law 'On Medications' regulates development, production, manufacturing, pre-clinical and clinical research of drugs, their quality control, efficiency, safety, trade in drugs and other activities in the field of circulation of drugs and medicines.

Chapter IX lists funding sources for the development of new medications; establishes patent protection of developers; sets procedures for pre-clinical research. According to Article 37 the objective of clinical research of medications is to assess and prove their efficiency and safety via scientific methods; obtain data about expected side-effects and interaction with other drugs.

The law establishes that clinical testing of a specific drug must be authorized by the federal executive agency responsible for supervision and control in the field of drug circulation, that is the Federal Service for Supervision of Healthcare and Social Development of the RF Ministry of Health and Social Development (Roszdravnadzor).

It should be noted that the list of documents to be presented includes a written approval by the Roszdravnadzor's Ethics Committee. Clinical testing of drugs must be conducted in medical institutions accredited by Roszdravnadzor. The same body prepares and publishes list of medical institutions authorized to conduct such research.

Unfortunately, it must also be noted that some of the most important provisions are only declared in valid laws, and lack appropriate subordinate legislation (bylaws). Also, there is a mismatch between various valid laws. On the one hand this hinders participation of Russian R&D and medical institutions in such research; on the other hand it creates opportunities for abuse in the field being discussed here.

According to Article 39 of this document, the head of the medical institution conducting clinical research of a drug must approve the programme of the experiment and appoint a responsible person (manager) who must be a physician with at least two years of medical research experience. The programme must be developed jointly with the institution's ethics committee. The programme manager must be informed about results of relevant pre-clinical research and has a right to access any other additional information about the drug in question. At the same time the law does not specify establishment procedure, status, functions, responsibilities or membership of the above-mentioned ethics committees, which makes observing this provision rather difficult and leads to conflict situations. However, the Basic RF Legislation on Public Health Care stipulates that procedures for establishment and operation of ethics committees (commissions) regarding public health care must be approved by the RF State Duma (Article 16). Since the Duma had never approved any such legislation, none of the ethics committees actually operating in the country has an official status.

Clinical research may be discontinued at any time if it turns out to be dangerous to the patients' health. Decision to discontinue the experiment is made by the programme manager.

Violation of clinical practice rules as well as falsifying results of clinical testing of drugs are punishable according to the RF legislation.

As to the patients' rights, they are established by Article 40 of this federal law which says that all human participation in clinical research must be voluntary. Patients must provide written consent to participate in medical experiments, and they must be informed about the following:

- the medication, and the nature of clinical research of the said medication;
- expected efficiency, safety of the medication, potential risks to the patient;
- what the patient should do in case the medication has unexpected effect on their health;
- terms and conditions of the patient's medical insurance.

Patients have the right to discontinue their participation in clinical research at any time.

The legal basis for clinical testing of drugs is provided by Roszdravnadzor's authorization to conduct the testing and a contract to clinically test the drug. The latter, apart from everything else, must include information about the terms and conditions of the patients' medical insurance and liability insurance of persons conducting the research in question.

Article 40 of the law 'On Medications' describes the rights of patients participating in clinical testing in great detail. These include voluntary participation, written consent, full disclosure of information about the drug and expected results of the research to the patient. The article includes a list of groups who cannot participate in clinical research.

However, the vast majority of clinical research projects are conducted without any registration whatsoever, since going by the book significantly increases the costs due to insurance payments, taxation pressure, the need to prepare official contracts between R&D organizations and their sponsors. Registration means transparent cash flows and certain participants of clinical research are not usually very keen on that.

Normally the patient signs a brief document confirming they've been informed by the researchers about goals, objectives, advantages and risks of the testing. Such a document cannot be considered a fully-fledged informed consent required by the law, and that is a serious violation. Despite the active administrative reform including creation of a specialized agency (Roszdravnadzor), an efficient monitoring and control system to ensure that clinical testing complies with all legal requirements is yet to be developed and implemented.

At the same time one should keep in mind that modern biomedical science becomes increasingly sophisticated, which significantly widens its potential. Creation of genetic databases, introduction of genetic passports, identification of genetic deviations, gene therapy, development of cloning techniques and much more – all these phenomena present the modern society with new challenges and make ever more important the need for monitoring, control and strict adherence to principles of quality, safety and ethics, their legalization and government regulation.

Lack of coordinated legal regulation leads to insufficient monitoring and control – which translates into insufficient protection of patients' and experiment participants' interests – and furthermore, creates legal uncertainty and makes shaky the positions of all parties involved in this sphere.

The increasingly complex nature of research, the need for cooperation, and the globalization processes demand harmonization of methodologies, approaches and principles. Lack of mechanisms for ethical evaluation and quality control on the one hand hinders promotion of Russian biomedical products on international markets, and on the other hand contributes to the growing imports of technologies, products and medications harmful to the health and well-being of the Russian people.

Legal and methodological uniformity can be achieved by developing a system of laws and regulations. All this calls not just for special laws but for preparing appropriate regulations and bylaws, amending existing legislation accordingly as well as effectuation of law-enforcement practice.

## Legal Provision of Protection of Intellectual Property and Licence Trade

General aspects of legal protection of intellectual property and licence trade, including biotech inventions – such as nature, content and types of licence agreements, procedures for making them, registration in the Federal Service for Intellectual Property, Patents and Trademarks (Rospatent) (for deals concluded in Russia), use of intellectual property objects are described in Part IV of the RF Civil Code: 'Rights to Results of Intellectual Activity and Individualisation Means' (Articles 1232, 1235–1239). Besides consolidation of the acting Intellectual Property Rights (IPR) legislation in the Civil Code of the RF, certain measures are taken for its advancement:

- The Civil Code of the RF involves not only traditional, but also new legal IPR institutions. Some of them are related to widely used objects of legal protection, such as know-how and trade names, which were, nevertheless, not covered by unified legal regulation. Other institutions are totally new for the Russian legislation, for instance, related rights (right to keeping database, and right of publisher of works of science, literature and art), as well as right in means of individualization (brand and domain name).
- With the enactment of the Civil Code of the RF in 2008 the legal protection of rights of results of intellectual activity (authors,

performers, inventors, and so on) was reinforced. The Law pays special attention to the rights of inventors/creators, as their interests may contradict both the societal demands for a wider and free usage of results, and the interests of more powerful commercial entities.

- Amendments and additions to the current legislation were put forward, aimed at eliminating the deficiencies of legal regulation in the sphere of intellectual property.
- Codification tasks were also solved, satisfying the need for provision of full and exact correspondence of the Russian IPR legislation with existing international obligations of the Russian Federation in this sphere, as well as its accession to Agreement on Trade-related Aspects of Intellectual Property Rights (TRIPs), in respect to Russia's expected accession to the WTO. This is related, *inter alia*, to the principle of advanced protection of authors' interests, liability rules for violation of exclusive rights, and so on
- Chapter 77 of the Part IV of the Civil Code of the RF for the first time regulated the right to use intellectual activity results as a single technology. The chapter's provisions are applied to relations in the sphere of establishing, transfer and disposal of rights to technologies of civil, military and special or double purpose, created fully or partially with the use of budgetary (public) funds, allocated through governmental contracts, income and expenditure budget, as well as subsidies.
- The enactment of Part IV of the Civil Code of the RF assures legislative regulation of asserting and distribution of rights in Russia, as well as disposal of these rights, including conditions of encumbrance, to technologies, created fully or partially with the use of budgetary (public) funds. The law regulates relations of various economic entities on holding intellectual property rights, and thus creates the legal basis for activating technology transfer.

According to the RF Civil Code, licence agreements must specify their subject matter and uses of intellectual activity results or means of individualization. In case a licence agreement which involves payment of a fee doesn't specify the amount or a procedure for calculating it, the agreement is considered invalid.

According to the RF Civil Code, the following types of licences can be used in Russia: non-exclusive (simple), exclusive, open, compulsory and sub-licence.

According to Article 1237 of the RF Civil Code, the licensee must present to the licensor reports on their use of intellectual activity results if the licence agreement doesn't provide otherwise. If the licence agreement does stipulate reports on use of intellectual activity results but doesn't specify appropriate terms and procedures, the licensee must present such reports to the licensor as may be required by the latter.

While the licence agreement remains valid, the licensor must abstain from any actions which could hinder the licensee's use of intellectual activity results in accordance with the licence agreement.

Use of intellectual activity results not specified by the licence agreement or beyond the agreement's validity, or in any other way exceeding the rights granted to the licensee by the agreement entails responsibility for violating exclusive rights to intellectual activity results or means of individualization as provided in this Code, in other laws or in the agreement.

If the licensee doesn't carry out their obligation to pay the licensor the fee for the right to use the scientific product according to the terms of the licence agreement, the licensor can unilaterally cancel the licence and demand compensation of damage caused by such cancellation.

If the licensee obtains the licensor's written consent they can grant the right to use the intellectual activity result to another party (via a sub-licence agreement).

In certain cases (described in the RF Civil Code) the court of law, following the request of an interested party, may decide to grant them the right to use the intellectual activity result, the exclusive right for which is owned by another person, on conditions defined by the court (compulsory licence).

The patent holder can inform Rospatent (the federal service for property, patents and trademarks) that any person or organization may acquire the right to use the invention (open licence). The terms of open licence are decided by the patent holder and communicated to Rospatent, which in turn publishes (at the patent holder's cost) appropriate information about the open licence. Patent holders must sign a simple (non-exclusive) licence agreement with persons willing to use the invention in question.

The law 'On Single Technology Transfer' is now valid; it sets procedures for trading (exchanging) R&D results.

It should be stressed that according to existing Russian legislation the following cannot be patented in the country:

1. Human cloning techniques;
2. Techniques for modifying genetic integrity of human germ line cells;
3. Use of human embryos for industrial or commercial purposes;
4. Other solutions contradictory to public interests, principles of humanity and morals (RF Civil Code, Article 1249 'Subjects of Patenting', para. 4).

Development of procedures for safe production, use and transfer of genetically modified organisms, their fragments and genetic engineering technologies is the responsibility of the Inter-Departmental Commission on Genetic Engineering.

**Commercialization of Technology**

It would seem that a thorough framework for commercialization of technologies, including biotechnologies, through licensing, is in place. However, the problem of practical application of intellectual property rights in the economy remains among the most acute in Russia. Actual implementation of R&D results developed with public funds in production depends on many factors. One of them is a legal framework matching the requirements of the Russian economy.

A fact to take into account is that the vast majority of R&D results in the country are still created by research organizations and higher education institutes incorporated as state-owned federal-level institutions, whose rights and responsibilities are defined in special documents and who are subject not just to civil but also to budget laws. Such organizations by definition are not interested in claiming rights to intellectual property they create and in entering such property in their books, because this would involve additional costs (including property taxes), and yet would not allow them to make profits through exercising those very rights. According to the budget laws, such profits (licence fees, royalties and so on) must be fully paid in the federal budget. This provision significantly reduces the Russian technology market's major players' interest in its development and extension, eliminating necessary motivation and incentives. Such conditions hinder a technological modernization of the Russian economy and gets in the way of the country's strategic development objectives.

It should be noted that one of the major players on the Russian technology market is the state, represented by executive authorities: about 95 per cent of all industrial property was created within R&D funded by federal, regional and municipal budgets of the Russian Federation. Approval of Section IV of the RF Civil Code legalized the principle of allocating the rights to results of intellectual activity created with the help of public funds. The general rule is: rights to intellectual activity results belong to the executor.

At the same time such legal provisions, despite being a motivating factor to implement research results in industrial production, still cannot ensure commercialization of intellectual property created with the help of public funds if the creator of that intellectual property (and holder of the rights to it) does not have a legal opportunity to implement appropriate R&D results in production.

It is clear that most of this kind of research results are created by state R&D and educational organizations. There are two kinds of state-owned institutions in Russia: budgetary organizations and federal unitary enterprises. If the latter have at least some opportunities to implement their R&D results, directly or by creating innovative business enterprises, up to August 2009 budgetary organizations did not have any such opportunities at all.

Research and educational institutions, due to strictly targeted funding and severely limited legal competence, can not use their results for application purposes; also, they were not allowed to set up firms (start-ups) for such purposes.

The only thing budgetary institutions can do with the intellectual property they own is to invest it in the chartered capital of business companies. However, due to provisions of budgetary legislation[4], they cannot provide production space for newly created business companies for free; they cannot provide even original circulating capital out of the money earned by legally authorized entrepreneurial and other income-generating activities; and they cannot freely use funds earned as dividends.

The reasoning of legislators who have created a whole host of limitations for budgetary institutions regarding their use of funds, was that such institutions should be funded specifically and exclusively to perform the social functions they have been established to perform in the first place. Otherwise budgetary institutions should change their legal status (for example be transformed into autonomous institutions).

However, Russian realities are such that on the one hand R&D and educational institutions cannot afford to refuse targeted government funding (and it can be provided only to budgetary institutions), and on the other hand they account for the majority of R&D results created with the help of public funds, own the rights to this intellectual property as executors and thus must have opportunities to actually implement these results in production.

The Federal Law from 2 August 2009 N 217-FZ 'On amendments to selected legislative acts of the Russian Federation on creation of firms by budgetary research organizations and higher education establishments for application of their intellectual activity results', which allowed public research organizations and universities creating small enterprises for commercialization of their intellectual property, may be considered a compromise (or preliminary solution) to the existing problem. According to this law, public research organizations and universities are allowed to make investments in the authorized capital of the newly created companies other than intellectual property, including financial investment, equipment and other property which is under their operational administration. Furthermore, they are allowed to obtain earnings from disposition of shares in the authorized capital of small enterprises, of which they are founders (participants), and respective shares of the business profits (including dividends), provided they put these profits in a separate account and use them only for the legal protection of intellectual activity results, fees and remunerations to the authors, as well as statutory activity.

It should be noted that this law is not yet fully operational, as it was adopted in a restricted format[5], that is no respective amendments were introduced to the budgetary legislation, and no by-laws (sublegislative acts) were adopted (on sources of funding, methodological recommendations for evaluation of intellectual property assets, which are transferred to firms, and so on).

So one of the pressing tasks regarding promoting application of new technologies in production is the development of appropriate government policies, including for example adoption of sublegislative acts to replace existing legislation and create legal opportunities for R&D and universities which have the status of budgetary institutions to set up small innovative enterprises in order to commercialize technologies, including biotechnology.

Obviously, a major area for improving government policy, aimed at creating a civilized technology market and ensuring efficient use of R&D results, is also the elimination of the most important mismatches between requirements of valid legislation (primarily budget, tax and civil laws, including legislation on non-profit organizations) and the legal basis of intellectual property rights integrated in Section 4 of the RF Civil Code.

This will provide wider opportunities for national R&D and educational institutions to participate in innovation activities. However, note that it is not just about creating small innovative enterprises; equally important is creating favourable conditions for commercialization of intellectual property generated within R&D funded by the state budget; stimulating cooperation between R&D and industrial production sectors; establishing and developing stable alliances between R&D organizations and businesses; participating in targeted investment foundations; widening the range of available funding sources. Of particular importance is providing support to innovative entrepreneurship with the help of credit and other financing tools. This is especially relevant during the post-crisis period and steps should be taken to get ready for it well in advance.

## CONCLUSIONS

Practical application of approaches to technological modernization and innovation development of the Russian economy is of the utmost importance. This study of the current state of development of the Russian biotechnology sphere is based on: results of interviews with respondents representing R&D and industrial organizations; opinions of leading experts in the area of biotechnology, obtained via expert polling conducted in the course of priority-setting and critical technologies selection exercises; and analysis of the legislation regulating the biotechnology sphere.

The study allowed for identification of the following issues:

- identification and discussion of areas within the Live Systems priority field where Russian researchers conduct R&D most actively, and areas where, according to the respondents, a significant increase of activities is expected in the next few years;
- major, promising biotechnology areas being developed in Russia; areas with high and low technological development level compared with the best international results;
- factors hindering the work of organizations developing new biotechnologies;
- specialization of Russian enterprises manufacturing products with the help of biotechnologies;
- innovation potential of biotechnologies, determined by the rate of emergence of new markets and growth of the existing ones; a list of the most important innovative products manufactured with the help of biotechnologies;
- factors hindering application of biotechnologies in production;
- socio-economic effects of biotechnologies' application.

This study has paid particular attention to analysing legislation and national policies in the biotechnology sphere, primarily laws and bylaws establishing safety requirements regarding use of DNA in industry, agriculture and the environment; quality assurance during molecular genetics-related testing; and commercialization of technologies.

The results of the survey indicate an urgent need to improve economic, legal and organizational mechanisms in the S&T sphere, focusing on current plans to develop the economy in terms of technological modernization and the creation of an effective national innovation system. The emphasis should be placed primarily on the promotion of innovation activities of the enterprise sector, combined with measures to increase the efficiency of the R&D sector.

Large-scale support of R&D organizations, monitoring of research in even a small number of promising technological areas, and radical restructuring of the economy to promote research-intensive sectors would allow an increase of the GDP growth rate, as well as the country's share in international technology trade. An urgent need for and the inevitability of bringing Russian S&T and innovation policy in line with the international mainstream is determined largely by the objectives of the country's socio-economic development and the S&T and innovation complex's role in accomplishing them. However, turning major international trends into practical steps to improve the said policy currently depends on the success of improving the existing models of S&T and innovation development.

These models must meet not only internal but also external interests of the Russian Federation. The quality and relevance of these models should relate also to the sustainability of the national economy in a globalizing world and to processes of its integration into the international economy. The initiation of processes of full participation of the Russian Federation in the OECD can be seen as a reflection of the national S&T and innovation policy strategies.

## NOTES

1. 'On Approval of the Rules for Defining, Adjusting and Implementing Priority Development Areas for Science and Technology in the Russian Federation and the List of Critical Technologies for the Russian Federation', available at http://www.government.ru/content/governmentactivity/rfgovernmentdecisions/archive/2009/04/22/535853.htm
2. 40% of Russian R&D organizations which were interviewed in 2007, employed up to 100 staff, and 60% employed more than 100 people.
3. The surveyed organizations can be broken down as follows: Number of staff: up to 100 employees: 50% of the organisations; 100–300 employees: 33%; more than 300 employees: 17% of the organizations. Share of products (services) produced with the help of biotechnologies in the total output: up to 35% of the total output: 16.7% of the enterprises; 35–65%: 33.3%; 66–95%–16.7%; over 95% of total output: 33.3% of the enterprises.
4. In particular the federal law of 26 April, 2007 N 63-FZ 'On Amendments to the RF Budgetary Code Concerning Regulation of Budgeting and Amendment of Certain RF Laws to Bring Them in Line With the RF Budgetary Legislation' and the federal law of 24 July, 2007 N 198-FZ 'On the Federal Budget for the Year 2008 and Planned Estimates for the Years 2009 and 2010'.
5. Changes are introduced in Article 5 of the Federal law from 23 August 1996 No. 127 'On science and state S&T policy' (Legislative Acts of the RF, 1996, N 35, Article 4137; 1998, N 30, Article 3607; 2000, N 2, Article 162; 2004, N 35, Article 3607; 2005, N 27, Article 2715; 2006, N 50, Article 5280; 2007, N 49, Article 6069; 2008, N 30, Article 3616; 2009, N 7, Article 786); Article 27 of the Federal law from 22 August 1996 N 125 'On higher and postgraduate vocational education' (Legislative Acts of the RF, 1996, N 35, Article 4135; 2004, N 35, Article 3607; 2006, N 1, Article 10; 2007, N 49, Article 6069; 2008, N 52, Article 6241; Article 15 of the Federal law from 8 February 1998 N 14 'On limited liability companies' and Article 4 of the Federal law from 24 July 2007 N 209 'On

development of small and medium enterprises in the RF' (Legislative Acts of the RF, 2007, N 31, Article 4006).

## LAWS AND REGULATIONS

1. The Constitution of the Russian Federation was adopted by national referendum on 12 December 1993. The Constitution came into force on the day of its official publication. The text of the Constitution was published in *Rossiiskaya Gazeta* newspaper on 25 December 1993, available at http://www.constitution.ru/.
2. Section 4 of the RF Civil Code 'Rights to intellectual activity results and individualisation means' (adopted on 18 December 2006, N 230-FZ), available at the RF Ministry of Education and Science website, http://mon.gov.ru.
3. FGP 'R&D in priority areas for developing Russian S&T complex in 2007–2012' (approved by the RF government regulation of 17 October 2006, N 613 as amended by the RF government regulation of 18 August 2007, N 511), available at the RF Ministry of Education and Science website, http://mon.gov.ru.
4. The RF Government Regulation 'On Approval of the Rules for Defining, Adjusting and Implementing Priority Development Areas for Science and Technology in the Russian Federation and the List of Critical Technologies for the Russian Federation', N 340 of 22 April 2009, available at http://www.government.ru/content/government activity/rfgovernmentdecisions/archive/2009/04/22/535853.htm.
5. The Federal Law 'On government regulation of genetic engineering activities' N 86-FZ of 5 July 1996 (adopted on 12 July 2000, N 96-FZ), available at http://ntc.duma.gov.ru/bpa/docview.phtml?bpaid=1&code=36276&rdk=1&nv=gd.
6. The Federal Law 'On sanitary and epidemiological wellbeing of population' N 52-FZ of 30 March 1999 (adopted on 30 December 2001 N 196-FZ, 10 January 2003 N 15-FZ, 30 June 2003 N 86-FZ, 22 August 2004 N 122-FZ), available at http://www.garant.ru/law/12015118-000.htm.
7. The Federal Law 'On quality and safety of food stuffs' N 29-FZ of 2 January 2000 (adopted on 30 December 2001 N 196-FZ, 10 January 2003 N 15-FZ, 30 June 2003 N 86-FZ, 22 August 2004 N 122-FZ, 9 May 2005 N 45-FZ, 5 December 2005 N 151-FZ, 31 December 2005 N 199-FZ, 31 March 2006 N 45-FZ, 30 December 2006 N 266-FZ, 12 June 2008 N 88-FZ), available at. http://base.consultant.ru/cons/cgi/online.cgi?req=doc;base=LAW;n=65016.

8. The RF government regulation 'On government supervision and control in quality assurance and safety of food stuffs', N 987, 21 December 2000, available at http://faolex.fao.org/docs/texts/rus41196.doc.
9. Federal Law 'Basic RF legislation on public health care' N 5487-1 of 22 July 1993 (adopted on 24 December 1993, 2 March 1998, 20 December 1999, 2 December 2000, 10 January, 27 February, 30 June 2003, 29 June, 22 August, 1, 29 December 2004, 7 March, 21, 31 December 2005, 2 February, 29 December 2006, 24 July, 18 October 2007), available at http://old.rospotrebnadzor.ru/docs/fedlaw.
10 The Federal Law 'On medicines', N 86-FZ of June 22 1998 (adopted on 2 January 2000 N 5-FZ, 30 December 2001 N 196-FZ, 10 January 2003 N 15-FZ, 30 June 2003 N 86-FZ, 22 August2004 N 122-FZ, 16 October 2006 N 160-FZ, 18 December 2006 N 231-FZ), available at http://base.consultant.ru/cons/cgi/online.cgi?req=doc;base=LAW;n=70874.
11. The Federal Law 'Bases of Legislation of the Russian Federation on Public Health Care' N 5487-1 of 22 June 1993 (adopted on 24 December 1993, 2 March 1998, 20 December 1999, 2 December 2000, 10 January, 27 February, 30 June 2003, 29 June, 22 August, 1, 29 December 2004, 7 March, 21, 31 December 2005, 2 February, 29 December 2006, 24 July, 18 October 2007), available at http://www.niiot.ru/doc/doc024/doc.htm.
12. The Federal Law 'On amendments to selected legislative acts of the Russian Federation on creation of firms by budgetary research organizations and higher education establishments for application of their intellectual activity results' N 217-FZ of 2 August 2009, available at http://base.consultant.ru/cons/cgi/online.cgi?req=doc;base=law;n=90201.

# PART V

# Implications for public policy and industry development

# 14. Implications for public policy and industry development[1]

**Bo Göransson and Carl Magnus Pålsson**

The twelve case countries examined in this book differ substantially in terms of level of economic development, industry structure and market orientation, but they all share the trait of attaching high importance to the development of the biotechnology sector. The preceding chapters have illustrated the diversity and heterogeneity of the biotechnology sector, and that the *raison d'être* for national policies in this area varies substantially. For industrialized countries, the building up of national competence in biotechnology is perceived as an integral part of maintaining and enhancing a competitive advantage for domestic industry. Consequently, national policies in this category of countries are aimed at supporting and complementing the national biotech industry's ability to compete in targeted segments. Measures include stimulating university–industry linkages and technology transfer through support programmes as well as joint R&D schemes.

For other countries, with a less developed industrial structure and fewer resources to spend on R&D, the biotech industry is nevertheless regarded as a promising tool for addressing urgent social needs and for holding the promise of future industrial growth and development. As the case of Cuba illustrates, national health care considerations can provide a powerful incentive for developing the biotech sector. The successful development of a synthesized vaccine against meningitis enabled the country to embark on a nationwide vaccination campaign at considerably lower cost than for imported, cultured alternatives. Also, countries like Mozambique and Tanzania consider biotechnology to hold the key to transforming indigenous knowledge into products for domestic use as well as for export and, thus, supporting the socio-economic development of the country.

The basic prerequisites for building up a robust and sustainable biotech industry obviously differ considerably between the countries, as do the present size and quality of the industry. The governmental balancing act of meeting societal needs while, at the same time, fostering national domestic technological capability in biotech, is evident in several of the chapters. In countries such as Vietnam and Tanzania there is a mismatch between the research performed at universities and the technology demanded by industry. In China as well as in Brazil, a contributing reason for this is that the quality of university research in biotech is generally higher than the research performed by domestic industry. The absorptive capacity of industry is thus not sufficiently high to accommodate the high-tech products being researched at the universities.

Such discrepancies between the performer of research and the users of research results can further reinforce the ever-present danger of brain drain in developing countries. In the global competition for human resources, foreign-trained researchers have little or no incentive to embark on a career in an environment where his or her area of competence is not demanded, as the case of Tanzania bears out.

For the transition economies among the case countries, the fostering of the biotech industry has, additionally, been burdened by a sometimes painful shift from a centrally planned market where technology was provided free of charge for the productive sector, to a competitive market-oriented system. This has in some cases left the research institute sector in a 'sink-or-swim' situation where competition for dwindling state support and low demand from industry has weeded out large parts of the research institution sector. Not least, these countries have had to struggle with the building up of a viable university research system to complement the earlier pre-eminent technology provider, the research institute sector.

It would appear that the path of development of the biotech sector in the case studies has followed a trajectory based on the historic strength of national industry. Strong national interests have influenced national policies in the field of biotech and the resulting policies have contributed to the present particular form of the biotech industry in each country. This seems to have occurred to a higher degree than in the case of more disruptive technologies, such as information and communication technologies – perhaps a consequence of the combined properties of the underlying skill-bases in biotechnology.

Path dependency in the knowledge base plays a crucial role when we look at the national specialization patterns in biotech. A case in point is the evolution of biotechnology in Latin America. Life sciences are particularly strong all over Latin America, in comparison to other research and knowledge-intensive fields. It is striking that the most famous and well-known contribution of Latin American science to world science are in health sciences: the Argentinean Nobel Prizes in science, all related to health and biosciences; the Uruguayan invention of a method to measure the uterus contraction at the moment of birth, giving rise to a new paediatric speciality, perinatology; the Brazilian major figures of public health, like Carlos Chagas; the Cuban medical doctor Carlos Finlay and his pioneering research on yellow fever. These strengths have accumulated over decades, having been sped up in recent years by a comprehensive system of national fellowships, as well as improved access to international fellowships. It is fairly safe to assert that the development of a biotechnology-based industry in Latin America in general, and in Uruguay and Cuba in particular, is not constrained by the knowledge supply-side. It should not be constrained by the demand-side either, because the applications of biotechnology fit perfectly well with the needs stemming from the important and diverse natural resources-based productive sectors that are so important in the region and in both countries. But the whole issue is to find the niches where one may nurture a productive development in the midst of a sector that is generally very expensive and highly regulated at a global level. The examples of Uruguay and Cuba show that it is possible to succeed in this endeavour, but they also show that to make biotechnology a real tool for development, a truly systemic policy approach is needed, something that has not yet been achieved in these countries.

Also in transition economies, such as Russia, Latvia, China and Vietnam, the expansion of the biotech sector is greatly influenced by their social and economic history. The development of S&T policies in these countries cannot be understood without taking into account the various overlapping socio-historical contexts and rapid changes that they have gone through in the transition towards market economies.

The path dependency can also be seen in the cases of the Danish and Swedish welfare systems (especially health care and environmental protection) and the production structure with important parts of the Danish biotech related to agriculture and food, and the Swedish to raw materials (wood, pulp and paper). The shaping of the Danish and Swedish biotech (especially life sciences) is based on the co-evolution of several key factors such as: a demanding home market and path dependency in the knowledge base; education; research and access to qualified employees; networks and triple helix collaborations; access to financial resources; and an attractive setting for clinical trials. These cases illustrate that a broad palette of policy initiatives are necessary in order to stimulate and *maintain* international strongholds within biotech (life sciences). It is a mixture of science, technology and innovation policy, industrial policy and regional policy; in practice, the different policy implications overlap.

## S&T POLICY IMPLICATIONS

Turning to key S&T policy implications following from experiences of the case countries, it would appear that the following factors have been

*Table 14.1 Characteristics of R&D performance, 2000 and 2007*

| | GERD as % of GDP | GERD (m. USD PPP) | GERD/cap (USD PPP) | Performed GERD/cap by sector 2007 | | |
|---|---|---|---|---|---|---|
| | 2007 | 2007 | 2007 | BUS | GOV | HEI |
| Sweden | 3.63 | 12451 | 1361 | 989 | 83 | 287 |
| Denmark | 2.54 | 5105 | 934 | 606 | 65 | 257 |
| Germany | 2.53 | 72300 | 879 | 615 | 120 | 143 |
| China | 1.49 | 106542 | 81 | 59 | 15 | 7 |
| Russia | 1.12 | 22809 | 165 | 106 | 48 | 11 |
| Brazil | 1.11 | 19709 | 103 | 41 | 22 | 40 |
| Latvia | 0.59 | 158 | 69 | 22 | 17 | 30 |
| Tanzania | 0.48 | 235 | 6 | 0.2 | 2.6 | 3.2 |
| Uruguay | 0.44 | 161 | 48 | 14 | 16 | 18 |
| Cuba | 0.41 | 317 | 28 | 14[a] | | |
| Vietnam | 0.41 | 883 | 10 | 1 | 7 | 2 |
| Mozambique | 0.25 | 43 | 2 | 0 | 1.7[b] | |
| Memo: USA | 2.68 | 370569 | 1229 | 884 | 132 | 163 |

*Notes:* Countries ranked after GERD as % of GDP in 2007. [a](BUS+GOV) [b](GOV+HEI)

*Sources:* UIS Database, UNESCO, Montreal: Unesco Institute of Statistics, available at (http://stats.uis.unesco.org/unesco/ReportFolders/ReportFolders.aspx), and *African Innovation Outlook 2010,* African Union/Nepad (for Tanzania and Mozambique).

universally essential in providing a framework conducive for development of the biotech sector:

- high levels of investment in R&D;
- high levels of investment in human resources;
- stability and a long-term perspective;
- intellectual property rights;
- public procurement.

## High Level of Investment in R&D

As Table 14.1 indicates, the resources allocated to all research and development (R&D) the case countries vary in relative terms between 3.63 per cent of GDP (Sweden) and 0.25 per cent (Mozambique). This means in real terms that the gross expenditure on R&D (GERD) per capita in Tanzania and Mozambique is less than a 150th of what Denmark and Sweden invests in R&D, with obvious consequences for the range of options open to them.

From Table 14.1 we can also see that there is a general tendency for the business sector to assume the role of the main performer of R&D the more developed the economy is. Conversely, the role of government in supporting and nurturing R&D is more pronounced in less-developed countries, where the business community is fragmented and generally holds a lower technological level as well as absorptive capacity.

Looking more specifically at biotechnology, R&D activities in this sector are limited by a lack of adequate financial resources in a number of the case countries. Tanzania and Mozambique, as indeed many other African countries, are facing declining growth rates of public investment in agricultural R&D, and consistently limited investment from the private sector. Without instituting innovative sources of funding it is difficult to imagine how these countries will be able to improve their performance within the biotech field. To ensure adequate availability of funds from sustainable sources for biotechnology research and development programmes, the range of policy options should be considered, such as promoting and attracting the participation of the private sector investors in biotechnology, providing motivation and incentives to local financial support as well as improving the leverage of international contributions. For example, in Tanzania the government has recently committed itself to raising the R&D budget to 1 per cent of GDP. Given the modest contribution by the business community, this would imply a substantial increase in public investment in R&D and would bring the country on par with many European countries that are struggling to meet the Barcelona target of 1 per cent public R&D spending. For Latvia, for instance, the economic and financial crisis since 2008 has resulted in considerable cuts in the R&D funding from the state budget, with GERD having dropped from 0.61 per cent of GDP in 2008 to 0.45 per cent in 2009; to a certain extent this is envisaged to be compensated by the EU SF funding. For other countries, such as China, Germany, Sweden and Denmark, the present levels of funding for biotech R&D are generally not seen as bottlenecks; rather, the comparatively high levels of funding are recognized as a prerequisite for the early growth and subsequent development of the industry.

## High Level of Investment in Human Resources

The importance of skilled human resources in biotechnology can not be overstated. Human resources play a dual role: on the one hand they are key drivers of the expansion of the knowledge frontier, in the process of scanning and searching for new opportunities; on the other hand, they often play a bridging role, strengthening the link between the firm and the university, for instance. Brazil is a case in point, proving that long-term continuous investment in human resources brings increases in capabilities that are crucial for development. However, the case also illustrates the need for comprehensive and coordinated policies in tune with human resource development. In Brazil there are significant pitfalls as the capabilities generated in the biotech sector have not been translated into economically useful knowledge, at least not on the same level as scientific developments. Policy mechanisms that have been put into practice since the mid-1980s to stimulate spin-off firms from university research, to support venture capital initiatives and to encourage joint university–industry research projects, did not succeed, as potential start-up firms have to compete in industrial structures with barriers to entry that are too high.

Several of the case countries face a severe shortage of scientific and technical staff qualified to carry out biotechnology research in universities and research institutes. Likewise, specialized laboratories for advanced biotechnological research, such as DNA sequencing, molecular cloning and genetic engineering, are very limited in countries such as Tanzania, Mozambique and Vietnam. Considering the potential of biotechnology in development, these countries should focus on building human capacity and other capacities, particularly in science and technology, entrepreneurship and technology transfer. This can be achieved by addressing the following policy issues:

- develop and sustain a critical mass of institutional and skilled human resource capacities;
- improve the current curriculum in biological/basic sciences at all levels of education in order to generate a pool of human resources who can work in the field of biotechnology;
- develop strong collaborative links with other R&D institutions nationally and internationally, including enrolling knowledge and expertise from the diasporas of the countries in question. Cooperation, collaboration and networking are vital to progress in science and technology innovations;
- devise effective strategies for technology transfer.

## Stability and a Long-term Perspective

One of the most significant S&T policy implications emanating from the case studies is the need for stability and a long-term perspective among the major actors in the sector. Contrary to this, the policy environment in many of the case studies is characterized as either underdeveloped or inefficient, or both. Perhaps surprisingly, this goes also for a country like China, which has experienced a positive development of the biotech sector for the last couple of decades. China has supported the development of the biotech industry through several national projects, including the national 863 plan focusing on applied research, the 973 programme emphasizing basic research, and the Torch Plan to promote the application and commercialization of research results in high-tech. Although the existing policies undoubtedly have resulted in important gains, China's government still has not put in place a national management organization for supporting the biotechnology industry R&D and industrialization, and still lacks an overall strategy. In order to play more effectively to the advantage of the existing policies, the government needs to strengthen the cooperation among industries, universities and research institutes, establish mechanisms and a policy environment to promote the development of the biotechnology industry, and give full play to the enterprise sector as the leading actor for biotechnology industry development.

In Russia, the current S&T policy in biotech is still ineffective despite the creation of new institutions and policy measures which try to officially match international standards, resulting mainly in an import of international practices into the local reality. Low R&D output, imbalances in the age structure of S&T staff, overall low social prestige of a scientific career and work are among numerous examples of actual policy shortcomings.

The present state is due not only to an inconsistent and fragmented policy, but also to persisting barriers and constraints that prevent the real implementation and adaptation of the chosen policy instruments. Among the factors which hinder the development of the Russian scientific and technological system one should mention an archaic organizational and institutional structure; weak integration of science, higher education and industry domains; contradictions and ambiguities in the interpretation of law provisions and legal regulations; violations of law enforcement practices; and the inertia of the governance system.

In neighbouring Latvia, there is no dedicated national S&T policy in the domain for biotech. However, this field can still be counted among those receiving particular attention by policy-makers. Over the last decade (2001–10) the biotechnology field has thus become one of the constitutive elements of the overall national S&T policy in Latvia. In terms of funding, this prioritization has been demonstrated by selected vertical as well as horizontal policy instruments (funding programmes supported by both national budget funding and allocations from the EU Structural Funds) that feature biotechnology-related topics among their thematic foci (for example, specific state research programmes in biomedicine and agricultural biotechnology). Support for the biotechnology field has thus been granted and/or envisaged for conducting research, upgrading the research infrastructure and improving the supply of human resources.

For countries with a more embryonic biotech industry there are a number of policy challenges that impact on the development and application of biotechnology. Progress is often limited due to lack of clear priorities and investment strategies in biotechnology. Investment plans need to be based on clearly articulated national priorities and goals. It is, for instance, because of this lack of priority that in Tanzania several institutions were found to be working on the same activities, thus leading to wasteful duplication of efforts and resources.

In Vietnam, to take another example, the current policies are unclear and do not appear to be specific enough to make enough of an impact on biotechnology development. The biotechnology sector has not developed as expected because it does not have a workable innovation system: the linkages between actors in the system are very weak, while the actors themselves have limited capabilities. This weak capability is aggravated by limited financial and human resources and insufficient physical facilities, such as equipment and laboratories.

Overall, the case studies included in this book indicate the importance of explicit STI policies which are not only conceived as stand-alone policies but also as transversally embedded across policy domains; this is critically important. An example from the Uruguay case may illustrate: environmental policy and public health policies, acting deliberately in consort with STI policies from the demand side, have been catalytic in speeding up innovation-related decisions of biotech firms.

## Intellectual Property Rights (IPR)

Connected to the need for stability and clear rules-of-the-game, the international laws on intellectual property rights have been a stabilizing factor, providing impetus for the evolution of the biotech sector in countries like Germany, Sweden and Denmark.

For Russia, the S&T sector in biotechnology does not demonstrate tangible results that lead to an improvement of the competitive advantage of the country. In particular, the number of patent applications for inventions in biotechnology, filed by the Russian applicants in the EPO, is almost insignificant (22 patents in 2007, or 0.2 per cent of the worldwide flow of EPO). And yet there are 57 research organizations in Russia with huge intellectual and technological potential.

The national intellectual property market in Russia is only now emerging, and the involvement of scientific and technological activities in the economic cycle on the basis of market principles is still very low. The patenting activity is much higher than the activity related to the exploitation of intellectual property rights. Thus, it is only during the process of commercialization that the real competitive advantages derived from the use of S&T activities in the production cycle are revealed.

For countries with a weaker knowledge base, intellectual property rights can act as major constraints for local firms' innovation processes. This is indeed true in the case of pharmaceutics in Uruguay, where local firms' innovation processes are strongly blocked by IPRs, creating boundaries of non-opportunities. For latecomers, IPR can thus amount to formidable roadblocks in the process of technological catching-up.

For these countries, the policy challenge is to position themselves so that they can strike a balance between, on the one hand, the need for building competence in acquisition, development and diffusion of biotechnology innovations and, on the other hand, the need for protection of research results from domestic R&D activities. Tanzania, for instance, has no IPR policy, and IP rights such as patents are still in their infancy, as only a few institutions have IP or patents offices. Key challenges here relate to the establishment of policies that will enable Tanzania and the region to regulate access to their genetic resources and share the benefits. In addressing IPR, the following policy issues could include:

- strengthen the institutions responsible for IPR issues;
- encourage R&D institutions to develop their institutional IPR frameworks;
- facilitate the development of policy guidelines on access to and on exchange of genetic resources.

**Public Procurement**

Public procurement can be a powerful tool for promoting domestic industry and for exploiting national competitive advantages. Strategic public procurement may be mentioned as a specific public policy instrument with a potential to combine innovation and improvement in a selected sector with domestic industrial job creation. Examples of this can be seen in the Nordic countries where the public demand from the health sector has provided a foundation for the development of a knowledge base, with long-term effects on the national specialization patterns in biotech. The clearest examples can be found within medical technology, illustrated for instance by the Danish hearing aid industry.

This interaction between public procurement and innovation activities appears to be important for all types of countries. Research agenda settings based on the actual needs of society, coupled with a supportive public procurement policy, can cultivate the knowledge base of the biotech sector, irrespective of the level of economic development of the country.

However, to be effective, public procurement policies need to be complemented by a range of other policy initiatives. A case in point is the experiences of the organizations examined in the Vietnamese case study. To varying extents Vietnam began to carry out R&D in biotechnology-related matters and tried to turn them into useful products for end users. Despite these efforts, so far, many research results remain in the lab. One of the main reasons has been that the research priorities that were set tended to rely on the existing expertise of scientists, and not always because of real needs. Moreover, there is no mechanism for funding the pilot phase to turn end-of-lab results into more fully developed final products for transfer. Financial regulations and systemic barriers have been identified as key obstacles. This reflects the fact that the financial policy is not conducive enough to research and innovation activities.

In the Uruguayan case, a specific biotechnology public procurement policy as part of the industrial policy could have major implications. Even in the most difficult of all biotechnology endeavours, human health, public procurement could speed up related research at the university, articulated from the start with business firms and with public–private platforms like the Instituto Pasteur Montevideo, a branch of the famous French institute. In the case of environmental biotechnology, a blend of regulation setting, regulation enforcement and industrial policy that foster the manufacturing of 'environmental hardware', can have a great impact. Even more than STI policy, industrial policy has the capacity of 'building a market' for biotech firms, and this is fundamental for the existing firms as well as for the new entrants that can be formed, particularly as spin-offs from the academic realm.

## PUBLIC POLICY IMPLICATIONS

As the case studies have illustrated, biotechnology is considered a strategic sector for developing countries. Placing it as a key component of a development strategy is shown to be fundamental in the analysis of biotechnology in countries as diverse as Cuba and Uruguay. In Cuba biotechnology can be characterized as a state affair: the ground for deciding what problems to focus on, how to conceive the innovation process, and how to establish international relationships around biotech projects and products, are part of an explicit national strategy. In Uruguay, biotechnology has been declared a high priority for the country, not so much in terms of health biotechnology but more for production and especially for agriculture. This technology is perceived as a great opportunity to add value to the overwhelming export's weight of primary commodities. However, the complex articulation of public policy and business firms within a market economy in a developing country has so far been much less systematic than needed.

Linkages between users and producers of knowledge, as well as between the local and the international community (academic institutions, firms, and so on), are a fundamental source of learning and innovation. In small countries like Uruguay and Cuba, the reliance on this type of linkage, to complement and cope with the lack of critical mass, has been acknowledged as crucial.

Looking at policy implications at the sector and firm level, some additional characteristics appear relevant to take into account. First, as pointed out earlier, biotech is a very heterogeneous sector, meaning that policies for stimulating innovation and learning should take sector specificity into consideration. For instance, drug development and production is typically dominated by large multinational players, while environmental biotechnology seems to be more domestically oriented and dependent on, for instance, public R&D.

Second, the large multinational companies are important drivers for the biotech sector when it comes to R&D activities, employment and growth. Countries at all levels of development have in place policies to attract multinationals. For countries with a weaker knowledge base in biotech the immediate goal is to attract industrial production capacity that hopefully can be linked to domestic R&D capacity. However, as the Brazilian case clearly demonstrates, the entry of multinational firms on the domestic market does not automatically lead to an improvement of the knowledge base and innovative capacity. The steady progress of the biotech sector that centred on a Brazilian pharmaceutical firm – Biobrás – was halted when the company was acquired by a multinational corporation. This led to a subsequent technological downgrading and dismantling of links with other national biotech firms. Large incumbent firms, mostly TNCs, have their own strategic agenda that most of the time does not match the intentions of policy programmes. Consequently, policies that do not take this into account are bound to fail.

The ambiguous role multinationals can play in technological upgrading is reflected in the efforts to regulate the sector and establish rules of the game for investments in biotech. Vietnam, for instance, is introducing a new Technology Transfer Law, and a High Technology Law recently passed by the National Assembly, which, taken together, are designed to contribute to a more adequate incentive structure, as well as to promote technological upgrading. The effort of Vietnam to restructure legal regulations, as well as policies for joining the WTO, are other factors that will affect the sector, and the recently enacted Law on Intellectual Property

Rights and other regulations on biosafety is expected to ensure greater reliability for stakeholders.

A third observation with public policy implications emanating from the case studies, is that universities hold the key for the development of the biotech sector. Typically, most biotech industries cluster and locate around larger universities in order to be able to tap into the newest knowledge and to recruit new graduates.

In China, universities, corporate R&D institutions and public research institutions are the three main executive institutions of biotechnology R&D covering at least 95 per cent of the biotechnology R&D funding and personnel in China. University R&D has received increasing attention from the Chinese government over the past ten years, following the popularization of the idea of research universities. Both the central government and local governments at all levels have issued a large number of policies aiming at promoting university R&D, and have also set up multiple R&D funds for universities. As a major support object of the Chinese government, biotechnology has particularly benefited from this development. In recent years, judging from the application and licensing of biotechnology patents in China, universities are also more active in the development of biotechnology than public research institutions, and second only to corporations.

For Vietnam, following a similar path towards research universities, the policy change to turn universities and R&D institutions into commercial-type entities or enterprise-type operations, without being accompanied by further specific plans of action, could be too much, and too soon. Most of the national biotechnology policies are only loosely linked to the university domain. Resources and capacity for performing biotechnology R&D are beyond the reach of many universities.

The role of public policy is appreciated as vital in advancing the biotechnology field in Latvia, though it relies to a great extent on international funding (for example, EU FP projects) and clientele (for example, contract research for foreign companies). Since most of the research units in the biotech field are public research organizations (independent research institutes or university agencies), the whole field so far strongly depends on the direction and overall framework of public policy. It can be argued that public policy has a considerable role to play in facilitating international research cooperation. The same goes for enhancing the role of universities as suppliers of highly skilled human resources for both the academic and industrial part of the biotechnology field and more generally for facilitating the attraction of human resources to science in Latvia.

## INDUSTRY POLICY IMPLICATIONS

Turning to industry policy implications, the level of generality tends to decrease as the specific national traits warrant more tailor-made actions. For Germany, the conclusion from the case study that the appropriation of biotechnology in industry until now did not perform satisfactorily, indicates that policy activities towards supporting entire value chains in certain sectors rather than focusing on individual levels (such as education, research, or technology transfer) are becoming more important. Along these lines, recent market-oriented initiatives are interesting. These include, for instance, the recent introduction of E10 fuel containing 10 per cent instead of 5 per cent ethanol. Thereby a certain push towards biotechnological ethanol production might be expected. Another example is the lead market initiative for biobased products (excluding fuels) of the European Commission. Currently there is a discussion within the German government as to whether this initiative would call for additional policy activities at a national level.

Currently Russia's share in world biotechnology production is very low (in 2010 approximately 0.2 per cent). The domestic biotech industry produces generic versions (accounting for 90 per cent of overall production) of mostly outdated drugs. Modern high-tech biotech products make up only 10 per cent. This points to the weak links between science and industry. A characteristic feature of the Russian innovation system is that the process of technology development is isolated from the real needs of manufacturing and service sectors. Intellectual property is often not created for a particular consumer, and therefore remains underutilized and unclaimed, both at production level and in the marketplace. In this regard, it is important to ensure integration of the science, education and production domains, as well as the joint creation of intellectual property.

In general, the focus of policy attention in designing and selecting tools for science and technology policy is gradually shifting toward the final stages of the innovation cycle (technological modernization of the economy; development of innovation infrastructure; creation of business venture; establishment of technological platforms; creation and development of clusters covering the scientific and educational institutions and manufacturing companies; promotion of innovation activities and demand-oriented innovation policy, and so on). In order to foster market development and capitalization of intellectual property, a number of measures have been implemented to expand the market segment of protected industrial property, especially know-how and patents, useful models (including the ones created with the involvement of the budgets of all levels), and to intensify the transfer of results of intellectual activity in the sphere of production.

Currently Russia is facing a choice of direction: either to remain a country of unrealized innovative features with a so-called 'high scientific potential' (an option which contains a range of S&T outcomes that would still be hard to utilize for innovation activities), or to focus on raising the technical level and quality of results and their commercialization and use in manufacturing and services.

The first option is a serious risk to the economy because it does not correct the existing imbalances and does not provide the necessary basis for innovation-based growth. Hence, given the international competitive pressure and the rapid development of modern science and technology trends, this scenario would inevitably lead to an insurmountable lag of the Russian economy, not only compared to developed but also to developing economies. Therefore, there is an urgent need to develop an alternative development scenario.

For Latvia, which has a weakly developed industrial sector (including only a few biotech companies) characterized by limited absorptive capacity, there are comparatively minor implications from biotechnology-related issues for the industrial policy in general. There are limited policy instruments specifically aimed at stimulating industrial applications of biotechnology. In contrast to research policy, biotechnology has not explicitly been put forward among the priority fields of economic relevance with a potential for boosting the economic development and enhancing the competitiveness of the national economy. While a certain development of spin-off companies can be identified in Latvia over the last decades, a need for establishing a transfer centre of biotechnology to bridge academic knowledge, applied research and production, is being voiced.

In Tanzania, the increasing public interest and participation in biotechnology have generated a variety of interrelated policy issues that the government needs to address. One of the most important policy issues is how to strengthen national competitiveness in biotechnology. Policy interventions should address key policy concerns such as redefining the nature and design of the public R&D institutions so as to achieve a system of innovation, determining the role of private sector/firms in biotechnology R&D and developing different mechanisms for sustaining coherent collaboration within firms and between firms and public research organizations. Whereas in most developed countries the private–public sector partnerships (PPPs), especially those between public universities and private firms, have become a prominent form of organizing and managing technological innovation, the investment of the private sector in biotechnology through PPP in Tanzania is negligible. Policies to attract private sector involvement are crucial because a significant portion of the body of scientific information on modern biotechnology and other cutting-edge sciences is in the private domain. Furthermore, the full value of R&D is only realized if it influences innovative activities in the productive and service sectors.

In addition to creating policies to enhance private–public partnerships, developing countries like Tanzania should put in place strategic measures that strengthen the role of universities as centres of training, research and technology diffusion. Universities should be restructured to catalyse innovation and to respond to the changing societal and environmental needs. Such changes may include, for example, creating incubators or information platforms for private and social enterprises to help expand economic opportunities. Biotechnology will benefit Tanzania only if knowledge and technologies developed through biotechnology can be transformed into goods and services through viable and sustainable business enterprises. The small and medium-sized enterprises (SMEs) could play a very significant role in the upscaling and dissemination of low to medium-level biotechnology, such as tissue culture of virus-free planting materials, disease diagnostic kits, biofertilizers, biopesticides and so on. The nature of biotechnology products and markets would require local industries and firms to learn new ways to improve their business by identifying new technologies ('technology prospecting') from the public R&D institutions, universities or other global sources of scientific and technical knowledge. Industrial and agricultural policies should stimulate innovation activities in firms and farms. Since to a large extent there are market failures in the area of innovation, industrial and agricultural policies should focus on the demand-side innovation policies. This can be in the form of taxation policies, financing via matching grants, public procurement and so on.

The case studies presented in this book emphasize the importance of a combined policy approach to STI: specific science, technology and innovation policies integrated with others that are pervasive to the different policy domains, including industrial policies. To turn innovation into an explicit and key driver of decision-making in the productive sector requires policy support as well as a cultural change that places innovation at the root of the decision process on how to go about solving knowledge-based problems. A case in point is the policies to foster university–industry links that in many countries implicitly still follow the linear vision of innovation, which assumes that the results of S&T developments will eventually and automatically lead to the creation of industrial capacity. Policies targeting university–industry links should, thus, include a deeper understanding of the industrial structure bound to be affected by them. A fully integrative and evolutionary perspective on education, research and industry policies is still, mostly, in the future; the cases delineated in this book prove that there are substantial processes of learning that may be transferred across national, organizational and institutional boundaries, to further such perspectives.

## NOTE

1. Parts of this chapter are based on contributions from the national teams.

# Index